Classics on Fractals

STUDIES IN NONLINEARITY
Robert L. Devaney, Boston University, Editor

Chaotic dynamics has been hailed as the third great scientific revolution of the twentieth century, along with relativity and quantum mechanics. The explosion of interest in nonlinear systems has led to the development of the Advanced Book Program's *Studies in Nonlinearity* series. This unique, interdisciplinary publishing program makes exciting new developments in nonlinear studies available to the scientific community.

Studies in Nonlinearity's textbooks, monographs, and reference books are targeted for advanced undergraduate and beginning graduate students, as well as researchers. The series aims to make new mathematical developments in this area accessible to a wide audience, including mathematicians, physicists, chemists, engineers, computer scientists, biologists, and economists.

Dynamics of Complex Systems
Yaneer Bar-Yam

Nonlinear Dynamics and Chaos
Steven Strogatz

Classics on Fractals
Gerald A. Edgar

A First Course in Chaotic Dynamical Systems
Theory and Experiment
Robert L. Devaney

Dynamical Systems Software
A First Course in Chaotic Dynamical Systems Software Labs: 1–6
James Georges and Del Johnson

An Experimental Approach to Nonlinear Dynamics and Chaos
With Macintosh disk
Nicholas B. Tufillaro, Tyler Abbott, and Jeremiah Reilly

An Eye for Fractals
Michael McGuire

Chaos, Fractals, and Dynamics
Computer Experiments in Mathematics
Robert L. Devaney

An Introduction to Chaotic Dynamical Systems
Second Edition
Robert L. Devaney

Exploring Chaos
Theory and Experiment
Brian Davies

Classics on Fractals

Edited by Gerald A. Edgar

Routledge
Taylor & Francis Group

LONDON AND NEW YORK

First published 2004 by Westview Press

Published 2018 by Routledge
52 Vanderbilt Avenue, New York, NY 10017
2 Park Square, Milton Park, Abingdon, Oxon OX14 4RN

Routledge is an imprint of the Taylor & Francis Group, an informa business

A Cataloging-in-Publication data record for this book is available from the Library of Congress.

ISBN 13: 978-0-367-00739-3 (hbk)
ISBN 13: 978-0-367-15726-5 (pbk)

Contents

Introduction

Read the masters! Experience has shown that this is good advice for the serious mathematics student. This book contains a selection of the classical mathematical papers related to fractal geometry. For the convenience of the student or scholar wishing to learn about fractal geometry, nineteen of these papers are collected here in one place. Twelve of the nineteen have been translated into English from German, French, or Russian.

In many branches of science, the work of previous generations is of interest only for historical reasons. This is much less so in mathematics.[1] Modern-day mathematicians can learn (and even find good ideas) by reading the best of the papers of bygone years. In preparing this volume, I was surprised by many of the ideas that come up.

I have arbitrarily fixed the end of the "classical" period of fractal geometry as 1975, when Benoit Mandelbrot coined the word "fractal". So, of course, none of the papers here uses that word. But the themes of fractal geometry—found for example in Mandelbrot's writing—can often be traced back to this earlier work. These themes include: nowhere-differentiable functions and curves; self-similar sets; fractional dimension. If Mandelbrot is the father of fractal geometry, the authors represented here are the grandfathers and uncles.[2]

The papers selected here are mathematics papers, of interest for one reason or another to the modern student of fractal geometry. There are no papers on applications outside of mathematics (except perhaps Mandelbrot's paper). In fact, during this "classical" period, any suggestion that a non-differentiable function or a Cantor set might have some application in the real world was not taken seriously. This situation has been completely reversed today. *Physics Briefs* for July to December, 1990, lists 37 references on fractals (including my own text). So physics papers on fractals are being published more frequently than once a week. I do not know whether the application of fractals will remain at this high level. I think it is even possible (but unlikely) that fractal geometry for applications will recede to the level that "catastrophe theory" has today. But, in any case, independent of any possible

---

[1]"Mathematical language is difficult but imperishable. I do not believe that any Greek scholar of to-day can understand the idiomatic undertones of Plato's dialogs, or the jokes of Aristophanes, as thoroughly as mathematicians can understand every shade of meaning in Archimedes' works." —M. H. A. Newman, *Mathematical Gazette* 43 (1959) 167.

[2]I regret that there are no women among these authors (no, not even Helge von Koch). I hope women who read this will take that as a challenge!

applications outside mathematics, fractal geometry (like catastrophe theory) will still be an interesting and viable chapter of mathematics. There is enough material for a volume twice this size. Consequently, many interesting or important papers have been omitted. Selections that are, in my opinion, more accessible or educational to a student have been preferred. The paper included is not always the original paper on the topic. For example: Jarnik's theorem is presented in a paper of Besicovitch; Kiesswetter's very elementary paper is included, rather than the earlier version due to Bolzano.

The selections are accompanied by my comments, before and/or after the text. The comments preceding a selection may point out things to watch for, and explain terminology and notation. The selections may be read independently of each other; so these explanations are repeated whenever appropriate. References are given separately for each selection; references for the selection itself followed by additional references for my comments. Footnotes on the selections are from the originals unless they are marked: *"Tr."* for translator, *"Ed."* for editor (or Edgar), *"Au."* for author.

Columbus, Ohio   G. A. Edgar
September 15, 1992

Classics on Fractals

J. D. Memory
Blake and Fractals

William Blake said he could see
Vistas of infinity
In the smallest speck of sand
Held in the hollow of his hand.

Models for this claim we've got
In the work of Mandelbrot:
Fractal diagrams partake
Of the essence sensed by Blake.

Basic forms will still prevail
Independent of the scale;
Viewed from far or viewed from near
Special signatures are clear.

When you magnify a spot,
What you had before, you've got.
Smaller, smaller, smaller yet,
Still the same details are set;

Finer than the finest hair
Blake's infinity is there,
Rich in structure all the way—
Just as the mystic poets say.

Presented at the Royal Academy of Sciences,
July 18, 1872

This paper was read in the Royal Prussian Academy of Sciences on July 18, 1872. But it was not published at that time. The example given in the paper was published in 1875 in the paper [5] of du Bois-Reymond. (Also apparently [11], p. 97, but I have not verified that.) The version translated here was only published in the collected works [10] of Weierstrass in 1895.

Karl Theodor Wilhelm Weierstrass (1815–1897) was born in Ostenfelde (near Münster); he studied law and finance before turning to mathematics. He taught at secondary schools before obtaining (at the age of 40) a university position. His best-known work is in function theory—for example the representation of a complex function by means of a power series. He clarified and extended the basic ideas of analysis: limits, continuity, and convergence. This clarification involved the "epsilon-delta" technique,[1] still used today. We owe to Weierstrass the concept of "uniform convergence" and the postulational definition of the determinant. [1], [4], [6]

---

[1]The branch of mathematics known as "analysis" is ill-defined. I sometimes wonder, in the midst of an argument that seems to be mostly symbol-manipulation, whether I am really doing analysis. But when the proof begins, "Let $\varepsilon > 0$ be given. . .", then I know it is analysis.

ONE

On Continuous Functions of a Real Argument that do not have a Well-defined Differential Quotient

Karl Weierstrass

In the last half of the Nineteenth Century, mathematicians were beginning to realize that the notion of a function of a real variable was more complicated than had been thought. Previously it had been believed that—although functions non-differentiable at isolated points could be defined—functions arising naturally must be differentiable, at least for most values of the argument. And certainly (it was thought) any function defined by an "analytic formula" would be among these naturally defined functions. So it was a surprise to find that functions defined by analytic formulas like

$$\sum_{n=1}^{\infty} \frac{\sin(n^2 x)}{n^2} \quad \text{and} \quad \sum_{n=0}^{\infty} b^n \cos(a^n x\pi),$$

might not be differentiable.

A function $f : R \to R$ is **differentiable** at the point x (in the language of the selection, "has a well-defined differential quotient") iff the limit of the difference quotient

$$\lim_{h \to 0} \frac{f(x + h) - f(x)}{h}$$

exists (as a finite real number). In order to show that his example is non-differentiable, Weierstrass shows that the lim sup of the difference quotient is $+\infty$ and the lim inf is $-\infty$. *–Ed.*

Until very recently it was universally assumed that a single valued continuous function of a real variable always had a first derivative whose value could become undefined or infinitely large only at isolated points. To my knowledge, even in the writings of Gauss, Cauchy, and Dirichlet there is no remark from which one can infer without doubt that these mathematicians, who in their science were accustomed to exercise the severest criticism, held any other opinion. As I heard from some of his audience, Riemann was the first to express explicitly in 1861, or perhaps even earlier, that the above assumption was inadmissible and that, for example, the infinite series

$$\sum_{n=1}^{\infty} \frac{\sin(n^2 x)}{n^2}$$

represented a function that did not verify this. Unfortunately, Riemann did not publish a proof and it does not appear anywhere in his papers nor was it preserved through an oral communication. This loss is compounded by the fact that I have not been able to find out exactly what Riemann stated to his audience on this subject. The mathematicians who worked on Riemann's claim after it became known in wider circles, seem (at least in the majority) to have been of the opinion that it sufficed to demonstrate the existence of functions which in any arbitrarily small interval of their arguments have points where they are not differentiable. That such functions exist is extraordinarily easy to show, and I thus believe that Riemann had only those functions in mind which do not have a well–defined differential quotient for any value of their arguments. The proof that the above indicated trigonometric series is an example of this kind appears to me to be somewhat difficult. One can, however, easily construct continuous functions of a real argument x, for which it is possible to show by the simplest of means that these functions do not have a well–defined differential quotient for any value of x.

This can, for instance, be accomplished as follows.

Let x be a real variable, a an odd positive integer, b a positive constant less than 1, and

$$f(x) = \sum_{n=0}^{\infty} b^n \cos(a^n x\pi),$$

so $f(x)$ is a continuous function which can be shown to have the property, provided that the product ab exceeds a certain limit, the function does not have a well-defined differential quotient anywhere.

Let x_0 be a fixed value of x, and m an arbitrarily chosen positive integer. Then there is a certain integer α_m for which the difference

$$a^m x_0 - \alpha_m,$$

denoted by x_{m+1}, is $> -1/2$ but $\leq 1/2$.

If one defines

$$x' = \frac{\alpha_m - 1}{a^m}, \qquad x'' = \frac{\alpha_m + 1}{a^m},$$

then

$$x' - x_0 = -\frac{1 + x_{m+1}}{a^m}, \qquad x'' - x_0 = \frac{1 - x_{m+1}}{a^m};$$

and consequently

$$x' < x_0 < x''.$$

But one can take m large enough so that x' and x'' are both as close to x_0 as one wishes.

Now one has

$$\frac{f(x') - f(x_0)}{x' - x_0} = \sum_{n=0}^{\infty} \left(b^n \frac{\cos(a^n x' \pi) - \cos(a^n x_0 \pi)}{x' - x_0} \right)$$

$$= \sum_{n=0}^{m-1} \left((ab)^n \frac{\cos(a^n x' \pi) - \cos(a^n x_0 \pi)}{a^n (x' - x_0)} \right)$$

$$+ \sum_{n=0}^{\infty} \left(b^{m+n} \frac{\cos(a^{m+n} x' \pi) - \cos(a^{m+n} x_0 \pi)}{x' - x_0} \right).$$

The first term of this expression is

$$\frac{\cos(a^n x' \pi) - \cos(a^n x_0 \pi)}{a^n (x' - x_0)} = -\pi \sin\left(a^n \frac{x' + x_0}{2} \pi \right) \frac{\sin\left(a^n \frac{x' - x_0}{2} \pi \right)}{a^n \frac{x' - x_0}{2} \pi}$$

and since the value of

$$\frac{\sin\left(a^n \frac{x' - x_0}{2} \pi \right)}{a^n \frac{x' - x_0}{2} \pi}$$

lies between 1 and -1, the absolute value of the first term is smaller than

$$\pi \sum_{n=0}^{m-1} (ab)^n,$$

and so is also smaller than

$$\frac{\pi}{ab-1}(ab)^m.$$

Furthermore, since a is odd, one has

$$\cos(a^{m+n}x'\pi) = \cos(a^n(\alpha_m - 1)\pi) = -(-1)^{\alpha_m},$$

$$\cos(a^{m+n}x_0\pi) = \cos(a^n\alpha_m\pi + a^n x_{m+1}\pi) = -(-1)^{\alpha_m}\cos(a^n x_{m+1}\pi),$$

hence

$$\sum_{n=0}^{\infty} b^{m+n}\left(\frac{\cos(a^{m+n}x'\pi) - \cos(a^{m+n}x_0\pi)}{x' - x_0}\right)$$

$$= (-1)^{\alpha_m}(ab)^m \sum_{n=0}^{\infty}\frac{1 + \cos(a^n x_{m+1}\pi)}{1 + x_{m+1}}b^n.$$

All terms of the sum

$$\sum_{n=0}^{\infty}\frac{1 + \cos(a^n x_{m+1}\pi)}{1 + x_{m+1}}b^n$$

are positive and the first term is not less than 2/3, since $\cos(x_{m+1}\pi)$ is non negative and $1 + x_{m+1}$ lies between 1/2 and 3/2.

It follows that

$$\frac{f(x') - f(x_0)}{x' - x_0} = (-1)^{\alpha_m}(ab)^m \eta\left(\frac{2}{3} + \varepsilon\frac{\pi}{ab-1}\right),$$

where η is a positive quantity > 1 and ε lies between -1 and $+1$.

Similarly, one finds

$$\frac{f(x'') - f(x_0)}{x'' - x_0} = -(-1)^{\alpha_m}(ab)^m \eta_1\left(\frac{2}{3} + \varepsilon_1\frac{\pi}{ab-1}\right),$$

where η_1 is also positive and > 1, and ε_1 lies between -1 and $+1$ as well.

One now takes a, b such that $ab > 1 + \frac{2}{3}\pi$, so that

$$\frac{2}{3} > \frac{\pi}{ab-1}$$

and get that

$$\frac{f(x') - f(x_0)}{x' - x_0}, \qquad \frac{f(x'') - f(x_0)}{x'' - x_0}$$

always have opposing signs but both become indefinitely large as m increases without bound.

From this it follows immediately that $f(x)$ at the point $(x = x_0)$ possesses neither a well–defined finite nor a well–defined infinite differential quotient.

Several later mathematicians, inspired by the Weierstrass example, examined continuous but nowhere differentiable functions. In this volume, Selection 3 (von Koch) arose from an attempt to give a simplified example of this kind; and Selection 18 (Kiesswetter) is one of many such examples that have been published over the years. Other references are in Kiesswetter's bibliography. An earlier example was found before 1830 by Bernard Bolzano (but not published until 1930 [3])—Bolzano constructed a continuous function and proved that it is not differentiable on a dense set of points. The Weierstrass example itself,

$$f(x) = \sum_{n=0}^{\infty} b^n \cos(a^n x\pi),$$

was analyzed thoroughly by G. H. Hardy in 1916 [7]. Hardy showed that the function is differentiable nowhere, provided

$$0 < b < 1, \quad a > 1, \quad ab \geq 1.$$

Hardy also analyzed the Lipschitz condition satisfied by $f(x)$. If $ab > 1$, let $\xi < 1$ be defined by

$$\xi = \frac{\log(1/b)}{\log a}.$$

Then, as $h \to 0$,

$$|f(x+h) - f(x)| = O\left(|h|^{\xi}\right) \qquad \text{for every } x$$

but

$$|f(x+h) - f(x)| = o\left(|h|^{\xi}\right) \qquad \text{for no } x.$$

This is much more than the assertion $f(x)$ is not differentiable, since in the case $ab > 1$, we have $\xi < 1$.

Let $G \subseteq R^2$ be the graph of f:

$$G = \left\{ (x, y) : y = f(x) \right\}.$$

The Lipschitz condition above, together with the result of Besicovitch and Ursell (Selection 11), shows that the graph of f has Hausdorff dimension $\dim G$ satisfying

$$\dim G \leq 2 - \xi = \frac{\log(a^2 b)}{\log a}.$$

It is believed by the experts that the Hausdorff dimension is exactly equal to $2 - \xi$. But that has not yet been proved. For recent work on this problem, see [9].

Mandelbrot proposed a variant of the function:

$$w(x) = \sum_{n=-\infty}^{\infty} b^n \left(1 - \cos(a^n x\pi)\right).$$

This function satisfies the ("self-affine") functional equation $w(x) = bw(ax)$. See [8], p. 388–390. Berry and Lewis [2] call this the "Weierstrass-Mandelbrot" function; their paper has lots of pictures.

As noted in the selection, Riemann proposed (in 1861 or earlier) that the function

$$g(x) = \sum_{n=1}^{\infty} \frac{\sin(n^2 x)}{n^2}$$

is non-differentiable on an everywhere dense set of values of x, but neither he nor Weierstrass proved it. This function is much more difficult to analyze than $f(x)$. Hardy [7] also provided proofs for this. In fact: if $\alpha < 5/2$, then the function

$$h(x) = \sum_{n=1}^{\infty} \frac{\sin(n^2 \pi x)}{n^{\alpha}}$$

is differentiable for no irrational value of x. –Ed.

Bibliography

[1] E. T. Bell, *Men of Mathematics*, Simon and Schuster, New York, 1937.

[2] M. V. Berry and Z. V. Lewis, *On the Weierstrass-Mandelbrot fractal function*, Proc. R. Soc. A **370** (1980), 459–484

[3] Bernard Bolzano, *Funktionenlehre*, Herausgegeben und mit Anmerkungen versehen von K. Rychlik, Prague, 1930.

[4] C. B. Boyer, *A History of Mathematics*, Second Edition (Revised by U. C. Merzbach) Wiley, New York, 1991.

[5] P. du Bois-Reymond, *Versuch einer Klassifikation der willkürlichen Funktionen reeler Argumente nach ihren Änderungen in den kleinsten Intervallen*, J. Reine Angew. Math. **79** (1875), 21–37.

[6] H. W. Eves, *An Introduction to the History of Mathematics*, Holt, Rinehart and Winston, New York, 1969.

[7] G. H. Hardy, *Weierstrass's non-differentiable function*, Trans. Amer. Math. Soc. **17** (1916), 301–325.

[8] B. Mandelbrot, *The Fractal Geometry of Nature*, Freeman, San Francisco, 1982.

[9] R. D. Mauldin and S. C. Williams, *On the Hausdorff dimension of some graphs*, Trans. Amer. Math. Soc. **298** (1986). 793–803

[10] K. Weierstrass, *Mathematische Werke*, Königlich Preussischen Akademie der Wissenschaften, Berlin, 1895.

[11] K. Weierstrass, *Abhandlungen aus der Functionenlehre*, Springer, Berlin, 1886.

Georg Ferdinand Ludwig Philip Cantor (1845–1918) was born in Russia, and educated in Germany, studying with Weierstrass and Kronecker. Most of his career was spent at the University of Halle. Cantor (along with Dedekind, Méray, Weierstrass, and Heine) established the foundations of analysis—the rigorous specification of the real number system—Dedekind using "cuts", Cantor using "fundamental sequences". C. B. Boyer [8], p. 560 calls this important development "the arithmetization of analysis". Cantor founded a new field—set theory—virtually single-handedly. His work on infinite sets was controversial at the time; but today the points of view considered controversial are those opposed to Cantor's. A discussion of Cantor's scientific work is contained in the book by J. W. Dauben [10]. Two of Dunham's twelve "great theorems" are results of Cantor [11]. A romantic's view of his life is in *Men of Mathematics* [7].

In this paper (extracted by the editors of the journal *Acta Mathematica* from a letter written to them by Cantor) we find a discussion of the cardinality of perfect sets in the line.

TWO

On the Power of
Perfect Sets of Points

Georg Cantor

A subset P of $\mathbb{R}$ is called **perfect** iff it is closed and has no isolated points. If $P \subseteq \mathbb{R}$ is a set, its **derived set** $P^{(1)}$ is the set of all limit points (accumulation points) of P. Thus, P is perfect if and only if $P = P^{(1)}$, and (as Cantor defines) P is closed if and only if $P^{(1)} \subseteq P$.

The important example of a nowhere dense perfect set with Lebesgue measure zero, today universally known as the **Cantor set**, is described here. There is an earlier description of the set in another paper of Cantor; I have chosen this selection because it foreshadows so much that came later.

An earlier paper [14] by H. J. S. Smith describes some sets related to the Cantor set. But Smith's sets seem to be only countable sets of endpoints, not the actual perfect sets. Of course, before "countable" and "uncountable" were clarified by Cantor, this distinction would not have seemed important.

Two sets A and B have the same cardinality (in the language of the selection, the same **power**) iff there is a one-to-one correspondence between them; Cantor writes $A \sim B$. Cantor shows here that all non-empty perfect subsets of the line have the same cardinality. This is easy to see for sets S with an interior point, but what about the remaining case? What if the set S has "magnitude" zero? This paper precedes Lebesgue, but the reader will not go wrong thinking of the "magnitude" $\mathfrak{J}(S)$ of a closed set S as the Lebesgue measure of S; see equation (11).

Other old-fashioned notation deserves comment. An open interval in the line:

$$(a \ldots b) \quad \text{means} \quad \{ x \in \mathbb{R} : a < x < b \}.$$

If ν runs through an index set (such as the positive integers), and a number a_ν is defined for each ν, then $\{a_\nu\}$ denotes the set of all those values a_ν. Similarly, if a variable z is specified by certain conditions, then $\{z\}$ denotes the set of all those values z. Union of (disjoint) sets is called "addition":

$$A + B \quad \text{means} \quad A \cup B.$$

If A is a subset of B, then Cantor says A is a **partial system** of B. Sometimes he says A is a **divisor** of B, and B is a **multiple** of A. The union of a family K_ν of sets is the **smallest multiple** of the family,

$$\mathfrak{M}(K_\nu) \quad \text{means} \quad \bigcup_\nu K_\nu.$$

–Ed.

———————————

... My theorem says that perfect sets of points all have the same power, the power of the continuum. As a first step in the proof, I will show it for perfect linear sets.[1] Let S be an arbitrary perfect set of points *which is not condensed in the span of any interval,* no matter how small.[2] We assume that S is contained in the interval $(0 \ldots 1)$ whose endpoints 0 and 1 belong to S and it is clear that all other cases in which the perfect set is not condensed in the span of any interval can be projected to this one.

However, by considerations in Acta Mathematica **2**, p. 378, there exists an infinite number of distinct intervals, each totally separated from the others, that can be represented by ordering them according to their size in such a way that smaller intervals come after larger ones. We denote them under this ordering by

$$(a_1 \ldots b_1), (a_2 \ldots b_2), \ldots, (a_\nu \ldots b_\nu), \ldots . \tag{1}$$

With respect to the set S, these intervals have the property that no point of S lies in any of their interiors while their endpoints a_ν and b_ν in conjunction with the other limit points of the set of points $\{a_\nu, b_\nu\}$ belong to S and determine it. We denote by g a limit point of $\{a_\nu, b_\nu\}$ and by $\{g\}$ the set of these. We then have

$$S \equiv \{a_\nu\} + \{b_\nu\} + \{g\}. \tag{2}$$

Moreover, the sequence (1) of intervals is such that the space between any two $(a_\nu \ldots b_\nu)$ and $(a_\mu \ldots b_\mu)$ always contains an infinite number

———————————

[1] That is, perfect subsets of the line.—*Ed.*

[2] That is, S is not dense in any interval; or S is *nowhere dense.* Since S is closed, this means S has no interior point.—*Ed.*

of others and that for any interval $(a_\varrho \ldots b_\varrho)$ there are others in the same sequence (1) that approach arbitrarily closely either the point a_ϱ or the point b_ϱ since a_ϱ and b_ϱ are *points* of the perfect set S and so are *limit points.*

This having been established, I will take an arbitrary set of the first power

$$\varphi_1, \varphi_2, \ldots, \varphi_\nu, \ldots, \tag{3}$$

a set of distinct points all lying on the interval $(0 \ldots 1)$, *in the whole span of which they are condensed,*[3] and I will suppose that the endpoints 0 and 1 are terms of the sequence φ_ν.

In order to give an appropriate example I recall the type of sequence that I used to enumerate all rational numbers ≥ 0 and ≤ 1 in Acta Mathematica 2, page 319, where, for our purposes, one must remove the first two terms which are 0 and 1.

But I insist that sequence (3) be left in full generality.

This is what I am claiming: *the set of points $\{\varphi_\nu\}$ and the set of intervals $\{(a_\nu \ldots b_\nu)\}$ can be associated to each other uniquely in such a way that for any pair of intervals $(a_\nu \ldots b_\nu)$, $(a_\mu \ldots b_\mu)$ belonging to sequence (1) and φ_{k_ν}, φ_{k_μ} the corresponding points of sequence (3) one always has that the number φ_{k_ν} is smaller or larger than φ_{k_μ} depending on whether in the segment $(0 \ldots 1)$ the interval $(a_\nu \ldots b_\nu)$ comes before or after the interval $(a_\mu \ldots b_\mu)$.*[4]

Such a correspondence between the two sets $\{\varphi_\nu\}$ and $\{(a_\nu \ldots b_\nu)\}$ can be realized, for example, by the following rule:

We associate to the interval $(a_1 \ldots b_1)$ the point φ_1, to the interval $(a_2 \ldots b_2)$ the term of φ_{k_2} of sequence (3) having the smallest index and the same order relation to φ_1 that the interval $(a_2 \ldots b_2)$ has to $(a_1 \ldots b_1)$ with respect to their location in the segment $(0 \ldots 1)$. Moreover, we associate to the interval $(a_3 \ldots b_3)$ the term with smallest index that has the same order relation with respect to φ_1 and φ_2 that the interval $(a_3 \ldots b_3)$ has to the intervals $(a_1 \ldots b_1)$ and $(a_2 \ldots b_2)$ with respect to their locations in the interval $(0 \ldots 1)$.

In general, we associate to the interval $(a_\nu \ldots b_\nu)$ the term φ_{k_ν} of sequence (3) having the smallest index and the same order relation with respect to the points $\varphi_1, \varphi_{k_2}, \ldots, \varphi_{k_{\nu-1}}$ already constructed that the interval $(a_\nu \ldots b_\nu)$ has to the corresponding intervals $(a_1 \ldots b_1)$, $(a_2 \ldots b_2), \ldots, (a_{\nu-1} \ldots b_{\nu-1})$ with respect to their locations in the interval $(0 \ldots 1)$.

---

[3] The "first power" is the cardinality or power of the integers, also called $\aleph_0$. So this is a countable set, dense in the interval $(0 \ldots 1)$.—*Ed.*

[4] Thus, this is not the order of ν and μ in sequence (1).

I claim that, by this rule, *the points $\varphi_1, \varphi_2, \ldots \varphi_\nu, \ldots$ of sequence (3) will be successively, even with a different ordering than the one given by sequence (3), all associated to distinct intervals of sequence (1);* since for each order relation between a finite number of points in sequence (3) there are, for the same number of intervals, relations with respect to location in the interval $(0 \ldots 1)$ that conform to this ordering. This is due to the fact that the set S is a perfect set that is not condensed in any interval, no matter how small.

To simplify matters we put

$$\varphi_1 = \psi_1, \quad \varphi_{k_2} = \psi_2, \ldots, \varphi_{k_\nu} = \psi_\nu, \ldots .$$

As a consequence the following sequence

$$\psi_1, \psi_2, \ldots, \psi_\nu, \ldots \tag{4}$$

consists of exactly the same numbers as sequence (3). *The two sequences (3) and (4) differ only by the ordering of their terms.*

The sequence ψ_ν of points in (4) thus has the remarkable relation with the sequence of intervals (1): whenever ψ_ν is larger or smaller than ψ_μ one has that a_ν and b_ν are respectively larger or smaller than a_μ and b_μ. Once more I recall that, since the set $\{\psi_\nu\}$ corresponds to the given set $\{\varphi_\nu\}$ except for order, it is condensed in the entire segment $(0 \ldots 1)$ and the endpoints 0 and 1 do not belong to this set.

The consequences of such a correspondence between the two sets $\{\psi_\nu\}$ and $\{(a_\nu \ldots b_\nu)\}$ are, as it is easy to show, the following:

If $(a_{\lambda_1} \ldots b_{\lambda_1}), (a_{\lambda_2} \ldots b_{\lambda_2}), \ldots, (a_{\lambda_\nu} \ldots b_{\lambda_\nu}), \ldots$ is an arbitrary sequence of intervals belonging to series (1) that converges either to the point a_ϱ or the point b_ϱ then the corresponding series of points $\psi_{\lambda_1}, \psi_{\lambda_2}, \ldots, \psi_{\lambda_\nu}, \ldots$ belonging to sequence (4) converges to the point ψ_ϱ and vice-versa.

If $(a_{\lambda_1} \ldots b_{\lambda_1}), (a_{\lambda_2} \ldots b_{\lambda_2}), \ldots, (a_{\lambda_\nu} \ldots b_{\lambda_\nu}), \ldots$ is an arbitrary sequence of the same type but its terms converge to a point g of the set S (see formula (2) and the meaning of g) then the corresponding sequence $\psi_{\lambda_1}, \psi_{\lambda_2}, \ldots, \psi_{\lambda_\nu}, \ldots$ in turn converges to a well defined point of the interval $(0 \ldots 1)$ not occurring in sequence (3) or (4) and completely determined by g. We denote this point corresponding to g by h. Conversely, an arbitrary point on the interval $(0 \ldots 1)$ not belonging to sequences (3) or (4) determines a point g in the set S unequal to the points a_ν and b_ν. The variables g and h are single valued functions of each other so the sets $\{g\}$ and $\{h\}$ certainly have the same power.

Our theorem now follows.

For we have, by formula (2),

$$S \equiv \{a_\nu\} + \{b_\nu\} + \{g\}.$$

And it is clear that

$$(0\ldots 1) \equiv \{\varphi_{2\nu}\} + \{\varphi_{2\nu-1}\} + \{h\}\,.$$

But, since we have the following formulas,

$$\{a_\nu\} \sim \{\varphi_{2\nu}\}, \quad \{b_\nu\} \sim \{\varphi_{2\nu-1}\}, \quad \text{and} \quad \{g\} \sim \{h\}\,,$$

one concludes, by Theorem (E) of Acta Mathematica 2, page 318, the formula:

$$S \sim (0\ldots 1)\,,$$

i.e., the perfect set S has the same power as the continuous segment $(0\ldots 1)$, which is what was claimed.

... This proof has the advantage of revealing a large and remarkable class of *continuous* functions of a real variable x whose properties give rise to interesting research, either by considering them according to their definition relating to the above, *or by putting them in the form of trigonometric series that certainly represent them since these functions do not have an infinite number of maxima and minima.*

In fact, we can find a function $\psi(x)$ on the interval $(0\ldots 1)$ satisfying the following conditions:

When x lies in an interval $(a_\nu \ldots b_\nu)$, i.e., $a_\nu \leq x \leq b_\nu$, one has that $\psi(x)$ is equal to ψ_ν and when x has the value g, the limit of a sequence of intervals $(a_{\lambda_1} \ldots b_{\lambda_1}), \ldots, (a_{\lambda_\nu} \ldots b_{\lambda_\nu}), \ldots$, then one defines

$$\psi(g) = h = \lim_{\nu=\infty} \psi_{\lambda_\nu}\,. \tag{5}$$

Certainly, the function $\psi(x)$, by what we have seen, is a continuous monotone[5] function of the continuous variable x. As x increases from 0 to 1, $\psi(x)$ varies in a continuous way without decreasing from 0 to 1. Its geometric image consists of a set of line segments each parallel to the x axis and of some interpolated points making this curve continuous. A special case of this type of function was already included in an example that I mentioned in Acta Mathematica 2, page 407. By putting

$$z = \frac{c_1}{3} + \frac{c_2}{3^2} + \cdots + \frac{c_\rho}{3^\rho} + \cdots, \tag{6}$$

where the coefficients c_μ can take any of the values 0 or 2 and where the series can have a finite or infinite number of terms, the set $\{z\}$ is a *perfect* set S lying in the interval $(0\ldots1)$ with the endpoints 0 and 1 belonging to the set $\{z\}$. Moreover, the set $\{z\} = S$ has the property that it is not condensed in any interval, no matter how small. Finally, one also has that the set $S = \{z\}$ has *magnitude* $\mathfrak{J}(S)$ (a *concept* that I will soon explain) equal to zero.

Here the points that we have denoted by b_ν arise as values of z from formula (6) by taking $c_\varrho = 0$ starting from a certain ϱ greater than 1 in such a way that all the b_ν are included in the formula

$$b_\nu = \frac{c_1}{3} + \frac{c_2}{3^2} + \cdots + \frac{c_{\mu-1}}{3^{\mu-1}} + \frac{2}{3^\mu}. \tag{7}$$

The points a_ν arise as z values from the same formula by taking c_ϱ starting with a certain ϱ always equal to 2 so that, by the equation

$$1 = \frac{2}{3} + \frac{2}{3^2} + \frac{2}{3^3} + \cdots,$$

one has, by taking $c_\mu = 0$, $c_{\mu+1} = c_{\mu+2} = \cdots = 2$,

$$a_\nu = \frac{c_1}{3} + \frac{c_2}{3^2} + \cdots + \frac{c_{\mu-1}}{3^{\mu-1}} + \frac{1}{3^\mu}. \tag{8}$$

We now relate the variable z to another variable y defined by the formula

$$y = \frac{1}{2}\left(\frac{c_1}{2} + \frac{c_2}{2^2} + \cdots + \frac{c_\rho}{2^\rho} + \cdots\right) \tag{9}$$

in which we agree that the coefficients c_ϱ have the same value as in (6).

This connection clearly implies that y is a function of z, denoted by $\psi(z)$. We note that the two values of $\psi(z)$ for $z = a_\nu$ and for $z = b_\nu$ now coincide and

$$\psi(a_\nu) = \psi(b_\nu) = \frac{1}{2}\left(\frac{c_1}{2} + \frac{c_2}{2^2} + \cdots + \frac{c_{\mu-1}}{2^{\mu-1}} + \frac{2}{2^\mu}\right).$$

From this one has a continuous and monotone function $\psi(x)$ of the continuous variable x defined as follows:

For $a_\nu < x < b_\nu$ one sets: $\psi(x) = \psi(a_\nu) = \psi(b_\nu)$, and for $x = z$ one has $\psi(x) = y = \psi(z)$.

L. Scheefer in Berlin has observed that this function $\psi(x)$, as well as many others, contradicts a theorem of of Harnack (see Math. Annalen **19**, page 241, Lehrs. 5). In fact, this function $\psi(x)$ has zero derivative $\psi'(x)$ for all x values except for those which we have denoted by z; these comprise a perfect set $\{z\}$ whose magnitude $\mathfrak{J}(\{z\})$ is equal to zero. But Mr. Scheefer has also informed me that he could replace this theorem with another free of any doubt. I hope that he will soon publish in Acta his researches on this subject as well as on other interesting questions that he is studying.

———

...I showed above that all perfect linear sets of points that are not condensed in any part of the segment on which they lie, no matter how small, have the same power as the linear continuum.

Let us now take an *arbitrary* perfect linear set of points S lying on the interval $(-\omega \ldots + \omega)$. I assert that S has the power of the continuum $(0 \ldots 1)$.

In fact, as we have already dealt with the case in which the set S is not condensed in any part of the segment $(-\omega \ldots + \omega)$, let us take an arbitrary interval $(c \ldots d)$ in whose interior S is everywhere condensed. All the points of $(c \ldots d)$ will also belong to S since S is a perfect set.

The set of points $(c \ldots d)$ is a partial system of S and S is a partial system of the segment $(-\omega \ldots + \omega)$. Since the set $(c \ldots d)$ has the same power as the set $(-\omega \ldots + \omega)$ one concludes that S has the same power as $(-\omega \ldots + \omega)$, i.e., the power of $(0 \ldots 1)$; for one has the general theorem:

"*Given a well defined set M of arbitrary power, a partial set M' contained in M and a partial set M'' contained in M', if the last system M'' has the same power as the first M, then the intermediate set M' also has the same power as M and M''.*" (see Acta Mathematica **2**, page 392).[6]

When a set P is such that its first derived set $P^{(1)}$ is a divisor, I will say that P is a *closed set.*[7]

Each *closed* set P of a larger power than the first decomposes, as we know, in a unique way into a set R of the first power and a perfect set S.[8] Using our theorems we conclude the following: "*All closed sets of*

---

[6] The Cantor–Bernstein Theorem, for example [13, p. 23].—*Ed.*

[7] If $P^{(1)} \subseteq P$, then P is a closed set.—*Ed.*

[8] The Cantor–Bendixson Theorem [13, p. 133].—*Ed.*

points fall into two classes, the ones of the first power, and the ones of the power of the arithmetic continuum." In an upcoming communication I will show that this dichotomy also holds for sets of points that are *not closed*. By this we will arrive at, with the help of the principles of paragraph 13 of my memoir in Acta Mathematica **2**, page 390, the determination of the *power* of the *arithmetic continuum*, by showing that it coincides with that of the *second number class (II)*.[9]

... There is a *notion* of *volume* or *magnitude* that relates to any set P lying in an n–dimensional plane space G_n, whether this set P is continuous or not.

In the case when P reduces to an n–dimensional continuous set or to a system of such sets, this notion coincides with ordinary volume.

When P is a certain continuum with a smaller number of dimensions than n, the value of the volume becomes zero. The same thing happens when P is such that $P^{(1)}$ has the first power and also in a number of other cases. But what at first glance seems surprising is that this volume, that I denote by $\mathfrak{J}(P)$, sometimes has a nonzero value for sets P contained in G_n of the kind that are not condensed in any continuous n–dimensional part of G_n, no matter how small.

I come to this general notion of *volume* or *magnitude* $\mathfrak{J}(P)$ of an *arbitrary* set P contained in[10] G_n by taking *each* point p that belongs to P or $P^{(1)}$ as center of a full n–dimensional sphere of radius ϱ that we will call $K(p, \varrho)$. The smallest multiple of all these full spheres $K(p, \varrho)$ (see the definition of smallest multiple, Acta mathematica **2**, page 357) written

$$\mathfrak{M}[K(p, \varrho)],$$

is, for each value of ϱ (ϱ is a constant), a set composed of n–dimensional pieces whose volume is determined by known rules by means of an n–fold integral.

Let $f(\varrho)$ be the value of this integral; then $f(\varrho)$ is a continuous function of ϱ that decreases with ϱ. The limit of $f(\varrho)$ as ϱ goes to zero will serve as the definition of the volume $\mathfrak{J}(P)$ so that we have

$$\mathfrak{J}(P) = \lim_{\varrho=0} f(\varrho). \tag{10}$$

---

[9]The "second number class" is the class of all countable ordinal numbers. It has cardinality or power $\aleph_1$.—*Ed.*

[10]Write G_n for an n-dimensional Euclidean space. Below, H_m is an m-dimensional Euclidean space.—*Ed.*

I stress the fact that this value of *volume* or *magnitude* of an arbitrary set P contained in an n-dimensional continuous plane space G_n is completely dependent on the plane space G_n itself of which P is a part and especially on the number n in such a way that, if one considers *the same set P* as a part of another continuous space H_m, the value of the volume of P with respect to the space H_m will in general be different than the one for the same set P considered as a part of G_n.

A square with unit side has *magnitude* equal to zero when considered as a part of three dimensional space but has magnitude equal to 1 when considered as a part of a two dimensional plane. This general notion of *volume* or *magnitude* is essential to my researches on *dimensions* of *continuous sets*, as promised in Acta Mathematica 2, page 407 (that I will submit later in your journal).

By limiting ourselves to *linear* sets of points contained in the interval $(0 \ldots 1)$, the *volume* or *magnitude* of such a set P is easily determined by following the method described in Acta Mathematica 2, page 378, where we considered intervals denoted by $(c_\nu \ldots d_\nu)$ subject to a condition on P and $P^{(1)}$ or, as I wrote there, to $\mathfrak{M}(P, P^{(1)})$. We set

$$\sum (d_\nu - c_\nu) = \sigma,$$

where σ is a constant ≤ 1. However, in our case it is easy to see that one has

$$\mathfrak{J}(P) = 1 - \sigma. \tag{11}$$

———

$\ldots$ Perfect linear sets of points S that are not condensed in any interval, no matter how small, have, in general, a magnitude $\mathfrak{J}(S)$ different from zero, but they can also have a magnitude $\mathfrak{J}(S)$ equal to zero.

Those for which $\mathfrak{J}(S)$ is different from zero can be reduced by composition (addition) to those for which $\mathfrak{J}(S) = 0$ and to perfect sets that not only have magnitude different from zero but for which all *perfect* parts obtained by taking partial intervals of $(0 \ldots 1)$ *in turn* have magnitude different from zero.[11]

---

[11] These sets meet every subinterval of $(0 \ldots 1)$ in a set of positive measure. Now there do exist perfect sets of positive measure with no such subsets (for example, a fat Cantor set [15, (2.82)]). So I do not know what reduction Cantor intends here.—Ed.

For *this last class* of perfect linear sets there is a very simple proof of the theorem shown above, that their power is that of the continuum.

In fact, take such a perfect set in the interval $(0 \ldots 1)$ and suppose that the endpoints 0 and 1 belong to S. We first establish the sequence (1) of intervals $(a_\nu \ldots b_\nu)$ belonging, in the sense explained above, to the perfect set S.

Let x be an arbitrary quantity > 0 and ≤ 1. We denote by S_x the set consisting of all points of S lying in the interval $(0 \ldots x)$ and define a function $\varphi(x)$ by the following conditions:

$$\varphi(0) = 0, \quad \varphi(x) = \mathfrak{J}(S_x) \quad \text{for all } x > 0 \text{ and } \leq 1.$$

This function $\varphi(x)$ is, as is easily seen, continuous and monotone in the interval $(0 \ldots 1)$. At the value $x = 1$ it takes the value $\varphi(1) = \mathfrak{J}(S) = c$, different from zero by the hypothesis on S. Moreover, in each of the intervals $(a_\nu \ldots b_\nu)$, that is, for $a_\nu \leq x \leq b_\nu$, it has the constant value $\varphi(x) = \varphi(a_\nu) = \varphi(b_\nu)$. When x is smaller than a_ν one always has $\varphi(x) < \varphi(a_\nu)$ and when x is larger than b_ν one has $\varphi(x) > \varphi(b_\nu)$. *This holds because* we have assumed that each partial perfect set of S obtained by restriction to partial intervals of $(0 \ldots 1)$ itself has nonzero *size*.

The continuous function $\varphi(x)$ takes on all values between 0 and c. It takes on each value $\varphi(a_\nu) = \varphi(b_\nu)$ an infinite number of times for all x that are[12] $\leq a_\nu$ and $\leq b_\nu$, but it takes each value h in the interval $(0 \ldots c)$ different from $\varphi(a_\nu) = \varphi(b_\nu)$ *exactly once* for a *distinct* value g of x, where g is unequal to any value belonging to the intervals $(a_\nu \ldots b_\nu)$, either to the endpoints a_ν and b_ν or to any intermediate value.

And since for each of these values of g and x there is a certain value $h = \varphi(g)$ different from the values $\varphi(a_\nu) = \varphi(b_\nu)$ and conversely, one has, as in our first proof, that

$$\{g\} \sim \{h\}$$

from which one concludes, as before, that the power of S is that of the continuum $(0 \ldots c)$.

———

$\ldots$ After obtaining these results I returned to my research on trigonometric series that appeared thirteen years ago and that I set aside

---

[12]Here, $\leq a_\nu$ should be $\geq a_\nu$. —Ed.

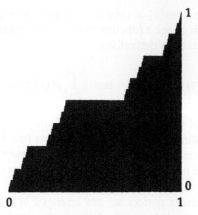

Figure 1. Cantor's singular function.

for a long time. Not only was I able to prove that the theorem Acta Mathematica 2, page 348 remains true when the system of points that I denote by P is such that its derived set $P^{(1)}$ has the first power, but I now have some results for the case when $P^{(1)}$ has power greater than the first. I will send you these at a later time.

In Cantor's investigations of set theory, one of the surprises was that the line $\mathbf{R}$ and the plane $\mathbf{R}^2$ have the same cardinality. Cantor himself said, "I see it but I don't believe it." [10], p. 55. There is a one-to-one correspondence between the line and the plane. So the plane can be "described by one parameter". Before Cantor, saying a set can be described by one parameter was considered the same thing as saying the set is 1-dimensional. Is there, then, any real sense that we can say the line is 1-dimensional and the plane is 2-dimensional? The study of topological dimension attempts to answer this question.

Another question investigated by Cantor is the evaluation of the cardinality of subsets of the line (or the plane). There are (of course) the countable sets, such as the integers, or the rational numbers. There are the subsets with cardinality of the continuum, such as the line itself, or any subset with an interior point. Are there any other possibilities? Cantor eventually came to believe that there are no other possibilities: every uncountable subset of the line has the cardinality of the continuum. This is known as the Continuum Hypothesis. This selection shows that the Continuum Hypothesis is true when restricted to closed sets. Cantor announces here that he will soon publish a proof of the Continuum Hypothesis without the restriction. But he did not publish that proof. In fact, the Continuum Hypothesis was never proved or refuted by Cantor. Much more recently it was found that proof and refutation are both impossible in the usual framework of set theory [12], [9] (Gödel, Cohen).

The second part of this selection concerns the Cantor singular function (or the "Devil's Staircase"). See Figure 1. It is a continuous non-constant function that has derivative zero except on a set {z} of Lebesgue measure zero (in fact, this exceptional set is the Cantor set). Today, such a function is called a **singular** function. This function is not equal to the integral of its derivative:

$$\psi(1) - \psi(0) = 1 \quad \text{but} \quad \int_0^1 \psi'(x)\, dx = 0.$$

The third section of the paper deals with n-dimensional volume. A completely satisfactory theory of n-dimensional volume in $\mathbb{R}^n$ was first formulated after 1900 by Lebesgue and his followers. But Cantor's remarks here deserve notice.

The Cantor set and its variants are the most important examples worked out by F. Hausdorff in his paper on fractional dimension (Selection 5). The complementary intervals of a nowhere-dense perfect set are arranged in (1) in order of decreasing length. Selection 15 (Besicovitch and Taylor) investigates how the lengths of these intervals are related to the Hausdorff dimension of the set. The definition (10) of the "magnitude" of a set is taken further in Selection 7 (Bouligand).

Questions. Note that the set

$$T(\rho) = \mathfrak{M}\Big[K(p, \rho)\Big]$$

appearing just before equation (10) is an open set, its volume $f(\rho)$ is defined by a multiple (Riemann) integral, so that no Lebesgue theory is required. Or is it? Is it important that the set $T(\rho)$ be a set with Jordan content—that is, that the characteristic function of $T(\rho)$ be a Riemann integrable function? The characteristic function of an open set is Riemann integrable if and only if the Lebesgue measure of the boundary is 0 [6], §10-5. Does $T(\rho)$ have boundary of measure zero? —*Ed.*

Bibliography

[1] G. Cantor, *Une contribution à la théorie des ensembles*, Acta Mathematica **2** (1883) 311–328.

[2] G. Cantor, *Sur les ensembles infinis et linéaires de points*, Acta Mathematica **2** (1883), 349–380.

[3] G. Cantor, *Fondaments d'une théorie générale des ensembles*, Acta Mathematica **2** (1883), 381–408.

[4] Ch. Neumann, *Über die nach Kreis-, Kugel- und Zylinderfunktionen fortschreitenden Entwickelungen*, Leipzig, 1881.

[5] A. Harnack, *Vereinfahrung der Beweise in der Theorie der Fourier'sche Reihe*, Math. Annalen **19** (1882), 235–279.

[6] T. Apostol, *Mathematical Analysis*, Addison-Wesley, Reading, MA, 1957.

[7] E. T. Bell, *Men of Mathematics*, Simon and Schuster, New York, 1937.

[8] C. B. Boyer, *A History of Mathematics*, Second Edition (Revised by U. C. Merzbach), Wiley, New York, 1991.

[9] P. J. Cohen, *Set Theory and the Continuum Hypothesis*, Benjamin, New York, 1966.

[10] J. W. Dauben, *Georg Cantor: His Mathematics and Philosophy of the Infinite*, Harvard University Press, Cambridge, Mass., 1979.

[11] William Dunham, *Journey Through Genius: The Great Theorems of Mathematics*, Wiley, 1990.

[12] K. Gödel, *What is Cantor's Continuum Problem?*, Amer. Math. Monthly **54** (1947), 515–525.

[13] E. Kamke, *Theory of Sets*, Dover, New York, 1950.

[14] H. J. S. Smith, *On the integration of discontinuous functions*, Proc. London Math. Soc. **6** (1875), 140–153.

[15] K. R. Stromberg, *An Introduction to Classical Real Analysis*, Wadsworth, 1981.

[16] J. F. Fleron, *A note on the history of the Cantor set and Cantor function*, Mathematics Magazine **67** (1994), 136–140.

The following paper has been communicated
on October 12, 1904 by G. Mittag–Leffler and Karl Bohlin

There is more than one way to look at mathematics. On the one hand is the verbal, symbolic, analytic approach. On the other hand is the visual, geometric, tactile approach. (The popular press refers to "left brain" and "right brain" thinking.) Contributions of both sides are valuable to mathematical thought. Some people do best thinking visually, some people do best thinking verbally. When teacher and student favor opposite approaches, learning is likely to be hindered, unless the teacher is trained to recognize the differences [9]. I have seen very bright students who were convinced that—even in geometry—drawing a picture is a Bad Thing. Presumably their teachers had trained them this way to prevent the opposite problem of depending too much on pictures.

The non-differentiable function defined by Weierstrass (Selection 1) is explained completely in terms of formulas, with no pictures. But Helge von Koch found that unsatisfying. So he constructed a non-differentiable curve that is explained by pictures, using geometric language.

Helge von Koch (1870–1924) was a Swedish mathematician, known principally for his work on the theory of infinitely many linear equations and the theory of matrices derived from such systems. He also did work in differential equations and in the theory of numbers. [11]

THREE

On a Continuous Curve without Tangents Constructible from Elementary Geometry

Helge von Koch

Almost immediately after von Koch published this paper, Ernesto Cesàro recognized that the Koch curve is **self-similar**: it is made up of four parts, each of which is similar to the whole [10]. Koch included his description of this curve with no tangents in a later paper [12] which deals with the use of geometric ideas in the study of plane curves. Koch's curve (and related self-similar constructions) was later examined carefully by Paul Lévy (Selection 12). Self-similarity can be seen implicitly in several other papers: for example Selection 5 (Hausdorff), Selection 6 (Menger), Selection 13 (Moran), and Selection 19 (Mandelbrot). Self-affine sets will be seen in Selection 16 (de Rham) and Selection 18 (Kiesswetter). —Ed.

Until Weierstrass constructed a continuous function not differentiable at any value of its argument,[1] it was widely believed in the scientific community that every continuous curve had a well determined tangent (except at some singular points). It is known that, from time

---

[1] See Journal f. Math., **79** (1875)

to time, some geometers had tried to establish this, no doubt based on the graphical representation of curves.[2]

Even though the example of Weierstrass has corrected this misconception once and for all, it seems to me that his example is not satisfactory from the geometrical point of view since the function is defined by an analytic expression that hides the geometrical nature of the corresponding curve and so from this point of view one does not see why the curve has no tangent. Rather it seems that the appearance is actually in *contradiction* with the factual reality established by Weierstrass in a purely analytic way.[3]

This is why I have asked myself–and I believe that this question is of importance also as a didactic point in analysis and geometry– whether one could find a curve without tangents for which the geometrical aspect is in agreement with the facts. The curve that I found and which is the subject of this paper is defined by a geometrical construction sufficiently simple, I believe, that anyone should be able to see through "naive intuition"[4] the impossibility of the existence of a tangent.

Introduction

Some basic facts about plane curves

We begin by recalling some concepts that will be needed later.

A set of points C in the plane is called an arc of a *curve* if one can set up a correspondence with a line segment AB in such a way that the points of AB correspond in a determined way to the points of C.

Consider such a set and denote by $K(X)$ the point on C corresponding to the point X on the segment AB. If X' is any point on AB, let $K(X')$ be the corresponding point on C. One says that the curve

---

[2]Among these attempts we cite those of Ampère (J. é. pol. cah. 13), of Bertrand (Traité de C. diff. et intég.: Vol. 1), and of Gilbert (Brux. mém. S., 23 (1872)). One can find historical references and a bibliography in the work of E. Pascal: Esercisi e note crit. di calcole infinitesimale, p. 85–128, Milano 1895. See also Encyklopädie der. Math. Wiss. II, A.2, p. 63 and the work of M. Dini (translation by Lüroth–Schepp): Grundlagen für eine Theorie der Funktionen einer veränderlichen reellen Grösse, p. 88, 205–229.

[3]Among the numerous analogs that were published after Weierstrass there is none to my knowledge for which this remark does not apply. A paper of C. Wiener (J. f. Math. 90, p. 221; also Weierstrass, Functionenlehre, p. 100) attempts to elucidate geometrically the function of Weierstrass but does not seem to remove adequately the difficulty in question.

[4]I borrow this expression from a lecture of Klein on the mathematical character of spatial intuition.

is *continuous* at the point $K(X)$ if the point $K(X')$ approaches $K(X)$ whenever X' approaches X. If this condition holds for every point on the arc in question then one says that this is an arc of a *continuous curve*.[5]

Let us consider such an arc of a continuous curve C. Let X_1, X_2 be two points lying on the segment AB and denote by x_1, x_2 their respective distances from the point A. One says that the point $K(X_1)$ of C *precedes* or *succeeds* the point $K(X_2)$ depending on whether $x_1 < x_2$ or $x_1 > x_2$. Given three points X_1, X_2, X_3 on AB, one says that the point $K(X_2)$ is *intermediate* to the points $K(X_1), K(X_3)$ or that this point is *between* the two others, if the point X_2 is between the points X_1 and X_3. The points $K(A), K(B)$ of the curve corresponding to the endpoints of AB are called the endpoints of the arc in question. If one lets X run on the segment AB in the direction taken to be the positive direction, then one says that the point K runs on the curve C in the positive direction.

If one joins two points K, K' on the curve by a line, then we call this a *secant* to the curve and the segment of the line lying between K and K' is called a *chord* of the curve.

Fixing the point K and letting K' tend to K in an arbitrary way, if the secant KK' tends to a well defined limiting position T, then the line T is called the tangent to C at the point K.[6] If this does not hold, one says that the curve at K has no well defined tangent or, more succinctly, is *without tangent* at K.

Let us suppose that the curve in question has a tangent T at the point K. Let L and M be two neighboring points on the curve such that K is *between* L and M. It follows that the secant LM necessarily approaches T in the limit as L and M approach K all the while staying at opposite sides with respect to K.[7]

Finally, recall the definition of the *length* of an arc of the curve KK'. On this arc, between K and K', select points $K_1, K_2, \ldots, K_n$ and consider the polygonal line given by the chords $KK_1, K_1K_2, \ldots, K_nK'$. Let the number of such intermediate points go to infinity in such a

---

[5]Analytically, this last condition says that the Cartesian coordinates u, v of a point on the curve are continuous with respect to a parameter.

[6]We consider the direction of K towards K' as the positive direction of the secant KK' if K' succeeds K on the curve. This determines the positive direction of the tangent T.

[7]This simple theorem (that we have not found elsewhere) will be very useful in what follows. The proof is immediate: in fact, if K is preceded by L and succeeded by M then LK and KM coincide in the limit with the positive direction of T, so the angle subtended by the these lines goes to zero. This angle is greater than KLM so this also goes to zero showing that LM coincides, in the limit, with the positive direction of T.

way that the length of each chord goes to zero. If the length of the polygonal line defined above tends to a finite value L, then we say that the arc KK' of the curve is *rectifiable* and has *length L*.

If this does not occur, then we say that the arc is not rectifiable. If the length of the polygonal line tends to infinity, then we say that the length of the arc is infinite.[8]

I. Definition of the curve P and the function $f(x)$. Continuity and nonexistence of the tangent.

1. Join by a line two points A and B in the plane (Figure 1). Divide the segment AB into three equal parts AC, CE, EB, and let CE be the base of an equilateral triangle CDE. We get a polygonal line $ACDEB$ consisting of four equal segments. To fix the side towards which this triangle has to be turned we fix a direction (for example, from A to B) as being positive and consider as positive the side to the left of the positive direction of the segment. To simplify our notation, we denote by Ω the operation by which the polygonal line $ACDEB$ on the positive side of the line segment AB given with direction from A to B is constructed.

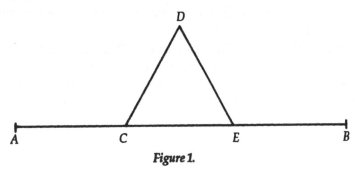

Figure 1.

2. Let us begin now with a given segment AB, the positive direction being from A to B (Figure 2). The operation Ω replaces the segment AB with the polygonal line $ACDEB$, the segments AC, CD, DE, EB being equal and their positive directions given by A to C, C to D, D to E, and E to B.

The operation Ω is now applied on each of the above segments so that the line $ACDEB$ is replaced by the polygonal line $AFGHCIKLD$

---

[8]It can be seen without too much difficulty that if this process does not converge, then it must diverge to infinity. —Tr.

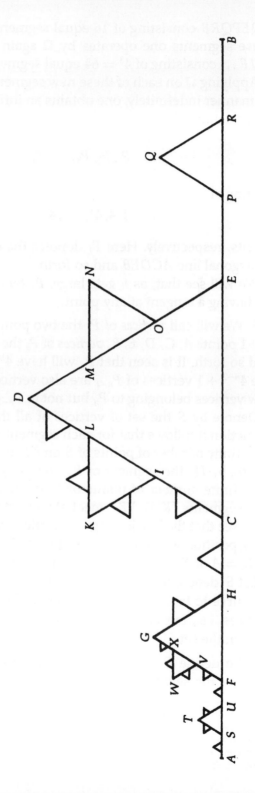

Figure 2.

$MNOEPQRB$ consisting of 16 equal segments AF, FG, etc. On each of these segments one operates by Ω again giving a polygonal line $ASTUF\ldots$ consisting of $4^3 = 64$ equal segments AS, ST, etc.

Applying Ω on each of these new segments and continuing in the same manner indefinitely, one obtains an infinite sequence of polygonal lines denoted by

(1) $$P_1, P_2, P_3, \ldots, P_n, \ldots$$

consisting of

$$1, 4, 4^2, \ldots, 4^{n-1}, \ldots$$

segments, respectively. Here P_1 denotes the original segment AB, P_2 the polygonal line $ACDEB$ and so forth.

We will see that, as n gets large, P_n tends to a continuous curve P not having a tangent at any point.

3. We will call *vertices* of P_1 the two points A and B, *vertices* of P_2 the $4+1$ points A, C, D, E, B, *vertices* of P_3 the 4^2+1 points $A, F, G, \ldots,$ B, and so forth. It is seen that P_n will have $4^{n-1} + 1$ vertices, and that all the $4^{n-2} + 1$ vertices of P_{n-1} are also vertices of P_n and the number of new vertices belonging to P_n but not to P_{n-1} is equal to $3 \cdot 4^{n-2}$.

Denote by S the set of vertices of all the lines in (1). From the construction it follows that for each segment KL of some P_α there will be an infinite number of points of S on KL in each neighborhood of K. Denoting by IK the segment of P_α preceding KL, there will similarly be an infinite number of points of S on IK in each neighborhood of K. The segments IK and KL meet at an angle of either 60^0 or 120^0, so one can see that the line joining two vertices K and K' cannot tend to a limiting position when the point K' (which is always a vertex) tends to K. (If $K = A$ or $K = B$ the above argument must be slightly modified.)

Let S' denote the set of limit points[9] of the points of S. Each point of the curve to be constructed will either be a vertex or a limit point of vertices, i.e., our curve will consist of a set of points P contained entirely in the set S'.[10]

4. To define P we will set up a correspondence between each point X of the segment AB to a well defined point $K(X)$ on P, and

---

[9]Following the terminology of Cantor, S' is the first derivative of S. The above discussion shows that each point of S is in S'. Thus, saying that a point K belongs to S' is equivalent to saying that K is a vertex or a limit point of vertices.

[10]Conversely, each point of S' belongs to P, that is $P = S'$, which follows easily from the results we are about to prove.

we will also introduce a continuous function $f(x)$ which will play a vital role in the study of the curve.

Denote by x the distance between A and the point X. If the point belongs to S' we take

$$K(X) = X$$

and

$$f(x) = 0.$$

Otherwise we draw a perpendicular XX_1 to AX (directed toward the positive side of AB).

Extending this perpendicular will lead to an intersection with one or more of the lines P_ν. Let P_α be the first such line and X_1 be the point of intersection. If X_1 is a point of S' we let

$$K(X) = X_1$$

and let $f(x)$ be the length of XX_1.

If X_1 is not a vertex of P_α then X_1 belongs to one the line segments comprising P_α, say $S_1 S_2$. If X_1 does not belong to the set S' draw a perpendicular $X_1 X_2$ to $S_1 S_2$ (directed toward the positive side of $S_1 S_2$). Let P_β be the first of the lines of (1) that intersects this perpendicular— say at X_2–by extending this line sufficiently.

If X_2 is a point in S' then we put

$$K(x) = X_2$$

and let $f(x)$ be the sum of the lengths of XX_1 and $X_1 X_2$.

If X_2 is not in S' then denote by $T_1 T_2$ the segment of P_β on which X_2 lies. We draw a perpendicular $X_2 X_3$ from $T_1 T_2$ in the positive direction and denote by X_3 the first intersection with a line P_n.

Continuing this process presents two possibilities. After constructing a certain number of perpendiculars

$$XX_1, \ X_1 X_2, \ldots, X_{k-1} X_k$$

whose lengths we denote by $f_1(x), f_2(x), \ldots, f_k(x)$, the point X_k will belong to S' and we can take $K(X) = X_k$ and let $f(x)$ be the sum of the $f_i(x)$. Otherwise, this process will never lead to an X_k in S'. In this case, we will have an infinite sequence of perpendiculars

(2) $$XX_1, \ X_1 X_2, \ X_2 X_3, \ldots$$

of lengths $f_1(x), f_2(x), f_3(x), \ldots$, the sum of which is finite, as will be shown.

In fact, taking the distance AB as the unit of length, the segment CD is equal to $1/3$ and the perpendicular from D to CE (see Figure 2) is equal to $\frac{1}{6}\sqrt{3}$. The triangle CDE is the largest one in the figure and one obviously has

$$f_1(x) = XX_1 \leq \frac{1}{6}\sqrt{3}$$

for any x in the interval

(3) $$0 \leq x \leq 1.$$

The triangles FGH, IKL, etc. lie on the sides of $P_2 = ACDEB$, so have sides equal to $1/9$. Similarly, the triangles constructed on the sides of P_3 have sides equal to $1/27$ and so forth. One thus has

$$f_2(x) = X_1X_2 \leq \frac{1}{2} \cdot \frac{1}{9}\sqrt{3}$$

$$f_3(x) = X_2X_3 \leq \frac{1}{2} \cdot \frac{1}{27}\sqrt{3}$$

(4)

$$\vdots \qquad \vdots \qquad \vdots$$

$$f_k(x) = X_{k-1}X_k \leq \frac{1}{2} \cdot \frac{1}{3^t}\sqrt{3}$$

for x in the interval (3).

The sum of the length in (2) cannot therefore be greater than the number

$$\frac{1}{2}\sqrt{3}\left(\frac{1}{3} + \frac{1}{3^2} + \frac{1}{3^3} + \cdots\right) = \frac{1}{4}\sqrt{3}$$

and so this sum converges to a given number which we denote by $f(x)$. Following the broken line $XX_1X_2X_3\ldots$ one approaches a given point denoted by $K(X)$ corresponding uniquely to X and a distance $f(x)$ away from X. It is immediately seen that $K(X)$ belongs to the set S'.

In case the series of perpendiculars in (2) contains only k terms let

$$f_{k+1}(x) = 0, \ f_{k+2}(x) = 0, \ldots$$

so that the function $f(x)$ is now defined for any x in the interval 0–1 by the formula

$$f(x) = f_1(x) + f_2(x) + \cdots.$$

All the points $K(X)$ so constructed comprise, by definition, our set P. For each point X on AB there corresponds a well defined point $K(x)$ of P whose position is given via $f(x)$.

We must now show that this set gives a *continuous curve* in the ordinary sense of the word.

5. Given two points K_1 and K_2 of P corresponding to values x_1 and x_2 of x, we say that K_1 *precedes* K_2 if $x_1 < x_2$, and otherwise that K_1 succeeds K_2.[11]

Given three points K_1, K_2, K_3 corresponding to x_1, x_2, x_3, where

$$x_1 < x_2 < x_3$$

then K_2 is *intermediate* to K_1, K_3. For example, between the two end points A and B of P we have three intermediate points C, D, and E belonging to the line P_2, 15 intermediate points F, G, H, etc. belonging to P_3, and so forth. Similarly, between two vertices S_1 and S_2 of the line P_α, which, by definition, belong to P, there are three intermediate points belonging to $P_{\alpha+1}$, 15 points belonging to $P_{\alpha+2}$ and so forth.

We have defined a point K on P by constructing perpendiculars

(2) $XX_1, X_1X_2, \ldots$

meeting respectively, P_α at X_1, P_β at X_2 and so on, where X_1 is between two vertices S_1, S_2 of P_α and X_2 is between two vertices T_1, T_2 of P_β, etc. The point K is thus, by its definition, intermediate to S_1 and S_2, intermediate to T_1 and T_2, and so on.

If the sequence (2) is infinite, then we will get a corresponding infinite sequence of decreasing segments

$$S_1S_2, \ T_1T_2, \ldots$$

that contain all points K that are intermediate between points an arbitrarily small distance apart.

6. We must prove that the function $f(x)$ which is well defined in the interval

(3) $0 \le x \le 1$

is also *continuous* in this interval. To do this we will first show that each of the functions

$$f_1(x), f_2(x), \ldots$$

---

[11] These are the definitions of the start of the paper.

is continuous on the interval in question and that the sum of these function converges uniformly there.

By definition, $f_1(x)$ is the distance from a point on a certain continuous line C_1 (composed of an infinite number of line segments) to the straight line AB, and this function is thus necessarily continuous. At the endpoints A and B this function is zero.

$f_2(x)$ is the distance from a point on a certain continuous line C_2 (similar to C_1) to a certain segment $S_1 S_2$ of the polygonal line P_α; considering $f_2(x)$ as a function on the segment $S_1 S_2$, one sees that the function is continuous on this segment, and so is a continuous function of x on the corresponding interval. Moreover, since $f_2(x)$ is zero at the endpoints S_1 and S_2, and similarly for the adjacent sides of P_α, one sees that $f_2(x)$ is continuous on the whole interval (3). The same argument holds for the functions $f_3(x)$, $f_4(x)$,....

All the functions $f_1(x)$, $f_2(x)$,... are thus continuous on the interval (3).

Now, since these functions satisfy the inequality (4) in this interval, it follows that their sum $\sum f_\nu(x)$ converges uniformly there. Therefore, by a classical result, *the function $f(x)$ given by this infinite series is continuous on this interval.*

7. Denote by $K(x)$ the point P corresponding to the value x in the interval $0 \ldots 1$. To see that P is the arc of a continuous curve at the point K, we must show that

$$\lim K(x') = K(x)$$

when

$$\lim x' = x,$$

that is, that the distance between the points $K(x')$ and $K(x)$ becomes arbitrarily small with $|x' - x|$.

We first consider the case when $K(x)$ is a point belonging to one of the lines P_ν, or, equivalently, that the point is given by a finite number of perpendiculars

$$XX_1, X_1 X_2, \ldots, X_{k-1} X_k$$

meeting the polygonal lines

$$P_{\alpha_1}, P_{\alpha_2}, \ldots, P_{\alpha_k}$$

at the points $X_1, X_2, \ldots, X_k$ of the sides

(5) $S_1 S_1', S_2 S_2', \ldots, S_k S_k'$

respectively, where the point X_i lies on P_{α_i} between the vertices S_i and S'_i. Let

$$X'X'_1, X'_1X'_2, \ldots$$

be the (finite or infinite) sequence of perpendiculars defining the point $K(x')$, where x' is a neighboring value of x. It follows by our construction that if one chooses $|x' - x|$ sufficiently small the points

$$X'_1, X'_2, \ldots, X'_k$$

belong respectively, to the sides in (5), and that the distance between X'_k and X_k are smaller than any given quantity δ specified in advance.

One goes from the point X'_k to the point $K(x')$ by a sequence of perpendiculars

$$X'_kX'_{k+1}, X'_{k+1}X'_{k+2}, \ldots$$

of length

$$f_{k+1}(x'), f_{k+2}(x'), \ldots,$$

respectively. Since

$$f_{k+1}(x) = 0, \quad f_{k+2}(x) = 0, \ldots$$

we have by the continuity of the sum $\sum f_\nu(x)$

$$\lim_{x'=x}(f_{k+1}(x') + f_{k+2}(x') + \cdots) = 0.$$

The absolute distance between the points X'_k and $K(x')$ cannot be greater than the length of the polygonal line

$$X'_kX'_{k+1}X'_{k+2}\cdots$$

which is

$$f_{k+1}(x') + f_{k+2}(x') + \cdots$$

so this distance goes to zero with $|x' - x|$. Since this is also true for the distance between X'_k and $X_k = K(x)$, we have shown that the distance between $K(x)$ and $K(x')$ decreases indefinitely with $|x' - x|$.

Let us consider the other case in which the given point $K(x)$ is defined by an unbounded number of perpendiculars

$$XX_1, X_1X_2, \ldots$$

meeting the lines

$$P_{\alpha_1}, P_{\alpha_2}, \ldots$$

at the points

$$X_1, X_2, \ldots$$

lying on the sides

$$S_1 S_1', S_2 S_2', \ldots$$

respectively, where we have kept our notation as in the previous case.

Let ε be a given quantity, and choose k sufficiently large so that the sum

$$f_{k+1}(x) + f_{k+2}(x) + \cdots$$

is less that $\varepsilon/3$ in the interval (3). It is seen that the distances between the X_k and $K(x)$ as well as the distances between the X_k' and $K(x')$ are less than $\varepsilon/3$. As well, take $|x' - x|$ sufficiently small so that the distance between X_k and X_k' is less than $\varepsilon/3$, as is always possible by the previous case. It is clear then that the distance between $K(x)$ and $K(x')$ is less than ε, in other words, that this distance can be made as small as desired by making $|x' - x|$ sufficiently small.

Thus the set P defines an arc of a curve continuous at each point.

8. In what follows we continue to use the letter P to denote the curve defined above.

Theorem. The curve P does not have a well defined tangent at any point.

Let us first consider a point K of the curve that is also a vertex of one of the polygonal lines P_α. In each neighborhood of K there are an infinite number of vertices K' and, by the above, we know that the line joining K to a point K' cannot tend to a well defined limit as K' approaches K arbitrarily. However, since the points K and K' are on the curve, the line KK' is a secant of P which would tend to a well determined limit if a tangent existed at K. Thus the curve cannot have a well determined tangent at K.

Let us consider the second case in which the point K lies on a polygonal line P_α, but is *not* a vertex. So K is necessarily a limit point of vertices and lies on all the lines

$$P_\alpha, P_{\alpha+1}, \ldots .$$

Thus, one can take the index α to be as large as one wants. Given this fact, let LM be the side of P_α on which K lies (Figure 3) and $LNRTM$ be the broken line given by applying Ω on LM. The vertices N, R, T are points on the curve P and one has

$$LN = NR = RT = TM .$$

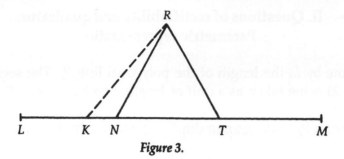

Figure 3.

From this it follows that the angle RKN is between 30^0 and 60^0.

However, K lies on the curve P *between* the points L and N (or *between* the points T and M), therefore,[12] if the curve had a well defined tangent at K, the secant KR would tend (for $\alpha = \infty$) towards the same limit as LN (or TM), which is impossible since the angle between these lines belongs to the interval $30^0 \ldots 60^0$.

We now consider the last case in which the point K of P does not lie on any of the P_α. In this case K is defined by an infinite sequence of perpendiculars

$$XX_1, X_1X_2, X_2X_3, \ldots$$

meeting the lines

$$P_\alpha, P_\beta, P_\gamma, \ldots$$

at the points $X_1, X_2, X_3, \ldots$ lying on the sides

(6) $$S_1S_2, T_1T_2, U_1U_2, \ldots$$

respectively. By the above, K is a point on the curve P lying *between* the points S_1 and S_2, *between* T_1 and T_2, *between* U_1 and U_2, and so forth. The sequence of points

$$S_1, T_1, U_1, \ldots$$

approaches K from the same side so all these points *precede* the point K and from the opposite side, the sequence of points

$$S_2, T_2, U_2, \ldots$$

approaches K. Thus, if the curve had a well defined tangent at K, the secants in (6) would have this tangent as a common limit. But this would be impossible since the angle made by two consecutive secants is either equal to 60^0 or 120^0.

The theorem has thus been proved for each point on the curve.

---

[12]See the introduction

II. Questions of rectifiability and quadrature.
Parametric representation

9. Denote by L_i the length of the polygonal line P_i. The segment AB (Figure 2) being taken as a unit of length, one has $L_1 = AB = 1$. The operation Ω takes P_1 to P_2 and this second line clearly has length 4/3. Going from P_2 to P_3 again multiplies the length by 4/3. In general one has

$$L_r = (4/3)^{r-1},$$

so that

$$\lim_{r=\infty} L_r = \infty.$$

It follows that the length of the arc of the curve P lying between A and B is infinite. In the same way, it can be shown that the same result holds for any arc lying between any two vertices so with little difficulty we conclude that *the length of the arc lying between any two points of the curve is infinite.*

It is also easy to evaluate the area contained between the curve and one of its chords. Let us take, for example, the chord AB. The area between $AB = P_1$ and P_2 is equal to the area of an equilateral triangle with base 1/3 which is

$$\frac{1}{36}\sqrt{3} = \frac{1}{16}\sqrt{3} \cdot \frac{4}{9}.$$

The area contained between P_2 and P_3 is equal to the sum of 4 equilateral triangles (see Figure 2) with base 1/9 so this area is equal to

$$4 \cdot \frac{1}{4}\left(\frac{1}{3}\right)^4 \sqrt{3} = \frac{1}{16}\sqrt{3} \cdot \left(\frac{4}{9}\right)^2.$$

To compute the area A_ν between P_ν and $P_{\nu+1}$ recall that P_ν is a polygonal line consisting of $4^{\nu-1}$ sides, each of which has length $1/3^{\nu-1}$. Going from P_ν to $P_{\nu+1}$ consists of constructing on each of these sides a little equilateral triangle with base $1/3^\nu$. One therefore has

$$A_\nu = 4^{\nu-1} \cdot \frac{1}{4} \cdot \frac{1}{3^{2\nu}} \cdot \sqrt{3} = \frac{1}{16}\sqrt{3} \cdot \left(\frac{4}{9}\right)^\nu.$$

The total area A is equal to the sum of the areas $A_1, A_2, \ldots$, so one gets

$$A = \sum_{\nu=1}^{\infty} A_\nu = \frac{1}{16}\sqrt{3} \sum_{\nu=1}^{\infty}\left(\frac{4}{9}\right)^\nu$$

and finally

$$A = \frac{1}{20}\sqrt{3}.$$

10. We now indicate how one can express the Cartesian coordinates u, v of a point on the curve P as uniform functions of a parameter.

We take the u axis to be the line AB (Figure 2), and the v axis to be the line perpendicular to AB passing through A (with the positive direction being taken from down to up). Let x be the distance of point X on AB to the origin A, and $K(x)$ the corresponding point on the curve (defined, as above, by a certain sequence of perpendiculars $XX_1, X_1X_2, \ldots$), and $u = u(x)$ and $v = v(x)$ the rectangular coordinates of the point K.

We specified above that

$$XX_1 = f_1(x), \ X_1X_2 = f_2(x), \ldots$$

and

$$f(x) = f_1(x) + f_2(x) + \cdots.$$

The line XX_1 is perpendicular to the u axis so projects to a point on this axis, while the line X_1X_2 forms an angle of either 30^0 or 150^0 with the u axis. We denote by $\{f_2(x)\}$ the projection of X_1X_2 on the u axis. Similarly, we denote by

$$\{f_3(x)\}, \{f_4(x)\}, \ldots$$

the projections of $X_2X_3, X_3X_4, \ldots$ on the u axis.

Finally, we let

$$\{\{f_1(x)\}\}, \{\{f_2(x)\}\}, \{\{f_3(x)\}\}, \ldots$$

be the projections of $XX_1, X_1X_2, X_2X_3, \ldots$ to the v axis. (It is clear that $\{\{f_1(x)\}\} = f_1(x)$.

It follows from these definitions that $\{f_i(x)\}$ and $\{\{f_i(x)\}\}$ are continuous functions of x in the interval $0 \ldots 1$ and that the modulus of these functions is at most equal to $f_i(x)$. Since $u(x) - x$ and $v(x)$ are respectively projections on the u axis and v axis of the broken line

$$XX_1X_2X_3\ldots$$

(the endpoints of this line being the points X and K), we can write

(7)
$$u = x + \{f_2(x)\} + \{f_3(x)\} + \cdots$$
$$v = f_1(x) + \{\{f_2(x)\}\} + \{\{f_3(x)\}\} + \cdots.$$

Since the series $\sum f_\nu(x)$ converges uniformly on the whole interval $0 \ldots 1$, it follows that the same holds for the new series defining the u, v coordinates of a point on our curve. These series thus represent *continuous* functions in the interval in question.

The formulas in (7) thus express the coordinates u, v in terms of uniform and continuous functions of a parameter x on the whole curve.

In the following section it will be seen how a simple transformation allows us to go from the curve P to a curve P' for which the abscissa u itself can be taken as the parameter and for which the ordinate v is a continuous and uniform function with respect to u on the whole curve.

III. Transformation of P into a curve P' where the ordinate is a uniform function of the abscissa

11. Consider in the x, y plane a line segment AB forming an arbitrary angle with the x axis (Figure 5). Split AB into three equal parts AC, CE, EB and construct, with CE as base, an equilateral triangle whose *median* MD (M being the point dividing the base CE into two equal parts CM and ME) is parallel to the y axis, directed in the positive y direction, and has length

$$\frac{CE}{2}\sqrt{3}.$$

We know that this median is equal to the median of an *equilateral* triangle constructed on the same base CE.

We will call Ω' the operation that takes a line segment AB to the broken line $ACDEB$.

12. Let us now consider a segment AB on the x axis, with A being the origin and the distance AB being fixed as the unit of length (Figure 4). We apply our operation Ω' (which reduces to the operation Ω defined in paragraph 1). Thus AB is replaced by a polygonal line $ACDEB$ made up of 4 segments that we will denote by P_2'. Applying the operation Ω on each of these segments, we get a polygonal line P_3' made up of 4^2 segments, on which the same operation will be applied and so on. More symmetrically, denote AB by P_1', so that an infinite sequence of polygonal lines

$$P - 1', P_2', P_3',$$

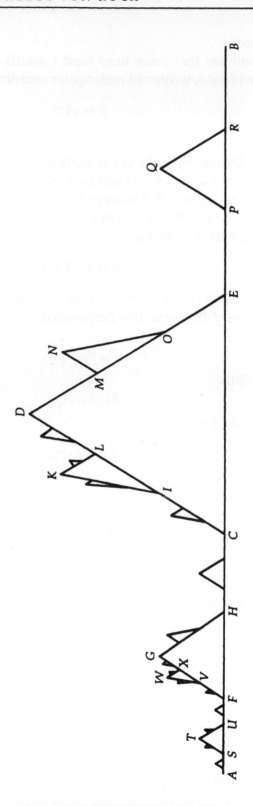

Figure 4.

is defined.

I will say that these lines tend towards a continuous curve P' whose equation, written in rectangular coordinates, can be written as

$$y = \varphi(x),$$

where $\varphi(x)$ is a continuous uniform function of x in the interval $0 \ldots 1$.

13. Denote by S the set of vertices (that is, points at which two segments of a line P'_ν meet) and by S' the set of limit points of S. (It is seen that each point of S belongs to S'.)

Let x be the distance from a given point X on AB to the origin A. If X is a point of S' we take

$$y = \varphi(x) = 0.$$

Otherwise, construct from X a line perpendicular to AB directed in the positive y direction. This perpendicular will successively meet certain lines

$$P'_\alpha, P'_\beta, P'_\gamma, \ldots$$

at the points

$$X_1, X_2, X_3, \ldots$$

respectively. Put

$$XX_1 = \varphi_1(x), \ X_1X_2 = \varphi_2(x), \ X_2X_3 = \varphi_3(x), \ldots$$

with the convention that

$$\varphi_{k+1}(x) = 0, \ \varphi_{k+2}(x) = 0, \ldots$$

when X_k belongs to S'.

By the same reasoning used in paragraph 6 we see that the functions $\varphi_\nu(x)$ are uniform and continuous in the interval $0 \ldots 1$ and that their sum converges uniformly there. Thus, if we set

(8) $$y = \varphi(x) \quad (0 \le x \le 1)$$

We now have a curve P' whose ordinate is expressible uniformly and continuously (8) with respect to its abscissa in the whole interval under consideration.

14. I assert that *the function $\varphi(x)$ does not admit, for any value of x, a finite well defined derivative.*[13]

---

[13] We leave unresolved in what follows whether there are values of x where the derivative exists but is *infinite.*

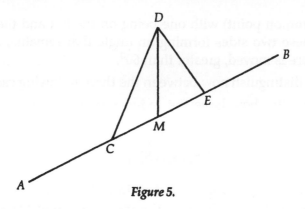

Figure 5.

If K is a point of P' that also belongs to one of the lines P'_v, the proof is analogous to the one used above for the curve P.

Let us now consider the other case in which the point K is a limit of an infinite sequence of points $X, X_1, X_2, \ldots$, and whose distance y to the x axis is equal to the infinite series

$$y = \varphi_(x) + \varphi_2(x) + \varphi_3(x) + \cdots .$$

Suppose that the perpendicular XK successively meets the polygonal lines

$$P'_{\alpha_1}, P'_{\alpha_2}, \ldots$$

at the points

$$X_1, X_2, \ldots$$

lying respectively on the sides

$$S_1 S'_1, S_2 S'_2, \ldots$$

of these lines. All the triangles constructed in our figure have their medians parallel to the y axis so we can distinguish (see Figure 5) the side CD of such a triangle which lies on the left of the median of the side DE located on the right. For ease of notation we call sides like CD *left sides* and sides like DE *right sides*.

With this convention, note first that in a triangle CDE of our figure (Figure 5) constructed on a *left* side AB, CD is a *left* side forming an angle DCB with AB *less* than 60^0 and that DE is a *right* side forming an angle DEC with AB *greater* than 60^0.

The other case would present itself if AB were a right side. Thus, in the constructed figure we consider two successive sides (that is, having a common point) with one being on the left and the other on the right, these two sides forming an angle that remains, no matter how the figure is moved, greater than 60^0.

15. We distinguish now between the three following cases.

1. No matter how large the index k is chosen, there is in the sequence

(9) $$S_k S'_k, S_{k+1} S'_{k+1}, \ldots$$

an infinite number of left sides and an infinite number of right sides. From what we have just said about the angle formed by two successive sides, it follows that the lines (9) cannot tend towards a well determined limiting position; and therefore, since the point K lies *between* the two points S_v and S'_v for all large v, there cannot be a well determined tangent at the point K.

2. From some index k on, all the sides (9) are on the left.

In this case it is easy to see that the line $S_v S'_v$ in the limit (for $v = \infty$) coincides with a line parallel to the y axis. In fact, let AB, CD be two consecutive left sides (see Figure 5) and let DE be the corresponding right side (by our construction we have $CM = ME$ and the median MD is parallel to the y axis). Denote by β the angle DMB, i.e., the angle of AB to the vertical, and by β' the angle of CD to the vertical. By construction, DM is larger than CM so the angle β' (or CDM) is smaller than the angle DCM and one gets that

$$\beta' < \frac{1}{2}\beta.$$

Consider a left side consecutive to CD and denote by β'' the angle it makes to the vertical. We similarly have

$$\beta'' < \frac{1}{2}\beta',$$

and so on. It is seen that the angles

$$\beta, \beta', \beta'', \ldots$$

decrease indefinitely and tend to zero.

However, the point K is between S_v and S'_v so we know that if there were a well defined tangent T at the point K, the secant $S_v S'_v$ would tend to T in the limit. Thus T would have to be parallel to the

y axis, and so the derivative of $\varphi(x)$ at the point in question would be infinite.

3. From some index k on, all the sides (9) are on the right.

As in the previous case, we come to the conclusion that if $\varphi(x)$ had a well defined derivative at x this derivative would be infinite.

The theorem thus holds in all cases.

Bibliography

[1] P. du Bois-Reymond, *Versuch einer Klassifikation der willkürlichen Funktionen reeler Argumente nach ihren Änderungen in den kleinsten Intervallen*, J. Reine Angew. Math. 79 (1875), 21–37.

[2] A. Ampère, Journal de l'Ecole Polytechnique 13 (1872).

[3] J. Bertrand, *Traité de Calcul Differentiel et de Calcul Intégral*, 1864.

[4] P. Gilbert, Nouveau Mémoirs, Acad. R. Sci. Bruxelles 23 (1872).

[5] M. E. Pascal, *Esercisi e Note Crit. di Calcole Infinitesimale*, Milano, 1895.

[6] H. Burkhardt, W. Wertinger, and R. Fricke, *Encyklopädie der Mathematischen Wissenschaften*, Leipzig, 1916.

[7] C. Wiener, *Geometrische und analytische Untersuchung der Weierstrassschen Funktion*, J. Reine Angew. Math. 90 (1881), 221–252.

[8] K. Weierstrass, *Abhandlungen aus der Functionenlehre*, Springer, Berlin, 1886.

[9] W. L. Akey, *Personality Type and Mathematics Anxiety Factors Affecting Remedial College Freshman*, Ph. D. Dissertation, The Ohio State University, 1991.

[10] E. Cesàro, *Remarques sur la courbe de von Koch*, Atti della R. Accad. Sc. Fis. Mat. Napoli 12 (1905), 1–12.

[11] C. C. Gillespie, editor, *Dictionary of Scientific Biography*, Scribners, New York, 1970.

[12] Helge von Koch, *Une méthode géometrique élémentaire pour l'étude de la théorie des courbes planes*, Acta Mathematica 30 (1906), 145–174.

Introduced by F. Klein
at the meeting of October 24, 1914

In this selection, Constantin Carathéodory studies the "length" of sets in Euclidean spaces $\mathbb{R}^n$ (denoted in the selection $\mathfrak{R}_n$). The length of a set in the line $\mathbb{R}$ had been covered by Lebesgue's theory of measure; Carathéodory extends and adapts the theory to generalize arc length in other spaces. The generalization to p-dimensional measure in q-dimensional space is then easily defined in the same way.

The first chapter is a fine introduction to the theory of metric outer measure, an approach that is often used to this day (for example, [9], Chapter I, [7], §1.3, [10], §12.7, [8], p. 126–130).

Note the notation $A \prec B$ for A is a subset of B; multiplication AB for intersection of sets; and addition $A + B$ for union of sets. The distance between two sets A, B is written $E(A, B)$; the letter E is for German *Entfernung*.

Constantin Carathéodory (1873–1950) was a German mathematician of Greek descent. He received his doctorate at Göttingen in 1904 working under Minkowski. He held positions at universities in Bonn, Hanover, Breslau, Göttingen, and Berlin. After a short time in Greece (Smyrna and Athens) from 1920 to 1924, he returned to Germany (Munich). His main work dealt with the calculus of variations, function theory, measure theory, and applied mathematics. [6]

FOUR

On the Linear Measure of Point Sets—a Generalization of the Concept of Length

Constantin Carathéodory

Introduction

The idea of applying the path–breaking and extremely fruitful theories that Mr. *Lebesgue* has developed for the volume of point sets[1] to the concept of length is very tempting: It suffices to find an additive set function, whose value for every rectifiable curve is exactly equal to the conventional length of the curve, and that also retains the properties of ordinary Lebesgue measure for unions and intersections of sets.

Not only is it apparent that carrying through the above program presents no difficulties worth mentioning, but also that one obtains a theory that, despite its great generality, is just as simple as the previous ones. The only difference is that one has to use more general properties of point sets than before in order to carry the proofs over.

I have therefore deemed it useful to begin with a purely formal theory of measurability. In this way, a definition of measurability will be laid down which is more general than the conventional one, insofar

---

[1] An excellent exposition of this theory can be found in Ch. *J. de la Vallée Poussin*, Cours d'Analyse Infinitésimale (Louvain & Paris T. 1 3ᵉ éd. 1914, T. II 2ᵉ éd. 1912), while a complete treatment of the newest fundamental results of *H. Lebesgue* is given in Sur l'intégration des fonctions discontinues (Ann. Éc. Norm. sup. (3) T. 27, 1910, p. 361–450).

as it extends to point sets of infinite outer measure, though in other aspects it appears much more constrained. This definition is much more intuitively appealing than the previous one: It allows all relevant theorems to be proved without subtle tricks and it is also *completely equivalent* to the normal definition, as will be shown in § 21.

In the second chapter I treat the concept of linear measurability for which it suffices to show that the quantity, denoted there as linear measure, which without exception is uniquely defined for any arbitrary point set, possesses the five properties of an outer measure as required by the formal theory. One could have proceeded in exactly the same way for the concept of conventional Lebesgue measurability.

Finally, it will be shown that the linear measure of a rectifiable curve equals the length of the curve, and the concept of a p–dimensional measure in a q–dimensional space will be defined.

Chapter One: Formal Theory of Measurability

1. In many theories of analysis "set functions" generally denoted as *outer measure* play a leading role. For each example of such an outer measure there exists a class of point sets, which one calls *measurable* with respect to the outer measure. Measurability is thus a *relative concept*. The outer measures are characterized by five properties. The last two will be postponed, as one can already deduce a number of important conclusions from the first three. Hence we suppose:

I. To each point set A of a q–dimensional space $\mathfrak{R}_q$ there is assigned a unique number μ^*A which can be zero, positive, or positively infinite, and which is called *the outer measure of A*.

II. For any subset B of A the following always holds

$$\mu^*B \leq \mu^*A.$$

III. If A is the set union of a finite or countably infinite sequence of point sets $A_1, A_2, \ldots$, the following always holds

$$\mu^*A \leq \mu^*A_1 + \mu^*A_2 + \cdots.$$

Obviously, this last inequality is meaningful only if the sum on the right hand side converges.

Definition. *A point set A is said to be measurable if for any arbitrary point set W of finite outer measure the following relation holds*[2]

(1) $$\mu^* W = \mu^* AW + \mu^*(W - AW).$$

The measure μA of A is then defined by

$$\mu A = \mu^* A.$$

Remark 1. In defining measurability, we could have dropped the requirement that W should have finite outer measure without changing the definition. Namely, if $\mu^* W = \infty$, then, by property III, one of the numbers $\mu^* AW$ or $\mu^*(W - AW)$ must be infinite, and (1) always reduces to an identity.

Remark 2. Since our definition does not require the measure μA of a measurable point set to be finite, one can draw conclusions about the measurability of point sets without having to determine whether these sets have finite outer measure.

2. Let A' be the complement of A, then the following holds for an arbitrary point set W

$$A'W = W - AW, \qquad W - A'W = AW.$$

From this immediately follows

Theorem 1. *The complement of a measurable point set is measurable.*

3. Let A, B be two measurable point sets and $D = AB$ their intersection. By the measurability of A, our definition of finite outer measure for an arbitrary point set W gives

(2) $$\mu^* W = \mu^* AW + \mu^*(W - AW),$$

and furthermore, if one sets $AW = W_1$,

$$BW_1 = DW, \qquad W_1 - BW_1 = AW - DW.$$

---

[2] We will denote here and in what follows the intersection of two point sets A and W by AW.

Since B is also assumed to be measurable, one can write

(3) $$\mu^* AW = \mu^* DW + \mu^*(AW - DW).$$

If, on the other hand, one sets $(W - DW) = W_2$, then

$$AW_2 = AW - DW, \qquad W_2 - AW_2 = W - AW,$$

and, using the measurability of A,

(4) $$\mu^*(W - DW) = \mu^*(AW - DW) + \mu^*(W - AW).$$

Comparison of (2), (3), and (4) gives

$$\mu^* W = \mu^* DW + \mu^*(W - DW).$$

It follows that the point set D is measurable. By using this conclusion to go from n to $n + 1$, one finally obtains the theorem

Theorem 2. *The intersection of finitely many measurable point sets is measurable.*

4. Let A, B be two measurable point sets and denote the complement of B by B', then Theorem 1 implies that the point set B' is also measurable and Theorem 2 gives that the point set

$$AB' = A - AB$$

is measurable as well. One obtains the theorem

Theorem 3. *If A, B are measurable point sets, then the point set $(A - AB)$ is also measurable.*

5. In the following theorem essential use will be made of properties II and III of the outer measure.

Given a countably infinite nested sequence of *measurable* point sets $A_1, A_2, \ldots$, so that

$$A_1 \succ A_2 \succ \cdots,$$

we shall denote their intersection by Ω. We wish to show that Ω is measurable.

Let W be a point set of finite outer measure; we set

$$W_n = A_n W, \qquad W_0 = \Omega W,$$

and observe that all the W_n's as well as W_0 have finite measure since they are subsets of W. Moreover, since for any n

$$W_n \succ W_{n+1} \quad \text{and } W_n \succ W_0,$$

one has

$$\mu^* W_n \geq \mu^* W_{n+1} \quad \text{and } \mu^* W_n \geq \mu^* W_0,$$

which immediately implies the existence of the limit

$$(5) \qquad \lim_{n=\infty} \mu^* W_n = \lambda,$$

having the property that

$$(6) \qquad \mu^* W_0 \leq \lambda.$$

Now observe that

$$(7) \qquad W - W_0 = (W - W_1) + (W_1 - W_2) + (W_2 - W_3) + \cdots$$

or, after rearranging,

$$W = W_0 + (W - W_1) + (W_1 - W_2) + (W_2 - W_3) + \cdots,$$

so that, by Property III

$$(8) \quad \mu^* W \leq \mu^* W_0 + \mu^*(W - W_1) + \mu^*(W_1 - W_2) + \mu^*(W_2 - W_3) + \cdots.$$

Now $A_1, A_2, \ldots,$ are measurable point sets, so one can write

$$(9) \qquad \mu^*(W - W_1) = \mu^*(W - A_1 W) = \mu^* W - \mu^* W_1,$$

$$(10) \quad \mu^*(W_{n-1} - W_n) = \mu^*(W_{n-1} - A_n W_{n-1}) = \mu^* W_{n-1} - \mu^* W_n,$$

and inserting this into (8) gives

$$\mu^* W \leq \mu^* W_0 + \mu^* W - \lambda,$$

from which it follows, using (6), that

$$(11) \qquad \mu^* W_0 = \lambda.$$

On the other hand, using (9) and (10), (7) yields

$$\mu^*(W - W_0) \leq \mu^*(W - W_1) + \mu^*(W_1 - W_2) + \cdots$$

$$\leq \mu^* W - \lambda,$$

whereby one concludes from $W - W_0 \succ W - W_n$ that

$$\mu^*(W - W_0) \geq \mu^*(W - W_n)$$

$$\geq \mu^*W - \mu^*W_n,$$

and since this holds for any n,

$$\mu^*(W - W_0) \geq \mu^*W - \lambda.$$

Using (11), one also has

$$\mu^*W = \lambda + \mu^*(W - W_0)$$

$$= \mu^*W_0 + \mu^*(W - W_0)$$

$$= \mu^*\Omega W + \mu^*(W - \Omega W).$$

It follows that the point set Ω is measurable.

6. Let $A_1, A_2, \ldots$, be an arbitrary countable sequence of measurable sets and let D be their intersection. The point sets

$$B_1 = A_1, \quad B_2 = A_1 A_2, \quad B_3 = A_1 A_2 A_3, \ldots,$$

form a nested nested sequence and are measurable by Theorem 2. The intersection of the above point sets, which we have seen is measurable, coincides with D.

We therefore have the following theorem

Theorem 4. *The intersection D of countably many measurable point sets $A_1, A_2, \ldots$ is measurable.*

7. Once again let $A_1, A_2, \ldots$, be an arbitrary finite or countable sequence of measurable point sets. We let V denote their union set.

Denote the complements of $A_1, A_2, \ldots$, by $A', A_2', \ldots$, and the complement of V by V'. Then V' is equal to the intersection of the point sets $A_1', A_2', \ldots$. By Theorem 1 all the A_n' are measurable, so V' is measurable by one of Theorems 2 or 4. Thus V is also measurable.

Theorem 5. *The union V of finitely many or countably infinitely many measurable point sets $A_1, A_2, \ldots$, is measurable.*

8. From Theorems 1 through 5 it follows that all point sets from a countable sequence

(12) $$A_1, A_2, A_3, \ldots,$$

of measurable point sets obtained by successive application of complements, intersections, or unions, are likewise measurable.

For example, let V_n be the union of $A_n, A_{n+1}, \ldots$, then their intersection

$$\bar{\alpha} = V_1 V_2 V_3 \ldots$$

is the so-called *limit superior* of the sequence (12), i.e., the set of points contained in infinitely many of the A_n.

On the other hand, let D_n be the intersection of $A_n, A_{n+1}, \ldots$, then the union set $\underline{\alpha}$ of all point sets D_n is called the *limit inferior* of the series (12). The point set $\underline{\alpha}$ consists of the points in the space which are contained in all but a finite number of the point sets (12). This yields the following theorem

Theorem 6. *Both the limit superior and the limit inferior of a sequence of countably many measurable point sets are measurable point sets.*

9. If the point sets in question have finite outer measure, then one can make the above theorems more precise.

Let A be a measurable point set and B an arbitrary point set both having finite outer measure. Denote their union set by V, then one has

$$V = A + (B - AB),$$

and, since A is measurable,

$$\mu^* V = \mu^* AV + \mu^* (V - AV)$$

$$= \mu A + \mu^* (B - AB)$$

$$= \mu A + \mu^* B - \mu^* AB.$$

If, in particular, $AB = 0$, then

$$\mu^* (A + B) = \mu^* A + \mu^* B$$

and we have a result that can easily be generalized to the sum of finitely many measurable point sets.

10.   Let $A_1, A_2, \ldots,$ be a countable sequence of measurable point sets that are pairwise disjoint. We denote the sum of these sets by S. Let

$$S_n = A_1 + A_2 + \cdots + A_n,$$

then S_n is measurable and by § 9

$$\mu S_n = \mu A_1 + \mu A_2 + \cdots + \mu A_n.$$

Now since $S \succ S_n$, for any n

(13) $\mu S \geq \sum_{k=1}^{n} \mu A_k.$

If, in the limit, the sum on the right hand side tends to infinity, then $\mu S = \infty$. However, if

$$\sum_{k=1}^{\infty} \mu A_k = \lambda$$

and λ is finite, then it follows by comparing (13) with property III of outer measure that $\mu S = \lambda$.

Theorem 7. *The measure of the sum of finitely many or countably many measurable pairwise disjoint point sets is equal to the sum of the measures of the sets.*

We can immediately state the following Theorems

Theorem 8. *If the measurable point sets*

$$A_1 \prec A_2 \prec \ldots$$

form a nested sequence and V is their union, then it is always true that

$$\mu V = \lim_{n=\infty} \mu A_n.$$

(Namely, one can write

$$V = A_1 + (A_2 - A_1) + (A_3 - A_2) + \cdots$$

and apply Theorems 3 and 7.)

Theorem 9. *Let the measurable point sets*

$$A_1 \succ A_2 \succ A_3 \cdots$$

be a nested sequence and let D be their intersection, then

$$\mu D = \lim_{n=\infty} \mu A_n,$$

whenever μA_1 is finite.

This holds since

$$D = A_1 - \sum_{k=1}^{\infty} (A_k - A_{k+1}).$$

However, this last theorem is not necessarily true when the equation $\mu A_k = \infty$ holds for every k, which can happen for special examples of measures.

11. From condition II of § 1 one can deduce that the measure of a union set V cannot be smaller than the measure of any of its components $A_1, A_2, \ldots$, and consequently that

$$\mu^* V \geq \text{upper bound of } (\mu^* A_1, \mu^* A_2, \ldots).$$

In exactly the same way one sees that, for the intersection

$$D = A_1 A_2 \ldots$$

the relation

$$\mu^* D \leq \text{lower bound of } (\mu^* A_1, \mu^* A_2, \ldots)$$

holds.

Using the notation of § 8 and assuming that the A_n are measurable, one has

$$\mu^* V_n \geq \text{upper bound of } (\mu^* A_n, \mu^* A_{n+1}, \ldots),$$

and by § 10, under the further constraint that at least one V_n have finite measure,

(14) $$\mu \bar{\alpha} \geq \lim \sup \mu A_n.$$

Similarly, one finds that if the A_n are measurable then

(15) $$\mu \underline{\alpha} \leq \lim \inf \mu A_n$$

must always hold.

If the limit inferior $\underline{\alpha}$ and the limit superior $\bar{\alpha}$ of a sequence coincide, then it is appropriate to speak of the *convergence* of the sequence and to introduce the notation

$$\lim_{n=\infty} A_n = \alpha .$$

If the union V of all the A_n has *finite* measure, then it follows from (14) and (15) that the sequence of numbers μA_n also converges and that

$$\lim_{n=\infty} A_n = \mu\alpha .$$

I have mentioned these last theorems merely for the sake of completeness, as they are important in the integration theory of Lebesgue.

12. The theory that we have developed up to now is so general that it is impossible, given a specific point set, to decide if it is measurable. To show the existence of measurable point sets, we introduce a fourth property of outer measure:

IV. *Let A_1, A_2 be two point sets a distance $\delta \neq 0$ apart, then*

$$\mu^*(A_1 + A_2) = \mu^*A_1 + \mu^*A_2 .$$

13. We shall make use of the following observation: Let H be an arbitrary point set in our space $\mathfrak{R}_q$, and let H consist solely of interior points and with complement K (which is necessarily closed) containing at least one point. Let n be a positive integer and let H_n denote all points in the space whose distance from the point set K is greater than $1/n$.

The point sets $H_1, H_2, \ldots$, and their complement sets $K_1, K_2, \ldots$, have the following properties:

a) H_n is a subset of H for all sufficiently large n and consists solely of interior points. The point set K_n, however, is closed and contains K.

b) The point sets $H_1, H_2, \ldots$ form a nested sequence, i.e.,

$$H_n \prec H_{n+1}.$$

Consequently

$$K_n \succ K_{n+1}$$

also holds.

c) Every point of H lies in at least one H_n, hence also in each successive one.

(Since K is closed, the distance between K and an arbitrary point of H is always $\neq 0$, so larger than one of the numbers $1/n$.)

d) The distance δ_n between K_n and H_{n-1} is non zero.

In particular, let P be a point in H_{n-1} and Q an arbitrary point which is closer to P than the distance $1/n(n-1)$. The distance $E(Q, K)$ between Q and the point set K is then given by

$$E(Q, K) \geq E(P, K) - E(P, Q)$$

$$> \frac{1}{n-1} - \frac{1}{n(n-1)} = \frac{1}{n}.$$

The point Q is then a point of H_n, and hence cannot be a point in K_n, i.e.,

$$\delta_n = E(H_{n-1}, K_n) \geq \frac{1}{n(n-1)}$$

is not zero.

14. Let B be an arbitrary point set of finite measure contained in H. We set

$$B_n = BH_n,$$

and I claim that

(17) $$\mu^* B = \lim_{n=\infty} \mu^* B_n.$$

First, it follows from $B_n \prec B_{n+1}$ and $B_n \prec B$ that the sequence of numbers $\mu^* B_n$ increases monotonically and stays less than $\mu^* B$. Consequently, the limit

(18) $$\lim_{n=\infty} \mu^* B_n = \lambda$$

exists, and

(19) $$\lambda \leq \mu^* B.$$

We set

(20) $$B = B_n + R_n,$$

(21) $$C_n = B_{n+1} - B_n,$$

and observe that

$$R_n = C_n + C_{n+1} + \cdots .$$

We then have

(22) $$\mu^* R_n \le \mu^* C_n + \mu^* C_{n+1} + \cdots .$$

Furthermore, by (21),

$$B_{n+1} = B_n + C_n \succ B_{n-1} + C_n$$

and so

(23) $$\mu^* B_{n+1} \ge \mu^* (B_{n-1} + C_n) .$$

Now one observes that B_{n-1} as a subset of H_{n-1} and C_n as a subset of K_n are at a distance which is at least equal to δ_n, hence $\neq 0$. One can then write (§ 12)

$$\mu^* (B_{n-1} + C_n) = \mu^* B_{n-1} + \mu^* C_n ,$$

and it follows from (23) that

$$\mu^* C_n \le \mu^* B_{n+1} - \mu^* B_{n-1} .$$

Substituting this into (22) gives

$$\mu^* R_n \le \sum_{p=0}^{\infty} (\mu^* B_{n+p+1} - \mu^* B_{n+p-1})$$

$$\le (\lambda - \mu^* B_{n-1}) + (\lambda - \mu^* B_n)$$

from which follows that

(24) $$\lim_{n=\infty} \mu^* R_n = 0 .$$

Now, Property III yields for all n

$$\mu^* B \le \mu^* B_n + \mu^* R_n$$

and, by (18) and (24), in the limit

$$\mu^* B \le \lambda .$$

Comparing this with (19) yields equation (17).

15. Let W be an arbitrary point set of finite measure. We set

$$HW = B, \quad H_n W = H_n B = B_n,$$

where H and H_n are defined as before. Since the point sets $(W - HW)$ and B_n are subsets of K and H_n, respectively, their distance is non zero so that

$$\mu^*(B_n + (W - HW)) = \mu^* B_n + \mu^*(W - HW).$$

Since

$$W \succ B_n + (W - HW)$$

it follows that

$$\mu^* W \geq \mu^* B_n + \mu^*(W - HW)$$

and, by (17), taking the limit $n = \infty$ yields

$$\mu^* W \geq \mu^* HW + \mu^*(W - HW).$$

On the other hand, by Property III, we have

$$\mu^* W \leq \mu^* HW + \mu^*(W - HW).$$

So

$$\mu^* W = \mu^* HW + \mu^*(W - HW).$$

We have thus proved that H is measurable, so the complement K of H is also measurable.

Theorem 10. *Point sets consisting solely of interior points are measurable, as are closed point sets.*

16. If one interprets the expression $q-dimensional$ $interval$ in the space $\Re_q$ to mean a point set consisting of all points with coordinates satisfying the conditions

$$|x_k - x_k^0| < h_k, \qquad (k = 1, 2, \ldots, q)$$

(where the x_k^0 denote arbitrary constants and the h_k denote arbitrary positive and non zero constants), then one can replace Condition IV by the following equivalent condition:

IVa. *Intervals are measurable point sets.*

Indeed, intervals consist solely of interior points, so Condition IVa is a consequence of IV, by Theorem 10.

On the other hand, it is known that one can consider an arbitrary point set consisting solely of interior points as the union set of countably many intervals. These point sets and their complement sets, the closed point sets, are measurable if IVa is assumed to hold.

Finally, if A_1, A_2 are a non zero distance apart, then there exists a closed point set B containing A_1 as a subset which has no point in common with A_2. However, since B is measurable, one has

$$\mu^*(A_1 + A_2) = \mu^*B(A_1 + A_2) + \mu^*((A_1 + A_2) - B(A_1 + A_2))$$

$$= \mu^*A_1 + \mu^*A_2.$$

Condition IV is then a consequence of IVa, and both conditions are equivalent, as was to be shown.

17. One can finally complete the theory if the following fifth property of outer measure is satisfied

V. *The outer measure μ^*A of an arbitrary point set is the lower bound of the measure μB of all measurable point sets B containing A as a subset.*

If A has finite outer measure, then, by Condition V there exists for each integer n at least one measurable point set B_n satisfying

$$A \prec B_n \quad \text{and} \quad \mu B_n \leq \mu^*A + \frac{1}{n}.$$

If one chooses such a point set for each n and sets

$$\bar{A} = B_1 B_2 \ldots,$$

then $\bar{A}$ is measurable (Theorem 4) and furthermore, since $A < \bar{A} < B_n$,

$$\mu^*A \leq \mu\bar{A} \leq \mu^*A + \frac{1}{n}$$

holds for all n, from which one concludes that $\mu\bar{A} = \mu^*A$. We have thus proved

Theorem 11. *If A is an arbitrary point set of finite outer measure, then there exist measurable point sets $\bar{A}$ which contain A and for which $\mu\bar{A} = \mu^*A$.*

18. Let C be a measurable point set contained in A, then $(\bar{A} - A)$ is a subset of the measurable point set $(\bar{A} - C)$. We then have

$$\mu^*(\bar{A} - A) \leq \mu(\bar{A} - C) = \mu\bar{A} - \mu C,$$

from which follows

(25) $$\mu C \leq \mu^* A - \mu^*(\bar{A} - A).$$

Furthermore, let B be a measurable point set satisfying the conditions

$$(\bar{A} - A) \prec B, \qquad \mu B = \mu^*(\bar{A} - A).$$

Such a point set must always exist by Theorem 11. The point set $\bar{A}B$ is measurable, and

$$(\bar{A} - A) \prec \bar{A}B \prec B,$$

hence also

$$\mu\bar{A}B = \mu^*(\bar{A} - A).$$

Now the point set $\underline{A} = \bar{A} - \bar{A}B$ is a measurable subset of A, and one has

(26) $$\mu\underline{A} = \mu^* A - \mu^*(\bar{A} - A).$$

The above quantity is, by (25), the upper bound of the measures of all measurable subsets of A. We shall call it the *inner measure* of A and denote it by $\mu_* A$.

Note that the quantity $\mu_*(\bar{A} - A)$ is completely independent of the particular construction which gave us the point set $\bar{A}$.

Finally, we see that there exist measurable subsets of A whose measure is equal to the inner measure of A.

19. If A is measurable, then it follows from

$$\mu\bar{A} = \mu A + \mu(\bar{A} - A)$$

that $\mu(\bar{A} - A) = 0$.

Conversely, if $\mu(\bar{A} - A) = 0$ holds and we let W be an arbitrary point set of finite outer measure, then

$$\bar{A}W = AW + (\bar{A} - A)W,$$

and consequently

$$\mu^* AW \leq \mu^* \bar{A} W \leq \mu^* AW + \mu^* (\bar{A} - A) W .$$

Since $(\bar{A} - A)$ has measure zero, the same is true of $(\bar{A} - A)W$, and one has

(27) $$\mu^* \bar{A} W = \mu^* AW .$$

In exactly the same way one proves from

$$(W - AW) = (W - \bar{A}W) + (\bar{A} - A)W$$

that

$$\mu^* (W - \bar{A}W) = \mu^* (W - AW) .$$

Now, by our construction, $\bar{A}$ is measurable, and so

$$\mu^* W = \mu^* \bar{A} W + \mu^* (W - \bar{A}W)$$

or, using (27) and (28)

$$\mu^* W = \mu^* AW + \mu^* (W - AW) ,$$

i.e., A is measurable.

Using definition (26) of the inner measure of A, one can express this result as

Theorem 12. *In order that a point set A of finite outer measure be measurable, it is necessary and sufficient that its outer and inner measures coincide.*

20. Let A, B denote two arbitrary disjoint point sets both having finite outer measure. We set $S = A + B$ and specify (by § 17, 18) four measurable point sets

$$A', B', \bar{S}, \underline{S},$$

with the following properties:

$$A' \succ A, \qquad \mu A' = \mu^* A,$$

$$\bar{S} \succ S, \qquad \mu \bar{S} = \mu^* S,$$

$$B' \prec B, \qquad \mu B' = \mu_* B,$$

$$\underline{S} \prec S, \qquad \mu \underline{S} = \mu_* S.$$

I will now construct two new point sets $\bar{A}, \underline{B}$ using the operations

$$\bar{A} = A'\bar{S} - A'B',$$

$$\underline{B} = B' + (\underline{S} - \underline{S}\bar{A} - \underline{S}B').$$

The point sets $\bar{A}, \underline{B}$ are measurable and have the properties

$$A \prec \bar{A} \prec A', \qquad B' \prec \underline{B} \prec B,$$

and so

$$\mu\bar{A} = \mu^*A, \qquad \mu\underline{B} = \mu_*B.$$

Furthermore,

$$\bar{A}\underline{B} = 0, \qquad \underline{S} \prec \bar{A} + \underline{B} \prec \bar{S},$$

from which one concludes that

$$\mu\underline{S} \leq \mu\bar{A} + \mu\underline{B} \leq \mu\bar{S},$$

or that

(29) $$\mu_*S \leq \mu^*A + \mu_*B \leq \mu^*S.$$

Relation (29) is a counterpart to the inequalities

(30) $$\mu^*A + \mu^*B \geq \mu^*S,$$

(31) $$\mu_*A + \mu_*B \leq \mu_*S,$$

the first of which was postulated in § 1 while the second can be inferred from the observation that the union of two measurable point sets contained in A, B respectively, must be measurable and a subset of S.

21. Let S be measurable, of finite measure, and subject to the equality

$$\mu^*A + \mu^*B = \mu S.$$

Then both A and B must be measurable.
    Indeed, it follows from (29) that

$$\mu^*A + \mu_*B = \mu S,$$

hence $\mu^*B = \mu_*B$, and exchanging B and A gives the analogous equation for A.
    This result shows the equivalence of the conventional definition of measurability with ours.

22.   We finally prove the following, often very useful generalization of our Theorem 8

Theorem 13. *Let $A_1, A_2, \ldots,$ be a nested sequence of sets*

$$A_1 \prec A_2 \prec \cdots,$$

and let A be their union, then

$$\mu^* A = \lim_{n=\infty} \mu^* A_n.$$

Since for any n

(32) $$\mu^* A \geq \mu^* A_n$$

a proof is only necessary when

$$\lim_{n=\infty} \mu^* A_n = \lambda$$

is finite.

In this case, we construct measurable point sets $A'_1, A'_2, \ldots,$ satisfying the conditions

$$A'_n \succ A_n, \qquad \mu A'_n = \mu^* A_n,$$

and then form the intersections

$$\bar{A}_n = A'_n A'_{n+1} \cdots.$$

One has $A_n \prec \bar{A}_n \prec A'_n$ and

$$\mu \bar{A}_n = \mu^* A_n.$$

The $\bar{A}_n$'s form a nested sequence and their union is a point set $\bar{A}$ which contains A as a subset.

We therefore have

$$\mu^* A \leq \mu \bar{A} = \lim_{n=\infty} \mu \bar{A}_n$$

and consequently,

$$\mu^* A \leq \lambda.$$

On the other hand, it follows from (32) that $\mu^* A \geq \lambda$, hence $\mu^* A = \lambda$ must hold. Q.E.D.

Chapter Two: Linear Measure

23.   Let A be an arbitrary point set in q–dimensional space ($q > 1$). We consider a finite or countable sequence of point sets $U_1, U_2, \ldots,$ satisfying the two conditions:

   a) The original set A is a subset of the union of the U_k's.

   b) The diameter[3] d_k of each U_k is smaller than a given positive number ϱ.

   We consider the *lower bound* of the sums

$$d_1 + d_2 + \cdots$$

of the diameters d_k for all sequences $U_1, U_2, \ldots,$ satisfying conditions a) and b) and we denote this lower bound, which, incidentally, can be $+\infty$, by $L_\varrho A$.

   If ϱ decreases, then the conditions which U_k must satisfy become more restrictive and the quantity $L_\varrho A$ cannot decrease. The following limit therefore *always* exists

$$\lim_{\varrho=0} L_\varrho A = L^* A .$$

   The quantity L^*A, which is uniquely defined for any point set, will be called the outer linear measure of A.

24.   One can impose constraints on the U_k by adding additional conditions to a) and b) without increasing the value of the linear outer measure L^*A.

   For example, if one replaces each U_k with the smallest closed convex point set containing it, then the diameters d_k and the quantities $L_\varrho A$ and L^*A remain unchanged.

   One can ask that the U_k be convex point sets consisting solely of interior points. To show that L^*A still remains unchanged in this case, I denote by $\Lambda_\varrho A$ the quantity corresponding to $L_\varrho A$ when the U_k are considered under these new conditions.

   Let $U_1, U_2, \ldots$ be an arbitrary sequence of point sets satisfying our previous conditions. Let each point of the Q_k be the center of a q–dimensional sphere of radius $\varrho/2^{k+1}$ and let C_k be the smallest convex

---

[3]It is well known that the diameter of a point set U_k is the upper bound of the distance between two points in U_k.

point set consisting solely of interior points and which contains the interior points of all of these spheres, hence also U_k.

The estimate

$$\delta_k \leq d_k + \frac{\varrho}{2^k} \leq 2\varrho$$

then holds for the diameter δ_k of the point set C_k, and one has

$$\sum_{k=1}^{\infty} \delta_k \leq \sum_{k=1}^{\infty} \left(d_k + \frac{\varrho}{2^k} \right),$$

from which it is easily concluded that

$$\Lambda_{2\varrho} A \leq L_\varrho A + \varrho.$$

However, by the definition of $\Lambda_\varrho A$, the inequality $\Lambda_\varrho A \geq L_\varrho A$ must hold, and one finally concludes that

$$\lim_{\varrho=0} \Lambda_{2\varrho} A = L^* A,$$

which is what was to be proved.

25. We now verify that L^*A has all the properties required of μ^*A in Chapter 1.

First, we observe that if B is a subset of A then every sequence $\{U_k\}$ covers the point set B whenever it covers A. One has that for each ϱ

$$L_\varrho B \leq L_\varrho A$$

and so, in the limit,

$$L^* B \leq L^* A.$$

Condition II is therefore satisfied.

26. Let $A_1, A_2, \ldots$, be a finite or countable sequence of point sets with union A, such that the infinite series

$$\sum_{k=1}^{\infty} L^* A_k$$

converges. By assumption, we can cover the point set A_k with a finite or countable number of point sets

$$U_{k1}, U_{k2}, \ldots$$

whose diameters $d_{k1}, d_{k2}, \ldots$ satisfy the conditions

$$d_{kn} \leq \varrho, \quad \sum_{n=1}^{\infty} d_{kn} \leq L_\varrho A_k + \frac{\varrho}{2^k} \leq L^* A_k + \frac{\varrho}{2^k}.$$

Now the union of all the U_{kn}'s covers our entire set A, and we have

$$L_\varrho A \leq \sum_{k,n} d_{kn} \leq \sum_{k=1}^{\infty} \left(L^* A_k + \frac{\varrho}{2^k} \right)$$

and finally

$$L_\varrho A \leq \sum_{k=1}^{\infty} L^* A_k + \varrho.$$

Therefore

$$L^* A = \lim_{\varrho=0} L_\varrho A$$

$$\leq \sum_{k=1}^{\infty} L^* A_k,$$

i.e., Condition III is satisfied.

27. Let A_1, A_2 be two point sets of finite measure a distance $\delta \neq 0$ apart.

We choose $\varrho \prec \delta/2$ and cover $(A_1 + A_2)$ with a finite or countable number of point sets U_k, whose diameters d_k satisfy the conditions

$$d_k < \varrho, \quad \sum_{k=1}^{\infty} d_k \leq L_\varrho(A_1 + A_2) + \varrho \leq L^*(A_1 + A_2) + \varrho.$$

Now let $U_1', U_2', \ldots$ be those of the sets U_k which contain a point of A_1 and so (due to the condition $\varrho \prec \delta/2$) contain *no* point of A_2. Let $U_1'', U_2'', \ldots$ be the remaining point sets. Denoting the diameters of U_k', U_k'' by d_k', d_k'', respectively, one observes that the union of the U_k' contains the point set A_1 and that the union of the U_k'' contains the point set A_2. This gives

$$L_\varrho A_1 \leq \sum_{k=1}^{\infty} d_k', \quad L_\varrho A_2 \leq \sum_{k=1}^{\infty} d_k'',$$

and consequently

$$L_\varrho A_1 + L_\varrho A_2 \leq \sum_{k=1}^{\infty} d'_k + \sum_{k=1}^{\infty} d''_k$$

$$\leq L^*(A_1 + A_2) + \varrho.$$

If ϱ converges to zero then one gets

$$L^*A_1 + L^*A_2 \leq L^*(A_1 + A_2),$$

and, by Condition III,

$$L^*A_1 + L^*A_2 \geq L^*(A_1 + A_2)$$

holds, so Condition IV is satisfied.

28.   Let L^*A once more be finite and let $\varrho_1, \varrho_2, \ldots$ denote an arbitrary sequence of positive numbers monotonically decreasing to zero.

For $k = 1, 2, \ldots$, we cover A with countably many point sets

$$U_{k1}, U_{k2}, \ldots$$

whose diameters $d_{k1}, d_{k2}, \ldots$ are all $\leq \varrho_k$, and which also have the property that

(33) $$\sum_{n=1}^{\infty} d_{kn} \leq L_\varrho A + \varrho_k \leq L^*A + \varrho_k.$$

As we saw in § 24, choosing closed point sets for the U_{kn}'s imposes no restrictions on generality. Thus, all the U_{kn}'s are linearly measurable (Theorem 10 of § 15).

As a consequence, the union V_k of $U_{k1}, U_{k2}, \ldots$ is, by Theorem 5, also linearly measurable, and the same applies to the intersection

$$\bar{A} = V_1 V_2 \ldots$$

of all the V_k's.

Now $A \prec \bar{A}$ so

(34) $$L^*A \leq L\bar{A},$$

where we have written $L\bar{A}$ instead of $L^*\bar{A}$ because $\bar{A}$ is linearly measurable. On the other hand, $\bar{A}$, as a subset of V_k, is covered by the sequence $U_{k1}, U_{k2}, \ldots$, so by (33) one has

$$L_{\varrho_k}\bar{A} \leq \sum_n d_{kn} \leq L^*A + \varrho_k$$

for every k, and in the limit

(35) $$L\bar{A} \leq L^*A.$$

Comparison of (34) and (35) shows that our Condition V is also satisfied.

The theory developed in Chapter One can be applied to linear measurability.

29. Now let γ be a *curve* without double points, i.e., a single valued continuous image of a closed line segment. The curve γ is closed and is therefore a measurable point set. I claim that the linear measure of γ is equal to the upper bound λ of the set of polygons inscribed in γ.

First we show that $L\gamma$ is not smaller than the length of the secant joining the endpoints P_1, P_2 of γ. We cover γ with a sequence of point sets $U_1', U_2', \ldots$, which we require to be *convex domains* such that every point of γ lies in the interior of at least one of these domains. The diameters d_k' of these domains satisfy the following conditions (§ 24)

$$d_k' \leq 2\varrho, \qquad \sum d_k' \leq L_\varrho\gamma + \varrho \leq L\gamma + \varrho.$$

Since γ is closed, one can assume by the well-known Borel–Lebesgue Theorem that one only needs a finite number of U_k'. We now specify a subset $\{U_1, U_2, \ldots, U_m\}$ of the U_k' in the following way: The domain U_1 contains the initial point P_1 of the curve in its interior. If Q_1 is the *last* point of the curve γ lying on the boundary of U_1, then U_2 will have this point in its interior. Likewise, U_3 will contain the last point Q_2 of γ lying on the boundary of U_2, etc. One obtains a polygon $P_1 Q_1 \ldots Q_{m-1} P_2$ consisting of finitely many segments inscribed in γ and whose length does not exceed the sum of the diameters $d_1, d_2, \ldots, d_m$ of the U_k, since the U_k's are different from each other by construction. In any case,

$$E(P_1, P_2) \leq \sum_{k=1}^m d_k \leq \sum d_k' \leq L\gamma + \rho,$$

and taking the limit $\varrho = 0$

$$E(P_1, P_2) \leq L\gamma.$$

One now denotes by $\mathfrak{P}$ an arbitrary polygon inscribed in γ, where $\gamma_1, \gamma_2, \ldots, \gamma_m$, are the successive segments of γ with endpoints the vertices of $\mathfrak{P}$. Since a finite number of points has linear measure zero, we have

$$L\gamma = L\gamma_1 + L\gamma_2 + \cdots + L\gamma_m,$$

and it follows from the above results that the length of the polygon $\mathfrak{P}$ does not exceed the quantity $L\gamma$. One therefore has

$$L\gamma \geq \lambda.$$

We now partition γ into finitely many closed subcurves $\gamma_1, \gamma_2, \ldots, \gamma_m$, with diameters $\leq \varrho$, such that any two neighboring ones have only one point in common, and use these γ_k instead of our point sets U_k. On each γ_k there exist two points whose distance equals the diameter d_k of γ_k. If one lets these points be the vertices of an inscribed polygon, then the length of this polygon is always less than the sum of all the diameters d_k, and we have that for every ϱ

$$L_\varrho \gamma \leq \lambda$$

so $L\gamma \leq \lambda$, and the equality of the linear measure and the length of γ is proved.

30. *A point set [in $\mathfrak{R}_n$ $(n > 1)$] of finite linear outer measure possesses ordinary Lebesgue measure zero.*

By assumption, one can cover the point set A with countably many point sets U_k (by assumption) whose diameters are all smaller than ϱ, so that

$$\sum d_k \leq L^*A + 1.$$

Now the diameter of the projection of each U_k onto *any* axis is never larger than d_k, and one can therefore enclose each U_k in the interior of an n–dimensional cube of edge–length d_k, so that the conventional outer measure m^*A satisfies the condition

$$m^*A \leq \sum_{k=1}^{\infty} d_k^n$$

$$\leq \varrho^{n-1} \sum_{k=1}^{\infty} d_k$$

$$\leq \varrho^{n-1}(L^*A + 1),$$

which is a quantity that goes to zero with ϱ.

31. One can develop a theory of p-dimensional measure in a q-dimensional space in a completely analogous way. *In order to do so, retain conditions a) and b) that the U_k must satisfy in the definition of § 23, but, with regard to lower bounds, replace the ordinary diameters d_k of the point sets U_k with the "p-dimensional diameters" $d_k^{(p)}$ of these point sets.*

To define $d_k^{(p)}$, observe that one obtains d_k by orthogonally projecting the smallest convex point set C_k which contains U_k onto an arbitrary line. In every case this produces a segment, and one specifies an upper bound of the length of these segments for all possible orientations of the line in the space.

However, it is possible to orthogonally project the point set C_k onto a p–dimensional linear manifold. One obtains a projection which is itself a convex point set whose *volume* depends on the "orientation" of our p–dimensional manifold in q–dimensional space. The upper bound of this volume for all possible p–dimensional manifolds will be the p–dimensional diameter $d_k^{(p)}$ of our point set U_k.

The definition of measurability that is given here is the essential ingredient that was new. Hewitt [8], p. 127 remarks: "How Carathéodory came to think of this definition seems mysterious, since it is not in the least intuitive." Carathéodory's criterion for measurability is the key to the treatment of measures like the Hausdorff measure, where every open set has infinite measure, so the outer measure of a set A cannot be obtained as the infimum of the measures of the open sets $U \supseteq A$.

We do have some of Carathéodry's own thoughts on this definition: on his copy of the preceding selection, Carathéodory added a handwritten note, which is a commentary on the new definition of measurable sets:

Borel and *Lebesgue* (as well as *Peano* and *Jordan*) assigned an outer measure m^*A and an inner measure m_*A to every point set A. If A is bounded and I is an interval containing A, then the following has to hold:

$$m^*A + m_*(I - A) = m_*A + m^*(I - A) = mI.$$

The point set A was called measurable if $m_*A = m^*A$, i.e., if one has

(1) $$m^*A + m^*(I - A) = mI.$$

Now I proved the following theorem in July 1914:

If A is measurable in the sense of Borel-Lebesgue then the following holds for any point set X, measurable or not:

(2) $$m^*X = m^*AX + m^*(X - AX).$$

If one takes (2) as the definition of measurability, then, according to the Borel-Lebesgue theory, no measurable sets are lost, although the class of measurable sets appears to be more constrained by condition (2) than by definition (1).

The new definition has great advantages:

1. It can be used for the linear measure.

2. It holds for the Lebesgue case, even if $m^*A = \infty$.

3. The proofs of the principal theorems of the theory are incomparably simpler and shorter than before.

4. The main advantage, however, is that the new definition is independent of the concept of an *inner measure*.

An independent theory of inner measure involves certain difficulties.

See *Carathéodory*, Vorlesungen über reele Funktionen, p. 338–341. A. *Rosenthal*, Beiträge zu Carathéodory's Meßbarkeitstheorie, Gött. Nachr. 1916. A thorough theory of inner measure was first provided by *J. Ridder*, Acta Mathematica [Volume 73 (1941), p. 131–173].

Outer measures are often used today. It is customary to add an axiom:

I'. The empty set has outer measure zero, $m^*\emptyset = 0$.

This will rule out the set function identically $+\infty$. Modern terminology uses the term **outer measure** for a set function satisfying I', II, III. An outer measure satisfying IV is called a **metric** outer measure. An outer measure satisfying V is called a **regular** outer measure. —Ed.

Bibliography

[1] Ch. J. de la Vallée Poussin, *Cours d'Analyse Infinitésimal*, Louvain & Paris, 1914.

[2] H. Lebesgue, *Sur l'intégration des fonctions discontinues*, Ann. Éc. Norm. sup. **27** (1910), 361–450.

[3] C. Carathéodory, *Vorlesungen über reele Funktionen*, Leipzig and Berlin, 1918.

[4] A. Rosenthal, *Beiträge zu Carathéodory's Meßbarkeitstheorie* Nachr. Ges. Wiss. Göttingen (1916).

[5] J. Ridder, *Maß- und Integrationstheorie in Strukturen*, Acta Mathematica **73** (1941), 131–173.

[6] David Abbott, editor, *The Biographical Dictionary of Scientists: Mathematicians*, Peter Bedrick Books, New York, 1985.

[7] D. L. Cohn, *Measure Theory*, Birkhäuser, Boston, 1980.

[8] E. Hewitt and K. Stromberg, *Real and Abstract Analysis*, Springer-Verlag, New York, 1965.

[9] C. A. Rogers, *Hausdorff Measures*, Cambridge University Press, 1970.

[10] H. L. Royden, *Real Analysis*, Second Edition, MacMillan, 1968.

This selection is the foundation of the theory of fractional dimension, now known as Hausdorff dimension. In Selection 4, Carathéodory shows how to define p-dimensional measure for a set in q-dimensional space. Hausdorff adapts that definition so that it makes sense even if p is not an integer. He goes on with what he calls "a small contribution".

Note the usual AB and $A + B$ for intersection and union of sets, respectively. If A is a subset of B, then he says B is **over** A and A is **under** B. A colon denotes division, $a : b = a/b$.

Felix Hausdorff (1868–1942) was born in Breslau, Germany (now Wroclaw, Poland). He graduated in 1891 from Leipzig University. His writings for the next seven years dealt with philosophy and literature, as well as mathematics. He had a considerable reputation as a poet, under the pseudonym Paul Mongré. Finally, in 1902 he turned permanently to mathematics; he held positions at Leipzig, Bonn, Greifswald, and Bonn again. The selection reprinted here was written in Greifswald, on the Baltic Sea. Hausdorff's work dealt mainly with set theory and topology. His book, *Grundzüge der Mengenlehre* [6], is a milestone in both of these fields. The abstract formulation of topology ("Hausdorff space") is one of the most important developments of Twentieth Century mathematics.[1] [7]

With the renewed interest in fractals today, it seems to me that Hausdorff is a good candidate for a full-fledged biography.[2]
—Ed.

---

[1] "If any one book marks the emergence of point set topology as a separate discipline, it is Hausdorff's *Grundzüge*. It is interesting to note that although it was the arithmetization of analysis that began the train of thought that led from Cantor to Hausdorff, in the end the concept of number is thoroughly submerged under a far more general point of view. Moreover, although the word 'point' is used in the title, the new subject has as little to do with the points of ordinary geometry as with the numbers of common arithmetic. Topology has emerged in the twentieth century as a subject that unifies almost the whole of mathematics, somewhat as philosophy seeks to coordinate all knowledge. Because of its primitiveness, topology lies at the basis of a very large part of mathematics, providing it with an unexpected cohesiveness." —C. B. Boyer [[8], p. 622]

[2] Addendum: [10], [11], [12].

FIVE

Dimension and Outer Measure

by Felix Hausdorff at Greifswald

Mr. Carathéodory has defined an exceptionally simple and general measure theory[1] that contains Lebesgue's theory as a special case, and which, in particular, defines the p–dimensional measure of a point set in a q–dimensional space. In this paper we shall add a small contribution to this work. After introductory considerations, which generalize the Carathéodory length measure in an obvious way and present an overview of the many analogous concepts of measure, we introduce an explanation of p–dimensional measure which can immediately be extended to non–integer values of p and suggests the possibility of the existence of sets of *fractional dimension,* and even of sets whose dimensions fill out the scale of positive integers to a more refined, e.g., logarithmic scale. In this way, the dimension becomes a sort of characteristic measure of graduality similar to the "order" of convergence to zero, the "strength" of convergence, and related concepts. As is well known, Mr. Fréchet[2] has defined, in an essentially different way, dimension-types that interpolate the sequence of natural numbers. The proof that there exist sets that have exactly a given dimension (i.e., have a corresponding measure which is neither zero nor infinite), is not, however, as natural as our concept of dimension. In this respect we will confine ourselves to the simplest examples, namely to certain perfect linear sets which are nowhere dense (these

---

[1]C. Carathéodory, Über das lineare Maß von Punktmengen–eine Verallgemeinerung des Längenbegriffs, Gött . Nachr. 1914, pp. 404-426. The somewhat modified presentation in Vorlesungen über reelle Funktionen (Teubner 1918) omits the specific portion of the theory which this work addresses, and hence does not need to be considered.

[2]M. Fréchet. Les dimensions d'un ensemble abstrait, Math. Ann. 68 (1910), p. 145–168

typically exhibit abnormal behavior and appear here from a new perspective) and to planar and spatial sets arising from them by taking products.

The assumptions of Carathéodory's measure theory are:

I. Every point set A of a q-dimensional Euclidean space is assigned a unique number $L(A)$ with $0 \leq L(A) \leq \infty$.

II. If B is a subset of A then $L(B) \leq L(A)$.

III. If A is the sum of finitely many or countably many sets A_1, $A_2, ...,$ then $L(A) \leq L(A_1) + L(A_2) + \cdots$.

IV. If A and B are two sets a positive distance apart, then

$$L(A + B) = L(A) + L(B).$$

V. $L(A)$ is the lower bound of $L(B)$ as B runs through the measurable sets over A, i.e., containing A.

A set A is said to be *measurable* with respect to L if for every set W the equation

$$L(W) = L(AW) + L(W - AW)$$

holds.

Every set function $L(A)$ satisfying these five conditions is called an *outer measure*. It corresponds to a system of measurable sets and to a measure theory having the properties developed by Mr. Carathéodory. On the basis of the fourth property, cubes in q-dimensional space, closed sets, open sets or regions, and *Borel sets* (arising from open sets by repeated application of unions and intersections over sequences of sets) are always measurable.

If $L(A)$ only satisfies the first four conditions, then one defines $M(A)$ or $N(A)$ to be the lower bound of $L(B)$ for measurable (relative to L) or Borel sets B containing A, respectively. Both these functions then satisfy all five conditions and are therefore outer measures. We omit the simple proof of this remark, as we shall make no further use of it in the following.

We will naturally confine ourselves to those outer measures which are identical for congruent sets. Moreover, if $L(B) = m^p L(A)$ whenever B is similar to A in proportion $m : 1$, then we call L an outer measure *of dimension p*.

If for any set A two outer measures $L(A), M(A)$ are simultaneously zero or positive or infinite, then we call them *measures of equal order*. In particular, this holds if there are two positive constants h, k such that

$$hL(A) \leq M(A) \leq kL(A)$$

always holds. If $M(A)$ is always zero whenever $L(A)$ is finite (or if $L(A)$ is always infinite, whenever $M(A)$ is different from zero), then L is said to be *of lower order* and M *of higher order*. In particular, this is the case if for any positive ε

$$M(A) \leq \varepsilon L(A).$$

2. Let $\mathfrak{U}$ be a system of bounded point sets U in a q-dimensional space having the property that one can cover any set A with a finite or countable number of the sets U *having arbitrarily small diameters* $d(U)$. Let a finite, nonnegative number $l(U)$ be assigned to every set U. Then, given a positive ϱ, we cover A by[3] $\mathfrak{S} U_n$ where $d(U_n) < \varrho$. The lower bound of the sums $\sum l(U_n)$ arising from all possible sequences of sets will be denoted by $L_\varrho(A)$, and it can be either zero, positive, or infinite. As ϱ decreases, the constraints become stricter. L_ϱ does not decrease and has a finite or infinite limit,

$$L(A) = \lim_{\varrho \to 0} L_\varrho(A)$$

$$= \liminf_{\varrho \to 0} \sum l(U_n)$$

$$\geq L_\varrho(A).$$

This set function satisfies the first four of Carathéodory's conditions without further constraints on $l(U)$. If the U are Borel sets then this set function satisfies the fifth condition as well.

The proof will be omitted as it is almost word for word the same as Carathéodory's proof for linear measure. As for the fifth condition, note that for any set A there exists a set W containing it with $L(W) = L(A)$ and of the form: W is the intersection of countably many sets V, which themselves are sums of countably many sets U. If the U's are Borel sets, then W is also a Borel set, in particular, it is a measurable Borel set, and the fifth condition is satisfied.

---

[3]Here $\mathfrak{S} U_n$ denotes the union $\bigcup U_n$, although elsewhere in the paper it is denoted $\sum U_n$.– *Ed.*

3. Let $\mathfrak{V}$ be a second system of sets such as $\mathfrak{U}$ and to each set V of this system assign a finite number $m(V) \geq 0$. As above, we define the numbers $M_\varrho(A)$, $M(A)$ and assume that the following relationship exists between these systems and the given functions l, m:

α) For every set U and for every positive ε there exists a set V over U such that

$$d(V) < rd(U) + \varepsilon, \qquad m(V) < kl(U) + \varepsilon,$$

where r, k are two positive constants.

One readily concludes that

$$M_{r\varrho}(A) \leq kL_\varrho(A), \qquad M(A) \leq kL(A)$$

for every set A. A special case of this (with $r = k = 1$) occurs when $\mathfrak{U}$ is a subsystem of $\mathfrak{V}$ and $l(U) = m(U)$. One then has $M(A) \leq L(A)$.

If *each* of the two systems has such a relationship with the other, then the following holds for two positive constants h, k

$$hL(A) \leq M(A) \leq kL(A),$$

and both set functions are of the same order. Moreover, if $h = k$, $M(A)$ is proportional to $L(A)$, and $\mathfrak{V}$ consists of Borel sets, then the set function $L(A)$ which is based on $\mathfrak{U}$ also satisfies the fifth Carathéodory condition and for any set A there exists a Borel set B containing it with $L(B) = L(A)$.

4. The function $l(U) \geq 0$, which, by the way, we can think of as defined for any bounded set, may be one of the following properties:

A) For the smallest closed set U_α containing U we have

$$l(U_\alpha) = l(U).$$

B) If one constructs a sphere of radius β around every point of U and if U_β denotes the set of interior points of these spheres, then one has for arbitrary $\varepsilon > 0$ and sufficiently small β

$$l(U_\beta) < l(U) + \varepsilon.$$

In case A) we say that the function $l(U)$ is *closeable*, and in case B) that it is *continuous*.

Let $\mathfrak{U}$ again be a set system of known type. We understand $\mathfrak{V}$ to be either the system of closed sets U_α or the system of open sets U_β (where β can take on all positive values or a set of numbers with accumulation point 0). Since the diameter $d(U)$ is also a closeable and continuous set function, one can assign to a given U and ε a $V = U_\alpha$ or $V = U_\beta$ such that either

$$d(V) = d(U), \qquad l(V) = l(U),$$

or at least

$$d(V) < d(U) + \varepsilon, \qquad l(V) < l(U) + \varepsilon.$$

This is a relationship of the form α) and it implies that

$$M(A) \leq L(A),$$

where the outer measure M refers to the system $\mathfrak{V}$ and the same function $l(V)$.

On the other hand, if $\mathfrak{U}$ is the system of all bounded sets, so that $\mathfrak{V}$ is a subsystem of $\mathfrak{U}$, then

$$L(A) \leq M(A)$$

and $L(A) = M(A)$ is an outer measure which also satisfies the fifth Carathéodory condition assuring that for any set A there exists a Borel set B over A with $L(B) = L(A)$. Consequently, we have the theorem

If $l(U) \geq 0$ is a closeable or continuous set function defined for bounded U and if A is enclosed in finitely many or countably many bounded sets U_n with diameters $< \varrho$, then

$$L(A) = \liminf_{\varrho \to 0} \sum l(U_n)$$

is an outer measure.

5. Let the function $l(U)$ be *continuous* and *monotonic*, i.e., if U is a subset of V, then $l(U) \leq l(V)$.

Then the function $l(U)$ is also closeable, as U_α is the intersection of sets U_β.

Let $\mathfrak{U}, \mathfrak{V}$ be two set systems of the following type:

Between U and U_β there always lies a V and between V and V_β there always lies a U.

Since the diameter $d(U)$ is continuous and monotone, for every U and positive ε one can find a V such that

$$d(V) < d(U) + \varepsilon, \qquad l(V) < l(U) + \varepsilon,$$

and consequently $M(A) \leq L(A)$. Conversely, $L(A) \leq M(A)$, hence the set functions $L(A), M(A)$ coincide (with respect to $\mathfrak{U}, \mathfrak{V}$ and the function l).

This holds, for example, if the U are closed and the V are open spheres in q–dimensional space. The $L(A), M(A)$ based on this construction are outer measures over the function l without constraint, since one is dealing with Borel sets and only $l \geq 0$ is required. However, it is not possible in general to infer their equivalence.

The same holds true if one replaces spheres with cubes, parallelopipeds, or q–dimensional simplices. For continuous monotonic functions $l(U)$ it makes no difference whether these figures are chosen to be closed or open (with or without boundaries).

6. We now give some examples of monotonic continuous functions $l = l(U)$.

A) The diameter d, i.e., the upper bound of the distances between pairs of points $u_0 u_1$ belonging to U.

B) The upper bound Δ of the surface area of a triangle $u_0 u_1 u_2$, whose vertices lie on U.

C) The upper bound Δ_p of the p–dimensional volume of a simplex $u_0 u_1 .... u_p$, whose vertices lie on U.

Of these examples, d and Δ are the special cases $p = 1, 2$. An outer measure with respect to $l = \Delta_p$ has dimension p. An outer measure with respect to Δ_q is not only invariant under length–preserving transformations, but also under affine volume preserving transformations of the q–dimensional space.

D) The lower bound of the elementary geometry q–dimensional volume (in the sense of elementary geometry) of a sphere (or cube, parallelopiped, or simplex), which contains the set U.

E) The lower bound of the volume sum of *finitely* many cubes (or the like), enclosing U, i.e., the outer *volume* of U (in the sense of Jordan–Peano, not the outer Lebesgue measure).

F) The q–dimensional volume (in the Jordan or Lebesgue sense) of the smallest *convex* set U_c over U.

G) The upper bound of the p-dimensional volume of the orthogonal projection of U_c onto a p-dimensional plane. Carathéodory calls this the p-dimensional diameter of U. It is then apparent that A) and F) are the special cases $p = 1, p = q$.

The proof that these set functions are continuous reduces to elementary geometry except for the last two cases, which require detailed considerations of convex sets. Moreover, it suffices to show the easily provable property of closeability.

Additionally, one notes that every nonnegative, continuous, or monotonic function $\lambda(t)$ of $t \geq 0$, combined with a nonnegative, continuous, or monotonic set function $l(U)$, gives rise to a new set function $\lambda\big(l(U)\big)$. One can also combine other set functions in an analogous way and, if the choice of the system $\mathfrak{U}$ is left open, then there is a wide range of possibilities for defining outer measures in the sense of Carathéodory. Besides the considerations below which confine themselves to functions of $d(U)$, we give a further example of case B) or D): If one covers the planar set A with triangles Δ_ϱ of diameters $< \varrho$, then an outer measure (of dimension 2/3) is obtained by

$$L(A) = \liminf_{\varrho \to 0} \sum \Delta_n^{1/3},$$

which, according to a comment of Mr. Blaschke [4] leads to an "affine length" for convex curve segments

$$\int k^{1/3} ds$$

(k curvature, s arc length).

7. For $p = 1, 2, \ldots$, one obtains the p-dimensional outer measure of Carathéodory by choosing the p-dimensional diameter given in 6.G) for the set function $l(U)$ of the Theorem at the end of Section 4. Since this definition requires a substantial amount of knowledge about the theory of convex sets, a simplification is desirable. If, for instance, one takes spheres instead of the set U, then one always obtains a p-dimensional outer measure L_p (in the sense of Section 1) as follows.

---

[4] W. Blaschke, Über affine Geometrie III. Eine Minimumeigenschaft der Ellipse (Leipz. Berichte 69 (1917), p. 12).

Definition 1. One covers A with finitely many or countably many spheres K_n of diameters $d_n < \varrho$ and constructs

$$L_p(A) = \liminf_{\varrho \to 0} c_p \sum d_n^p,$$

where[5]

$$c_p = \pi^{p/2} : 2^p \Pi\left(p/2\right)$$

is the volume of a p–dimensional sphere of diameter 1.

Actually, we do not need to consider the p–dimensional diameter at all as L_p is certainly an outer measure by the argument of Section 2, i.e., since spheres are Borel sets. L_p remains an outer measure when the spheres are replaced by arbitrary bounded sets, since $d(U)^p$ is a continuous set function. Furthermore, the definition could be modified by using cubes, parallelopipeds, or simplices instead of spheres but this might result in a modification of the factor c_p. Since $d(U)^p$ is also monotonic, whether these sets are open or closed is not an issue. In addition, all these outer measures are of the same order, since one can enclose any set of a given type with diameter d in a set of another type of diameter rd, where r is a constant factor. For example, one can enclose a bounded set of diameter d of a q–dimensional space inside a sphere of diameter $2d$, and even[6] inside one of diameter $\left(\frac{2q}{q+1}\right)^{1/2} d$.

Let us confine ourselves to Definition 1 and show that in the cases $p = 1$, $p = 2$, $p = q$, it gives the usual expressions for length, area, and volume, at least for the simplest sets.

α) For every set A of q–dimensional space, $L_q(A)$ is exactly outer Lebesgue measure $m(A)$.

On the one hand, every cover of A by spheres yields a sum of volumes $\geq m(A)$, hence $L_q(A) \geq m(A)$. On the other hand, it is possible to use the known covering theorems of Lebesgue and Vitali to enclose A in spheres of arbitrarily small diameters such that the sum of their volumes exceeds $m(A)$ by an arbitrarily small amount, hence $L_{q\varrho}(A) \leq m(A)$ and $L_q(A) \leq m(A)$.

β) For a simple curve A, $L_1(A)$ is exactly its length $\lambda(A)$, i.e., the upper bound of lengths of polygons inscribed in A.

One has to enclose A in spheres K_n of diameter $d_n < \varrho$ and to construct

$$L_1(A) = \liminf_{\varrho \to 0} \sum d_n.$$

---

[5]Here $\Pi(x)$ is the generalized factorial, $\Pi(x) = x! = \Gamma(x+1) = \int_0^\infty t^x e^{-t} dt$.—Ed.

[6]H. Jung, Über die kleinste Kugel, die eine räumliche Figur einschließt, Diss. Marburg 1899 = Journal f. Math. 123 (1901) p. 241–257.

The proof that $L_1 \geq \lambda$ follows Carathéodory's proof exactly. Actually, one can say that it is contained in Carathéodory's proof since his L_1, which uses all bounded sets U, is smaller or equal to our proof which uses spheres. To prove that $L_1 \leq \lambda$, where λ can be taken to be finite, one decomposes A into segments of lengths $\lambda_n < \varrho$ and encloses each of them in a sphere of diameter λ_n. (A segment of length λ is entirely contained in a sphere of radius $\lambda/2$, whose center is the center point of the segment.) We then have $L_{1\varrho} \leq \sum \lambda_n = \lambda, L_1 \leq \lambda$.

γ) For the surface element

$$F: \qquad a \leq x \leq A, \quad b \leq y \leq B, \quad x = f(x, y),$$

with *continuous* derivatives $p = f_x$, $q = f_y$, one has that $L_2(F)$ is given by the integral

$$\int_a^A \int_b^B \sqrt{1 + p^2 + q^2} \, dx \, dy.$$

We only give a sketch of the proof. One has $L_2(F) = L = \lim_{\varrho \to 0} L_\varrho$ and L_ϱ (actually $L_{2\varrho}$) is the lower bound of $\pi \sum \varrho_n^2$ if F is covered by spheres of radii $\varrho_n < \varrho$. One now partitions the rectangle

$$R: \qquad a \leq x \leq A, \qquad b \leq y \leq B,$$

into subrectangles where p, q have arbitrarily small oscillation. We first assume that R itself is such a subrectangle, draw the tangent plane at a point s_0 in F and project each point s of F vertically (parallel to the z–axis) to a point t of the tangent plane, so that a planar rectangle T is produced. We then let T^* be the vertical projection of F on this same surface. We can then assume (if we denote the elementary contents of these sets by the same letters):

that T and T^* differ from F (the above integral) by less than εR,

and that the distortion quotient $ss_1 : tt_1$ of the first projection is smaller than $1 + \varepsilon$.

One then covers T with circles having centers at t_n (in T itself) and radii $\tau_n < \frac{\varrho}{1+\varepsilon}$ such that

$$\pi \sum \tau_n^2 < T + \delta,$$

where $\delta > 0$ is arbitrary. The spheres with centers at s_n (the surface points corresponding to t_n) and radii $(1 + \varepsilon)\tau_n < \varrho$ cover the surface F and we find that

$$L_\varrho \leq \pi \sum (1 + \varepsilon)^2 \tau_n^2 < (1 + \varepsilon)^2 (T + \delta),$$

which becomes $L \leq (1 + \varepsilon)^2 T$ as one lets ϱ and δ go to 0.

On the other hand, if one covers F by spheres of radii $\varrho_n < \varrho$ with

$$\pi \sum \varrho_n^2 < L_\varrho + \delta,$$

then vertical projection onto the tangent surface gives circles of the same radii enclosing T^*, hence

$$T^* \leq \pi \sum \varrho_n^2,$$

i.e., $T^* \leq L$. Summarizing,

$$F - \varepsilon R < L < (1 + \varepsilon)^2 (F + \varepsilon R).$$

Furthermore, one observes that $L(F)$ does not change if one removes the four boundary lines corresponding to the edges of the rectangle, since they have Lebesgue area measure 0, being plane curves, and so also have L–measure 0 (see Section 7.α).

We now return to the original interpretation of R and partition it in subrectangles R_n as described above, and observe that, according to the last remark $L = \sum L_n$. The above inequality is also obtained for the entire rectangle, and therefore $L = F$.

8. Now that Definition 1 has justified its existence at least to some extent, it is tempting to extend it to noninteger positive values of p (which is excluded in Carathéodory's definition). The only question is whether this is trivial, in the sense that all sets have outer measure 0 or ∞. The problem therefore arises to construct sets A for which $0 < L_p(A) < \infty$ and that are exactly of dimension p in this sense. This problem can be extended further. In the definition of L_p we have seen that, apart from the numerical constant c_p, the pth power of the diameter d has played the role of the set function l: Can one possibly replace this with a function such as $d^p \log \frac{1}{d}$, which tends to 0 as $d \to 0$ slower than d^p, but faster than any smaller power, and in this sense construct sets whose dimensions fall in between the scale of positive numbers? We shall make use of the freedom emphasized in Section 6 and assign to every *function $\lambda(x)$ that is positive, continuous, increasing with x, and that converges to zero with x* an outer measure $L(A)$ in the following way.

Definition 2. Cover A with a finite or countable number of spheres K_n, whose diameters are $d_n < \delta$, then

$$L(A) = \lim_{\varrho \to 0} L_\varrho(A),$$

where $L_\varrho(A)$ is the lower bound of the sums $\sum \lambda(d_N)$.

The constraints that we have imposed on $\lambda(x)$ are not absolutely necessary, but in the sense of Sections 4 and 5 they are useful for the purpose of excluding zero or infinite $L(A)$.

If $0 < L(A) < \infty$, then we say that A is of dimension $[\lambda(x)]$. We will denote the dimension $[x^p]$ by (p). A set of this dimension is of dimension p in the sense of Section 1, but not vice–versa, see below.

Let $\lambda(x), \mu(x)$ be two functions corresponding to the outer measures $L(A), M(A)$ respectively, and let L, M be of the same order. We then say that the dimensions $[\lambda], [\mu]$ are equal. Every set of dimension $[\lambda]$ is also of dimension $[\mu]$ and vice–versa. If M is of higher order than L, then we call $[\mu]$ the higher and $[\lambda]$ the lower dimension. A set of a given dimension has measure 0 for any higher dimension, and it has measure ∞ for any lower dimension. Naturally it is not necessary for two dimensions to be comparable in this sense.

Obviously $L(A)$ only depends on the behavior of $\lambda(x)$ in an arbitrarily small neighborhood of the point $x = 0$, i.e., the dimension appears to be a characteristic of graduality[7] such as the "order" of a function tending to 0 and related concepts. If $\lambda(x) \leq \mu(x)$ for sufficiently small x, then $L(A) \leq M(A)$ holds for any set A. If $\mu = c\lambda$, $(c > 0)$, then $M = cL$. If μ/λ lies between two positive bounds, then the dimensions $[\lambda], [\mu]$ are equal. If $\mu/\lambda \to 0$ for $x \to 0$, then $M \leq \varepsilon L$ for every positive ε, then M vanishes for finite L so $[\mu]$ is the higher dimension and $[\lambda]$ is the lower dimension. In particular, (q) is the higher dimension and (p) the lower dimension whenever $q > p$. A function like $e^{-1/x}$, which goes to zero faster than any x^p, results in an infinitely large dimension, which cannot be realized by any Euclidean point set, whereas infinitely small dimensions can be realized, as will be seen in the following. The functions of the well known logarithmic scale[8]

$$\lambda(x) = x^{p_0} \left(\frac{1}{l_1 \frac{1}{x}} \right)^{p_1} \left(\frac{1}{l_2 \frac{1}{x}} \right)^{p_2} \cdots \left(\frac{1}{l_k \frac{1}{x}} \right)^{p_k}$$

where the first nonvanishing p_i is assumed to be positive, are positive together with their derivatives for sufficiently small x, as is required by our initial assumptions. They define dimensions that may be denoted by[9]

$$P = (p_0, p_1, \ldots, p_k).$$

---

[7]Abstufungsmerkmal. —*Ed.*

[8]Here, we have iterated natural logarithms: $l_1 t = \log t$, $l_2 t = \log \log t$, $l_3 t = \log \log \log t$, etc. —*Ed.*

[9]A set of this type always has dimension p_0 for any choice of the p_i's, where this last use of dimension refers to our previous definition of dimension. This type of dimension is therefore a refinement of the previous one.

These can be ordered lexicographically with Q higher than P if the first nonvanishing difference $q_i - p_i$ is positive. We divide these dimensions into two classes:

$$\text{Dimensions less than (1)}: \quad \frac{\lambda(x)}{x} \to \infty.$$

Here $\lambda'(x) \to \infty$, and λ'' is negative for small x. The function $\lambda(x)$ is *convex* upwards.[10] The following holds for the exponents

either $p_0 = 0$ and the first nonvanishing $p_i > 0$,

or $0 < p_0 < 1$,

or $p_0 = 1$ and the next nonvanishing $p_k < 0$.

$$\text{Dimensions greater than (1)}; \quad \frac{\lambda(x)}{x} \to 0.$$

Here $\lambda'(x) \to 0$, and λ'' is positive. The function $\lambda(x)$ is *concave* downwards.

either $p_0 = 1$ and the next nonvanishing $p_k > 0$,

or $p_0 > 1$.

The transition between the two regions occurs in dimension (1) when $\lambda(x) = x$.

9.   To construct *linear* sets of dimension $[\lambda(x)]$ as described above, the previous argument suggests that we take $\lambda(x)$ to be *convex*. Our assumptions about the continuous function $\lambda(x)$ of positive variable x are then

(α) $\lambda(x) > 0.$

(β) $\lambda(x_1) < \lambda(x_2)$ for $x_1 < x_2.$

(γ) $\begin{vmatrix} \lambda(x_1) & x_1 & 1 \\ \lambda(x_2) & x_2 & 1 \\ \lambda(x_3) & x_3 & 1 \end{vmatrix} > 0$ for $x_1 < x_2 < x_3.$

(δ) $\lambda(x) \to 0$ as $x \to 0$

(ε) $\lambda(x) \to \infty$ as $x \to \infty$

---

[10]The use of convex, concave (and so on) seems inconsistent. But the meaning should be clear from the context. —Ed.

Only the first four conditions in a neighborhood of the point $x = 0$ are essential. For the sake of simplicity we imagine $\lambda(x)$ to be specified in such a way that these conditions are satisfied for all positive x and that, in addition, condition (ε) applies.

The curve $y = \lambda(x)$ is upwardly convex by condition (γ), and point 2 of the curve lies above the secant 13, so if one sets

$$ik = \frac{\lambda(x_k) - \lambda(x_i)}{x_k - x_i},$$

then

$$12 > 13 > 23$$

holds for the quotients of the differences. If $x_1 < x_4 < x_3$ also holds, then

$$14 > 13 > 43,$$

hence

$$14 > 23, \qquad 12 > 43.$$

When the first relation holds, assume that $x_4 = x_1 - x_2 + x_3$ is symmetric to x_2 with respect to the midpoint of x_1, x_3. We then have

(ζ) $\qquad \lambda(x_1 - x_2 + x_3) > \lambda(x_1) - \lambda(x_2) + \lambda(x_3), \qquad (x_1 < x_2 < x_3).$

In the second relation, we take $x_4 = \alpha x_1$ and $x_3 = \alpha x_2$ where $\alpha > 1$ and obtain

(η) $\qquad \alpha\lambda(x_2) - \alpha\lambda(x_1) > \lambda(\alpha x_2) - \lambda(\alpha x_1), \qquad (x_1 < x_2, \alpha > 1).$

The function $\alpha\lambda(x) - \lambda(\alpha x)$ is therefore increasing with x and is greater than its limit 0 as $x \to 0$ so

(ϑ) $\qquad\qquad\qquad \alpha\lambda(x) > \lambda(\alpha x), \qquad (\alpha > 1).$

If one lets $x_3 = x_2 + h$ in (ζ), then

$$\lambda(x_1 + h) - \lambda(x_1) > \lambda(x_2 + h) - \lambda(x_2), \qquad (x_1 < x_2, h > 0),$$

and the function $\lambda(x + h) - \lambda(x)$ decreases with increasing x and is smaller than its limit $\lambda(h)$ as $x \to 0$ so

(ι) $\qquad\qquad\qquad \lambda(x_1 + x_2) < \lambda(x_1) + \lambda(x_2).$

After these initial considerations we will show that there exist linear closed sets A of dimension $[\lambda(x)]$, that is, sets A with finite positive measure $L(A)$ where $L(A) = \lim_{\varrho \to 0} L_\varrho(A)$, where $L_\varrho(A)$ is the lower bound of $\sum \lambda(\alpha_n)$ as one covers A by intervals of length $\alpha_n < \varrho$.

10.　We consider a sequence of positive numbers $\xi_0, \xi_1, \xi_2, \ldots$ with

$$\xi_0 > 2\xi_1, \qquad \xi_1 > 2\xi_2, \ldots .$$

From the closed interval $[0, \xi_0]$ on the x–axis we delete the central open interval of length $\xi_0 - 2\xi_1$ denoted by $\beta\,(1/2)$. From the remaining two closed intervals of length ξ_1, we delete the central open intervals of length $\xi_1 - 2\xi_2$ and denote these deleted intervals by $\beta\,(1/4)$ (left) and $\beta\,(3/4)$ (right). Out of each of the four remaining closed intervals of length ξ_2 we delete the central open interval of length $\xi_2 - 2\xi_3$ and denote these deleted intervals (from left to right) by $\beta\,(1/8)$, $\beta\,(3/8)$, $\beta\,(5/8)$, $\beta\,(7/8)$, etc.

In this fashion every rational dyadic number y between 0 and 1 will be associated to an open interval $\beta(y)$, and after deleting the open set $B = \sum \beta(y)$ out of the original interval $[0, \xi_0]$, there remains a perfect, nowhere dense set A, which can be covered by 2^n intervals of length ξ_n for $n = 0, 1, 2, \ldots$ and whose measure $L(A)$ is smaller or equal to $\liminf 2^n \lambda(\xi_n)$, since $\xi_n \to 0$. Now we choose ξ_n to satisfy the condition

$$(2) \qquad\qquad 2^n \lambda(\xi_n) = 1,$$

so that $L \leq 1$. This choice is admissible, since $\lambda(x)$ takes on each positive value once and only once,[11] and since (ϑ) gives

$$\lambda(\xi_{n-1}) = 2\,\lambda(\xi_n) > \lambda(2\xi_n), \quad \text{so that } \xi_{n-1} > 2\xi_n .$$

On the other hand, we must show that $L \geq 1$. Our construction relied on the fact that between two intervals $\beta\,((k-1)/2^n)$ and $\beta\,(k/2^n)$, corresponding to successive numbers with the denominator 2^n, an interval of length ξ_n remained free. If we thus call $u(y)$, $v(y)$ the left and right endpoints of $\beta(y)$, respectively (both points not belonging to the open interval), then

$$(3) \qquad\qquad u(k/2^n) - v((k-1)/2^n) = \xi_n .$$

---

[11] We note that this property serves only to simplify the argument and is unnecessary. Naturally, L only depends on the limiting properties of the ξ_n as $n \to \infty$. For example, divide A into two congruent halves A_1, A_2 to the left and to the right of $\beta(1/2)$, then $L(A) = L(A_1) + L(A_2) = 2\,L(A_1)$. This shows that it is independent of ξ_0, and so is independent of the value of any finite number of terms of $\xi_0, \xi_1, \ldots$.

This is valid for $n > 1$ and $k = 2, 3, \ldots, 2^n - 1$, but also for $k = 1, k = 2^n$ and $n = 1, n = 0$, if one makes the assumption that

$$v(0) = 0, \qquad u(1) = \xi_0.$$

Equation (3), which is simplified to

(4) $$u(y_2) - v(y_1) = \xi_n,$$

can serve to recursively determine the interval endpoints, since a dyadic number $y = (2k-1)/2^{n+1}$ with denominator 2^{n+1} is the arithmetic mean of its "neighbors," i.e., two successive numbers $y_1 = (k-1)/2^n$, $y_2 = k/2^n$ with the denominator 2^n. Equation (3) then gives

(5) $$u(y_2) - v(y) = u(y) - v(y_1) = \xi_{n+1}.$$

We now claim that for the remaining piece[12] α between two intervals $\beta(y), \beta(\eta)$ with $y < \eta$

(6) $$\alpha = u(\eta) - v(y),$$

satisfies the inequality

(7) $$\lambda(\alpha) \geq \eta - y.$$

We will prove this generally for $0 \leq y < \eta \leq 1$. Since this inequality is immediately verified in the lowest cases, for example, $y = 0, \eta = 1, \alpha = \xi_0, \lambda(\alpha) = 1$, we shall prove it recursively for dyadic numbers with denominators $\leq 2^{n+1}$ assuming it for dyadic numbers with denominators $\leq 2^n$. One must only distinguish whether y or η or both have the reduced denominator 2^{n+1}.

A) Let y have reduced denominator 2^{n+1}, and η a denominator $\leq 2^n$. Let y_1, y_2 be the neighbors of y as above. Then $y_1 < y_2 \leq \eta$ and

$$u(y_2) - v(y_2) = \xi_n,$$

$$u(y_2) - v(y) = \xi_{n+1},$$

$$v(y) = v(y_1) + \xi_n - \xi_{n+1},$$

$$\alpha = \alpha^* - \xi_n + \xi_{n+1},$$

where

$$\alpha^* = u(\eta) - v(y_1) \geq u(y_2) - v(y_1) = \xi_n.$$

---

[12] The letter α is used both for an interval and for the length of the interval. —Ed.

(One observes that for $y < \eta$, $\beta(y)$ lies to the left of $\beta(\eta)$.) By (ζ) of Section 9 we have

$$\lambda(\alpha) \geq \lambda(\alpha^*) - \lambda(\xi_n) + \lambda(\xi_{n+1}),$$

and, since inequality (7) already holds for α^*,

$$\lambda(\alpha) \geq (\eta - y_1) - \frac{1}{2^n} + \frac{1}{2^{n+1}} = \eta - y.$$

B) Let η have reduced denominator 2^{n+1}, and y a denominator $\leq 2^n$. Let η_1, η_2 be the neighbors of η, then $y \leq \eta_1 < \eta_2$ and

$$u(\eta_2) - v(\eta_2) = \xi_n,$$

$$u(\eta) - v(\eta_1) = \xi_{n+1},$$

$$u(\eta) = u(\eta_2) - \xi_n + \xi_{n+1},$$

$$\alpha = \alpha^* - \xi_n + \xi_{n+1},$$

where

$$\alpha^* = u(\eta_2) - v(y) \geq u(\eta_2) - v(\eta_1) = \xi_n.$$

Furthermore

$$\lambda(\alpha) \geq \lambda(\alpha^*) - \lambda(\xi_n) + \lambda(\xi_{n+1})$$

$$\geq (\eta_2 - y) - \frac{1}{2^n} + \frac{1}{2^{n+1}} = \eta - y.$$

C) Let y and η both have reduced denominator 2^{n+1}. Let y_1, y_2, η_1, η_2 be their neighbors so that $y_1 < y_2 \leq \eta_1 < \eta_2$. Then, as above, we have

$$v(y) = v(y_1) + \xi_n - \xi_{n+1},$$

$$u(\eta) = u(\eta_2) - \xi_n + \xi_{n+1},$$

$$\alpha = \alpha^* - 2\xi_n + 2\xi_{n+1},$$

where

$$\alpha^* = u(\eta_2) - v(y)$$

$$> u(\eta_2) - v(\eta_1) + u(y_2) - v(y_1)$$

$$= 2\xi_n.$$

By inequalities (ζ), (η) of the previous section we find (for $\alpha = 2$)

$$\lambda(\alpha) > \lambda(\alpha^*) - \lambda(2\xi_n) + \lambda(2\xi_{n+1})$$

$$> \lambda(\alpha^*) - 2\lambda(\xi_n) + 2\lambda(\xi_{n+1})$$

$$\geq (\eta_2 - y_1) - \frac{1}{2^{n-1}} + \frac{1}{2^n} = \eta - y.$$

We have thus proved inequality (7).

We now cover A by open intervals $\alpha_n < \varrho$ such that

$$\sum \lambda(\alpha_n) < L_\varrho + \varepsilon.$$

One can consider a finite number of α's without increasing the sum on the left (Borel's Theorem). These α can be replaced by closed intervals which can be taken to be segments of the original interval $[0, \xi_0]$. One can replace two adjacent or overlapping α with a single one (Section 9 ι), and then one can shrink an α that overlaps a $\beta(y)$ (without containing $\beta(y)$ entirely) to the overlapping piece. Finally, $n + 1$ intervals $\alpha_0, \alpha_1, \ldots, \alpha_n$ remain, and together with certain intervals $\beta_1, \beta_2, \ldots, \beta_n$ of the sequence $\beta(y)$ they cover the original interval. If the latter intervals correspond to the dyadic numbers $y_1, y_2, \ldots, y_n$, then by Equation (7)

$$\lambda(\alpha_0) \geq y_1,$$

$$\lambda(\alpha_1) \geq y_2 - y_1,$$

$$\cdots \qquad \cdots$$

$$\lambda(\alpha_{n-1}) \geq y_n - y_{n-1}$$

$$\lambda(\alpha_n) \geq 1 - y_n,$$

so

$$L_\varrho + \varepsilon > \sum \lambda(\alpha_i) \geq 1.$$

and it follows that

$$L_\varrho \geq 1 \geq L \geq L_\varrho, \quad \text{so } L_\varrho = L = 1.$$

The set A thus has dimension $[\lambda(x)]$.

Examples: To generate a set of dimension p, $(0 < p < 1)$, we have to set $\lambda(x) = x^p$, which satisfies our requirements for all positive x. By (2) we have

$$2^n \xi_n^p = 1$$

$$\xi_n = \xi^n \text{ with } \xi = (1/2)^p < \frac{1}{2}.$$

For instance, to get dimension $1/2$, we must choose $\xi = 1/4$, i.e., delete the central half of each remaining interval. For a given ξ where $0 < \xi < 1/2$ the dimension is

$$p = \log 2 : \log \frac{1}{\xi}.$$

For example, the classical Cantor set for which the central third of each interval is deleted ($\xi = 1/3$), the dimension is given by $p = \log 2 : \log 3 = 0,63903$.

To construct a set of "infinitely small" dimension $(0, p)$ in the logarithmic scale ($p > 0$), set

$$\lambda(x) = 1 : \left(\log \frac{1}{x} \right)^p,$$

which is convex only for sufficiently small x, and for n greater than a certain value

$$\left(\log \frac{1}{\xi_n} \right)^p = 2^n.$$

$$\xi_n = e^{-\eta^n}, \quad \text{where } \eta = 2^{1/p} > 1.$$

The condition (1) $\xi_{n-1} : \xi_n > 2$, $\eta^n - \eta^{n-1} > \log 2$, is also satisfied for n greater than a certain value.

If one replaces (2) with $2^n \lambda(\xi_n) = L$, then $L(A) = L$. If one inserts into the intervals $\beta(y)$ that have been left free in A sets like A with the same dimension $[\lambda(x)]$ such that the sums of their L–measures converges and then repeats with the resulting open intervals, all of which have finite L–measures, then one obtains a set which has dimension $[\lambda(x)]$ in every interval.

On the other hand, it is not possible to construct a set A whose intersection A_0^a with the interval $0 \leq x < a$ would have a monotonically (in a) growing dimension $[\lambda_a(x)]$ in the sense that $\lim_{x \to 0} \lambda_b(x)/\lambda_a(x) = 0$ for $b > a$ (Definition 2). For if a_n were to converge to b from below, then $L_b(A_0^{a_n}) = 0$, and, since the L–measure of increasing sets converges to the L–measure of their sum, (Carathéodory, Theorem 13) we would also have $L_b(A_0^b) = 0$, and A_0^b would not have dimension $[\lambda_b(x)]$. If the dimension $[\lambda_a(x)]$ exists and is a function of a, then incomparable dimensions appear in every arbitrarily small interval.

11.   Returning to the perfect set A discussed in Section 10, we define the piecewise constant function $\varphi(x)$ in the set $B = \sum \beta(y)$

$$\varphi(x) = y, \qquad \text{if } x \text{ lies in } \beta(y) \, .$$

If x and $\xi > x$ are two points in B, then they either belong to the same $\beta(y)$ and $\varphi(\xi) - \varphi(x) = 0$, or they belong to different intervals $\beta(y), \beta(\eta)$ with $y < \eta$ in which case

$$\xi - x > u(\eta) - v(y) = \alpha \, ,$$

so by (7)

$$\lambda(\xi - x) > \eta - y = \varphi(\xi) - \varphi(x) \, .$$

In any case we have

$$0 \le \varphi(\xi) - \varphi(x) < \lambda(\xi - x) \text{ for } x < \xi \, .$$

This function is uniformly continuous in B and can therefore be extended to a continuous function $\varphi(x)$ on the original interval $[0, \xi_0]$ in which B is dense. The function $\varphi(x)$ will in general satisfy

$$(8) \qquad\qquad 0 \le \varphi(\xi) - \varphi(x) \le \lambda(\xi - x) \text{ for } x < \xi \, .$$

It is monotonic, does not decrease as x grows, runs from 0 to 1, and takes on each of its values at least once for some point of A. The interval $0 \le y \le 1$ is the unique image (not inversely unique, however) of the interval $0 \le x \le \xi_0$ under the map $y = \varphi(x)$, and an interval α of the x–axis corresponds by (8) to an interval $\gamma \le \lambda(\alpha)$ of the y–axis. This map illustrates the last part of the proof in Section 10 even more clearly: Cover A by intervals α_n such that

$$\sum \lambda(\alpha_n) < L_\varrho + \varepsilon \, ,$$

and which we assume to belong to the original interval. Then the corresponding intervals γ_n must cover the whole interval $0 \le y \le 1$, since every y ($0 \le y \le 1$) is the image of at least one point x of A, and if x lies in α_n, then y lies in γ_n. Thus $\sum \lambda_n \ge 1$, and $\sum \lambda(\alpha_n) \ge 1$, the proof continuing as above.

The inverse function $x = \psi(y)$ is infinitely–valued at dyadic rational points between 0 and 1, meaning that it takes on all values between $u(y)$ and $v(y)$ (inclusive), while $\psi(0) = 0$ and $\psi(1) = \xi_0$. Incidentally, this function can be explicitly constructed in the following way: Let y be a *dyadic irrational* with dyadic expansion

$$y = \frac{y_1}{2} + \frac{y_2}{2^2} + \cdots, \qquad (y_n = 0, 1)$$

then

$$x = \psi(y) = y_1 \xi_0 + (y_2 - y_1)\xi_1 + (y_3 - y_2)\xi_2 + \cdots.$$

This is true for $y = 0, y = 1$. On the other hand, if y is a dyadic rational, $0 < y < 1$, then y has two dyadic expansions, the first ending with ones, the second with zeroes

$$y = \frac{u_1}{2} + \frac{u_2}{2^2} + \cdots, \qquad (u_n \to 1)$$

$$= \frac{v_1}{2} + \frac{v_2}{2^2} + \cdots, \qquad (u_n \to 0)$$

and these correspond exactly to the endpoints of the interval $\beta(y)$

$$u(y) = u_1 \xi_0 + (u_2 - u_1)\,\xi_1 + (u_3 - u_2)\,\xi_2 + \cdots,$$

$$v(y) = v_1 \xi_0 + (v_2 - v_1)\,\xi_1 + (v_3 - v_2)\,\xi_2 + \cdots.$$

The proof of these claims will be suppressed: The representation of $u(y)$, $v(y)$ can again be verified recursively by means of (4) and (5), and the conclusion for $\psi(y)$ can be extracted from this.

The function $\varphi(x)$ has the following simple interpretation: If A_a^b is the intersection of A with the interval J_a^b $(a \leq x < b)$, then

$$\varphi(x) = L(A_0^x),$$

which can be obtained first at the points of $B = \sum \beta(y)$, then extended by the monotonicity of both functions. For if x lies in $\beta(y)$ with $y = k/2^n$, then the β–intervals corresponding to the dyadic numbers of denominator 2^n divide A into 2^n congruent pieces, each of measure $1/2^n$, and since there are k such pieces on the left of $\beta(y)$, we have:

$$L(A_0^x) = \frac{k}{2^n} = y = \varphi(x).$$

12.   The function (9) can be viewed in a more general setting.[13] Let the outer measure L vanish for single points and over every set let there be a Borel set of the same measure. Assume that A lies in the interval J_0^a and has positive (finite) $L(A) = L$. Then $\varphi(x)$ runs monotonically from 0 to L as x runs from 0 to a. On account of the measurability of intervals we have

$$(10) \qquad \varphi(\xi) - \varphi(x) = L(A_x^\xi), \qquad (x < \xi),$$

in particular,

$$L - \varphi(x) = L(A_x^a).$$

$\varphi(x)$ is continuous, since by an already mentioned property of the outer measure we have

$$L(A_0^x) \to L(A_0^\xi), \qquad L(A_x^a) \to L(A_\xi^a),$$

as x converges to ξ from below and above, respectively.
    The map

$$x^* = \varphi(x)$$

uniquely assigns to any set X in the interval $[0, a]$ a set X^* in the interval $[0, L]$. In general X^* is not uniquely invertible, as $\varphi(x)$ can have (at most countably many) intervals where it is constant. But if X, Y do not have points in common, then X^*, Y^* have at most countably many points in common. Although the inverse–image of a set X^* is not uniquely determined, it has a *maximal* inverse–image X, i.e., the set of *all* points x whose images fall in X^*. (It thus contains no points or all points of a constant interval.) The image and the maximal inverse–image of a Borel set is again a Borel set.
    For an interval J we have by (10) that

$$m(J^*) = L(AJ),$$

where m denotes outer linear Lebesgue measure. For an arbitrary set X, only the *inequality*

$$(11) \qquad m(X^*) \geq L(AX),$$

---

[13] $f(X) = L(AX)$ is for our L and measurable sets X a completely additive set function, and $\varphi(x)$ the corresponding function on points. See the paper of J. Radon, Theorie und Anwendungen der absolut additiven Mengenfunktionen (Wiener Akad. Ber. 122 (1913), p. 1295–1438), which contains many new results about Carathéodory's measure theory. Instead of using $x^* = \varphi(x)$ (see Radon, p. 1343) one can also use $x^* = x + \varphi(x)$, as Mr. Carathéodory has remarked to me in a letter.

holds, for if one covers X^* with intervals J_n^* (in the interval $[0, L]$) satisfying

$$\sum m(J_n^*) < m(X^*) + \varepsilon,$$

then the maximal inverse–image intervals J_n cover the set X and we have

$$L(AX) \leq \sum L(AJ_n) = \sum m(J_n^*).$$

The *equation*

(12) $$m(X^*) = L(AX)$$

(whose left side represents the upper variation of $\varphi(x)$ in X or the upper Stieltjes–Lebesgue integral $\int_X d\varphi$) certainly holds in the following cases:

α) When X is a Borel set. For if Y is the complement of X with respect to an interval J, then the Borel sets X^*, Y^* have at most countably many points in common and

$$m(J^*) = m(X^*) + m(Y^*)$$

$$\geq L(AX) + L(AY) \geq L(AJ) = m(J^*),$$

so that for X and Y in (11) equality holds.

β) If X is a subset of A. For if V is a Borel set over X with $L(V) = L(X)$, then AV lies between X and V, so $L(AV) = L(X)$, and one has

$$m(V^*) \geq m(X^*) \geq L(AX) = L(AV) = m(V^*).$$

γ) The property holds for every X if and only if A is measurable with respect to L. For if B is the complement of A (with respect to the original interval $[0, a]$), and if (12) holds in general, then we have that, in particular,

(13) $$m(B^*) = 0,$$

and conversely by β) one has for arbitrary X that

$$m(X^*) = m((AX)^*) = L(AX).$$

From (13) it again follows that there exists a Borel set V^* over B^* such that $m(V^*) = 0$. Its maximal inverse–image V is a Borel set over B with $L(AV) = 0$, whose complement is also a Borel set *under* A with

$L(U) = L(A)$, so A is measurable with respect to L. This conclusion once more leads us back to (13).

If A is measurable relative to L, then one can easily conclude from (12) that AX is measurable with respect to L if X^* is linearly measurable.

If L comes from $\lambda(x)$ as in Definition 2, then the function

$$\varphi_\varrho(x) = L_\varrho(A_0^x)$$

is remarkable for reasons which will become clear. It goes monotonically from 0 to $L_\varrho(A)$, since

$$L_\varrho(X + Y) \leq L_\varrho(X) + L_\varrho(Y)$$

implies that

$$\varphi_\varrho(\xi) - \varphi_\varrho(x) \leq L_\varrho(A_x^\xi), \qquad (x < \xi)$$

and for $\xi - x < \rho$

$$L_\varrho(A_x^\xi) \leq \lambda(\xi - x),$$

which proves the continuity of $\varphi_\varrho(x)$. The (ϱ-dependent) map $x^* = \varphi_\varrho(x)$ assigns to every interval $\alpha < \varrho$ an image interval $\alpha^* \leq \lambda(\alpha)$. Incidentally, one can conclude the inequality

$$m(X^*) \leq L_\varrho(X)$$

between a set and its image.

13. The construction of a planar set of a prescribed dimension is not without difficulties. Let A be a set on the x-axis, B a set on the y-axis, and $C = A \times B$ the planar set of points (x, y), where x runs through set A and y runs through set B. We will prove, under specific constraints, that C is of dimension $[\lambda(t)\mu(t)]$, if A is of dimension $[\lambda(t)]$ and B is of dimension $[\mu(t)]$. Let the corresponding outer measures of A, B, C be L, M, N and let both functions $\lambda(t), \mu(t)$ be chosen as in Section 9 so that, in particular, they are convex.

Let A be the perfect set of Section 10, based on interval lengths satisfying

$$2^m \lambda(\xi_m) = L,$$

then we do not need to specify B any further. Cover B with intervals $\eta_n < \varrho$ satisfying

$$\sum \mu(\eta_n) < M + \varepsilon.$$

Then C will be covered by the sets $C_n = A \times \eta_n$. Assuming that $\varrho \leq \xi_0$, choose an index m for every n such that

$$\xi_{m-1} > \eta_n \geq \xi_m$$

and enclose A in 2^m intervals of length ξ_m. This will cover C_n with 2^m rectangles having sides ξ_m, η_n or in 2^m circles of diameters $d_{mn} = \sqrt{\xi_m^2 + \eta_n^2}$. Since $\xi_m < \frac{1}{2} \xi_{m-1}$ one has

$$d_{mn} \leq \sqrt{2}\, \eta_n < 2\varrho \,,$$

$$d_{mn} \leq \sqrt{\xi_m^2 + \xi_{m-1}^2} < \frac{\sqrt{5}}{2} \xi_{m-1} \,,$$

so by Section 9 (ϑ)

$$\lambda(d_{mn})\, \mu(d_{mn}) < \frac{\sqrt{5}}{2}\, \lambda(\xi_{m-1})\, \sqrt{2}\mu(\eta_n) \,.$$

It follows that

$$N_{2\varrho}(C) \leq \sum_n 2^m \lambda(d_{mn}) \mu(d_{mn})$$

$$< \sum_n \sqrt{10}\, 2^{m-1} \lambda(\xi_{m-1})\, \mu(\eta_n)$$

$$= \sqrt{10}\, L \sum \mu(\eta_n) < \sqrt{10}\, L\, (M + \varepsilon) \,,$$

and finally

$$N \leq \sqrt{10}\, LM.$$

To give a lower bound for N, we assume that A, B are closed sets in the intervals $[0, \xi_0], [0, \eta_0]$ and cover C by circles K_n of diameters $d_n < \varrho$ such that

$$\sum \lambda(d_n) \mu(d_n) < N_\varrho + \varepsilon.$$

To each circle is assigned a circumscribed square with sides parallel to the coordinate axes and its intersection R_n with the original rectangle $(0 \leq x \leq \xi_0, 0 \leq y \leq \eta_0)$. The rectangle R_n has sides $\alpha_n, \beta_n \leq d_n$. We now make use of the functions mentioned at the end of Section 12

$$\varphi_\varrho(x) = L_\varrho(A_0^x), \qquad \psi_\varrho(y) = M_\varrho(B_0^y) \,.$$

Since

$$u = \varphi_\varrho(x), \qquad v = \psi_\varrho(y),$$

we can project R_n onto a rectangle in the uv–plane having sides

$$\gamma_n \leq \lambda(\alpha_n), \qquad \delta_n \leq \mu(\beta_n).$$

These rectangles must cover the entire rectangle

$$0 \leq u \leq L_\varrho, \qquad 0 \leq v \leq M_\varrho$$

as $\varphi_\varrho(x)$ takes on every value at least once for some point of A and $\psi_\varrho(y)$ takes on every value at least once for some point of B. We then have $\sum \gamma_n \delta_n \geq L_\varrho M_\varrho$ and furthermore

$$\sum \lambda(d_n)\,\mu(d_n) \geq \sum \lambda(\alpha_n)\,\mu(\beta_n)$$

$$\geq \sum \gamma_n \delta_n \geq L_\varrho M_\varrho,$$

hence $N_\varrho \geq L_\varrho M_\varrho$ and $N \geq LM$.

If we combine the above inequalities then we see, for example, that C will be exactly of dimension $[\lambda\mu]$, whenever A, B are perfect sets in the sense of Section 10 having dimensions $[\lambda], [\mu]$, and whenever they belong to convex functions. One can then assign to the set C any dimension in the logarithmic scale which is less than 2. If one increases this dimension beyond 1 and enlarges the set C (which has no connected subset other than isolated points) by adding countably many line segments also having N–measure 0, to a simple curve,[14] then the resulting set has the same N–measure as C. It follows that there are Jordan curves of arbitrary dimension between 1 and 2. It is well known that there are also curves of dimension 2, and similarly to how one proceeds in this case or how is suggested for linear sets at the end of Section 10, one can generate Jordan curves for which every curve segment has the same dimension between 1 and 2.

With three convex functions λ, μ, ν and three specific perfect sets A, B, C (of the type in Section 10) one can construct a spatial set $A \times B \times C$ of dimension $[\lambda\mu\nu]$ and draw the corresponding conclusions.

Greifswald, March 1918.

This selection contains the foundation of the theory of Hausdorff measure. But it contains many other ideas, some merely mentioned as asides, many of which even today have not yet been fully explored.

[14] See for example my Grundzüge der Mengenlehre (Leipzig 1914) p. 374.

The use in §8 of a function λ of the diameter to define a Hausdorff measure is very interesting. It is sometimes known as a **Hausdorff function**. When $\lambda(x) = x^p$, we get the usual p-dimensional Hausdorff measure. But other functions $\lambda(x)$ may also be useful. Rogers [9] is a text devoted to this.

A lot of work related to the Hausdorff dimension was done by Besicovitch (for example, Selections 9, 10, 11, and 15). So the term "Hausdorff-Besicovitch dimension" is sometimes used. –Ed.

Bibliography

[1] C. Carathéodory, *Über das lineare Maß von Punktmengen–eine Verall-gemeinerung des Längenbegriffs*, Nachrichten K. Gesell. Wissensch. Gött. (1914), 404–426.

[2] M. Fréchet, *Les dimensions d'un ensemble abstrait*, Math. Annalen **68** (1910), 145–168.

[3] W. Blaschke, *Über affine Geometrie III: Eine Minimumeigenschaft der Ellipse*, Berichte Verh. K. Sächs. Gesell. Leipzig **69** (1917), 12.

[4] H. Jung, *Über die kleinste Kugel, die eine räumliche Figur einschließt*, J. Reine Angew. Math. **123** (1901), 241–257.

[5] J. Radon, *Theorie und Anwendungen der absolut additiven Mengenfunktionen*, Wiener Akad. Ber. **122** (1913), 1295–1438.

[6] F. Hausdorff, *Grundzüge der Mengenlehre*, Leipzig, 1914.

[7] David Abbott, editor, *The Biographical Dictionary of Scientists: Mathematicians*, Peter Bedrick Books, New York, 1985.

[8] C. B. Boyer, *A History of Mathematics*, Second Edition (Revised by U. C. Merzbach), Wiley, New York, 1991.

[9] C. A. Rogers, *Hausdorff Measures*, Cambridge University Press, 1970.

[10] E. Eichhorn, *Felix Hausdorff/Paul Mongré: Some aspects of his life and the meaning of his death*, Mathematical Research, **67**, *Recent Developments of General Topology and its Applications*, 85-117, Akademie-Verlag, Berlin, 1992.

[11] H. Mehrtens, *Felix Hausdorff: Ein Mathematiker seiner Zeit*, Universität Bonn, 1980.

[12] E. Eichhorn and E.-J. Thiele, editors, *Vorlesungen zum Gedenken an Felix Hausdorff*, Berliner Studienreihe zur Mathematik **5**, Helderman Verlag, Berlin, 1994.

Karl Menger was born in 1902 in Austria. He was educated in Vienna (a student of Hans Hahn), and later taught there. As a prominent member of the "Vienna Circle", he showed an interest in logic, philosophy, didactics, and economics, in addition to his work in mathematics. Later, he came to the Illinois Institute of Technology. Menger is one of the originators of the ideas of topological dimension (other, independent, originators are Brouwer, Urysohn, and Lebesgue). See [13] for a brief summary of Menger's work.

Menger's results are published as two "communications" to the Amsterdam Academy of Sciences. These results were later included as Chapter IX in Menger's book [11]. The book is a systematic exposition of set-theoretic geometry, especially the theory of topological dimension.

SIX

General Spaces and Cartesian Spaces

Karl Menger

The "First Communication" contains the proof that a compact metric space of topological dimension 1 is homeomorphic to a subset of three-dimensional Euclidean space (called here three-dimensional Cartesian number space, and denoted R_3).

A compact metric space A is *one-dimensional* iff it admits a cover $A = A_1 \cup A_2 \cup \cdots \cup A_n$ by finitely many closed sets with arbitrarily small diameter such that every point of A belongs to at most two of the sets (covering dimension); by subtracting the interior of each set A_i from $A_{i+1}, \cdots, A_n$, we may arrange that the sets have disjoint interiors. If A is one-dimensional, then it admits a cover by finitely many arbitrarily small closed sets with zero-dimensional boundaries (small inductive dimension). Combining the two: there exists a cover by finitely many arbitrarily small closed sets with disjoint interiors such that every point belongs to at most two of the sets, and the boundaries of the sets are zero-dimensional; thus the intersection of two of the sets (being a subset of the boundary of each) is zero-dimensional.

The "Second Communication" contains the construction of a fractal nowadays known as the *Menger sponge*, denoted here I or R_3^1. It is a self-similar set, like Koch's curve (Selection 3). But it is of interest not primarily for that reason, but as a *universal curve*. The sponge I is a compact metric space with topological dimension 1; and any compact metric space with topological dimension 1 is homeomorphic to a subset of I. This is proved from a modification of the argument in the First Communication.

Note the use of old-fashioned notation for union and intersection of sets:

$$\sum_{n=1}^{\infty} G_n = \bigcup_{n=1}^{\infty} G_n, \qquad \prod_{n=1}^{\infty} I_n = \bigcap_{n=1}^{\infty} I_n.$$

The term *curve* simply means a space with topological dimension 1; and *surface* means a space with topological dimension 2. *Cantor's intersection theorem* states that a decreasing sequence of non-empty closed subsets of a compact metric space has non-empty intersection. *Borel's covering theorem* states that an open cover of a compact metric space admits a finite subcover. Here, *compact* means: every sequence in the space has a cluster point in the space. Menger notes that the Borel covering theorem is not constructive (in the sense of intuitionism); perhaps this remark was added because L. E. J. Brouwer was editor of the journal that published the paper. —Ed.

First Communication

Conditions for a general topological or metric space to be homeomorphic to a subset of Cartesian number space have been almost completely unknown. Now dimension theory, which makes possible the formulation of the facts in this area, also supplies the methods for the proof.[1]

In this first communication on the topic, the following result will be proved:

Every one-dimensional compact metric space is homeomorphic to a subset of three-dimensional number space.

The method of proof (which in related work will give the solution of important homeomorphism problems for general n-dimensional sets) does not completely adhere to Brouwer's requirements for constructivity (in particular Borel's covering theorem is used essentially), but large parts, not only of the theorem, but also of the proof given here, can be taken over into intuitionistic mathematics. A one-dimensional set arising in a certain way will correspond to a set in R_3 so that their homeomorphism follows from their mutual construction. We begin with a few preliminaries.

Let A be a compact metric space. A *finite neighborhood system* in A is a (countable) set of closed neighborhoods $A_{i_1 i_2 ... i_n}$ in A satisfying:

1. There are finitely many sets $A_1, A_2, ... , A_n$ (called sets "of the first level"), no two of which have interior points in common.

2. If a set $A_{i_1 i_2 ... i_n}$ "of the nth level" is already defined, there are finitely many closed neighborhoods $A_{i_1 i_2 ... i_n 1}, ... , A_{i_1 i_2 ... i_n m}$

---

[1]In the discovery of the first results in this area, Sierpiński already found the sets that from a general viewpoint one should call zero-dimensional—he gave conditions for a totally disconnected set to be homeomorphic with a subset of the real line (Fund. Math. II, p. 89). The main problems do not occur with totally disconnected sets. Mazurkiewicz's result (FundMathII, p. 130) that every continuously parameterizable tree curve[2] is homeomorphic to a plane continuum is another result that I could derive from dimension and curve-theoretic results (Math. Annalen 96, 1926).

[2]A "tree curve" (*Baumkurve*) is a curve that has no subset homeomorphic to a circle. —Tr.

contained in $A_{i_1 i_2 \ldots i_n}$, no two of which have interior points in common.

3. There is a sequence ε_n of positive numbers with $\lim \varepsilon_n = 0$, so that the diameter of each set $A_{i_1 i_2 \ldots i_n}$ of the nth level is $< \varepsilon_n$.

The *kernel* of a finite neighborhood system $\{A_{i_1 i_2 \ldots i_n}\}$ is the set of all points that, for every n, belong to at least one of the sets of the nth level of the system. Evidently the kernel of a finite neighborhood system is always closed.

Two finite neighborhood systems (possibly taken from different metric spaces) $\{A_{i_1 i_2 \ldots i_n}\}$ and $\{B_{i_1 i_2 \ldots i_n}\}$, in which sets with equal indices correspond uniquely to each other, are called *homologous* if the following condition holds: If any sets $A_{i_1 i_2 \ldots i_n}, A_{j_1 j_2 \ldots j_n}, \ldots, A_{k_1 k_2 \ldots k_n}$ of the nth level have nonempty intersection, then the corresponding sets $B_{i_1 i_2 \ldots i_n}, B_{j_1 j_2 \ldots j_n}, \ldots, B_{k_1 k_2 \ldots k_n}$ have nonempty intersection, and conversely. Then the following holds:

Lemma. The kernels of homologous finite neighborhood systems are homeomorphic.[3]

One can first give a one-to-one mapping between the kernels A and B of the homologous systems $\{A_{i_1 i_2 \ldots i_n}\}$ and $\{B_{i_1 i_2 \ldots i_n}\}$: Let a be a point of A; in the nth level, say a belongs to sets $A_{i_1 i_2 \ldots i_n}, A_{j_1 j_2 \ldots j_n}, \ldots, A_{k_1 k_2 \ldots k_n}$. Denote them by $A_1^n(a), A_2^n(a), \ldots, A_{r_n}^n(a)$ and the corresponding sets $B_{i_1 i_2 \ldots i_n}, B_{j_1 j_2 \ldots j_n}, \ldots, B_{k_1 k_2 \ldots k_n}$ by $B_1^n, B_2^n, \ldots, B_{r_n}^n$. Consider the set

$$B(a) = \prod_{n=1}^{\infty} \sum_{i=1}^{r_n} B_i^n.$$

By Cantor's intersection theorem, $B(a)$ is not empty. On the other hand, $B(a)$ cannot contain more than one point: since the sets $A_1^n(a)$, $A_2^n(a), \ldots, A_{r_n}^n(a)$ have the point a in common, by the homology of the two systems the intersection of the sets $B_1^n, B_2^n, \ldots, B_{r_n}^n$ is also nonempty; since the diameters of the sets B_j^n converge to 0 as n grows, the diameter of the set $\sum_{i=1}^{r_n} B_i^n$ converges to 0 as n grows. Thus every point a of A is assigned a point $B(a)$ of B, where different points of A correspond to different points of B and every point of B corresponds to a point of A. One is convinced that this one-to-one mapping

---

[3] A strong generalization of this lemma will be included in future work on the solution of the homeomorphism problem regarding sets that are subsets of semicompact spaces. Homology will be defined for more general neighborhood systems and the idea of the kernel will be generalized. For investigation of homeomorphism regarding compact spaces, the above form of the lemma suffices.

of the two kernels is continuous in both directions, and is therefore topological.

The problem, given a one-dimensional compact set A, of finding a homeomorphic subset of R_3, is reduced by the Lemma to the problem of finding homologous finite neighborhood systems in A and in R_3. The solution of this problem is divided into three further lemmas:

Lemma 1 says essentially, *that given a decomposition of a compact space, one can give a corresponding system of polyhedra within a given polyhedron P, satisfying certain boundary conditions: that is, the polyhedra of the system meet the surface of P in given segments.*

Stated more precisely, the *hypothesis*: Let A be a compact space; let B_1, B_2, ..., B_m be m pairwise disjoint subsets of A; let P be a polyhedron in R_3; let Q_1, Q_2, ..., Q_m be pairwise disjoint segments on the surface of P.

Further: suppose A is the union of n closed neighborhoods A_1, A_2, ..., A_n, no two of which have interior points in common, satisfying the following conditions:

a. Every point of A belongs to at most two sets A_i;
b. Every point of a set B_k belongs to exactly one set A_i;
c. Every set A_i meets at most one set B_k.

The *conclusion*: There exist n polyhedra P_1, P_2, ..., $P_n < P$, no two of which have interior points in common, satisfying the following conditions:

1. P_i and P_j have points in common (and then the intersection is exactly a segment, that we will call Q_{ij}) if and only if A_i and A_j have points in common.

2. P_k has points in common with the surface of P (and then the intersection is exactly a segment contained in Q_k, that we will call Q'_{ik}) if and only if A_i and B_k have points in common.

3. The segments Q_{ik} and Q'_{ik} are pairwise disjoint.

Proof: We assign to every set A_i a point p_i of P, and choose p_i on the surface of P (and then in Q_k) if and only if A_i and B_k have points in common. [By assumption c such a choice of points is possible; of course several points p_i may lie on the same segment Q_k.] We connect p_i and p_j by a polygonal path $\overline{p_i p_j}$ if and only if A_i and A_j have points in common, and the polygon $\overline{p_i p_j}$ is chosen so that it contains no other point p_k and (except possibly for its endpoints p_i and p_j) it lies in the interior of P. We also specify the polygons $\overline{p_i p_j}$ so that any two of them have at most endpoints in common. Then we choose in each polygon $\overline{p_i p_j}$ a point q_{ij}, different from the endpoints, and a small segment Q_{ij} that meets $\overline{p_i p_j}$ only in the point q_{ij}. If the Q_{ij} are sufficiently small, one can surround the two subpolygons $\overline{p_i q_{ij}}$ and $\overline{q_{ij} p_j}$ with chains S_{ij} and S_{ji} of simplices $< P$, so that S_{ij} and S_{ji} intersect exactly in the segment Q_{ij}; if the point p_i lies in Q_k, then each of the chains S_{ij} of simplices from p_i to a point q_{ij} should be specified so that p_i is not in its interior, but on its boundary, and so that this boundary meets the surface of P exactly in a subsegment Q'_{ik} of Q_k.

If the simplices used to form the chains are chosen small enough, then each chain S_{ij} has points in common with the surface of P (and then the intersection is exactly a subsegment of Q_k) only if[4] p_i lies on Q_k; two chains S_{ij} and S_{ik} that come together at the same point p_i have in common only points in the neighborhood of p_i, while two chains that do not meet in a point p_i or in a segment Q_{ij}, are disjoint. If we denote by P_i the sum of the chains S_{ij} coming together at p_i, then the polyhedra P_i satisfy the conclusions of Lemma 1.

Lemma 2 says essentially, that given a chain of polyhedra, one can find a corresponding decomposition of a compact one-dimensional space satisfying specified boundary conditions.

More precisely: *hypothesis*: Let A be a compact one-dimensional space; let $B_1, B_2, \ldots, B_m$ be m pairwise disjoint closed zero-dimensional subsets of A; let P be a polyhedron in R_3; let $Q_1, Q_2, \ldots, Q_m$ be m pairwise disjoint segments on the surface of P.

Further: Let $P_1, P_2, \ldots, P_n$ be a chain of polyhedra $< P$ with the following properties:

1. Any two consecutive polyhedra P_i and P_{i+1} $(i = 1, 2, \ldots, n - 1)$ meet exactly in a segment; any other two polyhedra are disjoint.

2. The intersection of the chain of polyhedra with the surface of P consists of subsegments of the segments Q_k, each such subsegment belongs to exactly one P_i, and no P_i meets two Q_k.

The *conclusion*: There exist n closed neighborhoods $A_1, A_2, \ldots, A_n$ in A, any two of which have at most zero-dimensional intersection, satisfying the following conditions:

a. Any point of A belongs to at most two sets A_i; to any two disjoint polyhedra P_i and P_k correspond disjoint sets A_i and A_k.

---

[4]The original has p in place of p_i. –Tr.

b. Every point of a set B_k belongs to exactly the set A_i such that Q_k has points in common with P_i.

c. Any set A_i has points in common with at most one B_k.

Proof: If the polyhedron P_1 is disjoint from the surface of P, then we choose as A_1 any closed neighborhood in A that is disjoint from all B_k and whose boundary[5] C_1 is zero-dimensional. Otherwise, if the polyhedron P_1 contains a subsegment of Q_k, we construct a closed neighborhood A_1 of B_k, that is disjoint from all other B_r, and whose boundary C_1 is zero-dimensional. Suppose the sets $A_1, A_2, \ldots, A_{i-1}$ have been defined; let C_{i-1} be the zero-dimensional boundary of A_{i-1}. We distinguish two cases: If P_i is disjoint from the surface of P, let A_i be a neighborhood, closed in $A - \sum_{j=1}^{i-1} A_j + C_{i-1}$, disjoint from all sets B_k, whose boundary C_i is zero-dimensional, and such that C_{i-1} is contained in the interior of $\sum_{j=1}^{i} A_j$. Otherwise, if P_i contains a subsegment of Q_k, let A_i be a neighborhood, closed in $A - \sum_{j=1}^{i-1} A_j + C_{i-1}$, with zero-dimensional boundary, satisfying the following conditions: C_{i-1} is contained in the interior of $\sum_{j=1}^{i} A_j$, B_k in the interior of A_i, while A_i is disjoint from all other sets B_r. Finally, let $A_n = A - \sum_{i=1}^{n-1} + C_{n-1}$. The sets $A_1, A_2, \ldots, A_n$ satisfy the conclusions of Lemma 2.

Lemma 3 says essentially *that given a compact one-dimensional space A and a polyhedron P, one can find a decomposition of A into arbitrarily small neighborhoods, and corresponding system of arbitrarily small polyhedra in P, satisfying given two-sided boundary conditions.*

More precisely: *hypothesis*: Let A be a compact one-dimensional space; let $B_1, B_2, \ldots, B_m$ be m pairwise disjoint zero-dimensional subsets of A; let P be a polyhedron in R_3; let $Q_1, Q_2, \ldots, Q_m$ be m pairwise disjoint segments on the surface of P; let ε and η be two positive numbers.

The *conclusion*: There are finitely many closed neighborhoods A_1, $A_2, \ldots, A_n$ in A, any two of which have at most zero-dimensional intersection, with all $< \varepsilon$, such that

a. Any point of A belongs to at most two sets A_i;

b. Any point of a set B_k belongs to exactly one set A_i;

c. Any set A_i has points in common with at most one set B_k;

and there are polyhedra $P_1, P_2, \ldots, P_n < P$, all $< \eta$, so that

1. P_i and P_j have points in common (and then the intersection is exactly a segment, that we will call Q_{ij}) if and only if A_i and A_j have points in common;

---

[5]The original has C_i, not C_1. –*Tr.*

2. P_i has points in common with the surface of P (and then the intersection is exactly a subsegment of Q_{ik}, that we will call Q'_{ik}) if and only if A_i and B_k have points in common;

3. The segments Q_{ik} and Q'_{ik} are pairwise disjoint.

Proof: By the theory of dimension, one can first decompose A into closed neighborhoods $A_1^\times, A_2^\times, \ldots, A_n^\times$, any two of which have at most zero-dimensional intersection, are $< \varepsilon$, and are in fact so small that each set $A_i^\times$ has points in common with at most one set B_k, so that every point of A belongs to at most two of the sets $A_i^\times$, and every point of a set B_k belongs to exactly one set $A_i^\times$.[6] By Lemma 1 one can find polyhedra $P_1^\times, P_2^\times, \ldots, P_n^\times < P$ so that

1. $P_i^\times$ and $P_j^\times$ have points in common (and then the intersection is exactly a segment, which we will call $Q_{ij}^\times$) if and only if $A_i^\times$ and $A_j^\times$ have points in common;

2. $P_i^\times$ has points in common with the surface of P (and then the intersection is exactly a subsegment of Q_k, which we will call $Q_{ik}^{\times'}$) if and only if $A_i^\times$ and[7] B_k have points in common;

3. the segments $Q_{ij}^\times$ and $Q_{ik}^{\times'}$ are pairwise disjoint.

If the polyhedra $P_i^\times$ are all $< \eta$, then the claims of Lemma 3 are satisfied. But it is possible for $P_i^\times > \eta$. Then on each of the segments $Q_{ij}^\times$ and $Q_{ik}^{\times'}$ that lie on the surface of P we choose a subsegment $< \eta/2$, which we call (respectively) Q_{ij} or Q'_{ik}. Then we make a polygonal path, which has one point in common with each segment Q_{ij} and Q'_{ik} on the surface of $P_i^\times$, and otherwise lies completely in the interior of $P_i^\times$. By adjoining a suitable row of simplices, this polygon may be replaced by a chain $P_{i1}^\times, P_{i2}^\times, \ldots, P_{in_i}^\times$ of polyhedra $< P$ and $< \eta$ with the conditions specified in Lemma 2. By Lemma 2, there correspond $A_{i1}^\times, A_{i2}^\times, \ldots, A_{in_i}^\times$, subsets of $A_i^\times$. If consecutive sets $A_{ik}^\times$ and $A_{ik+1}^\times$ are disjoint, then we can replace the corresponding polyhedra by smaller disjoint polyhedra, that we will call $P_{ik}^\times$ and $P_{ik+1}^\times$ instead. If one chooses on $Q_{ij}^\times$ the same segment Q_{ij} for $P_i^\times$ and $P_j^\times$, then the neighborhoods $A_{ik}^\times$ and the polyhedra $P_{ik}^\times$ ($i = 1, 2, \ldots, n$, $k = 1, 2$,

---

[6]See for example Monatshefte f. Math. u. Phys. 34, p. 153; the decomposition can be chosen so that every point of one of the zero-dimensional sets B_k belongs to exactly one set $A_i^\times$, since one can carry out the construction with neighborhoods whose boundaries are disjoint from all B_k.

In this reference the Borel covering theorem is required.

[7]The original has $B_k^\times$ instead of B_k. –Tr.

..., n_i) satisfy all the conclusions of Lemma 3, when these sets are de-noted $A_1, A_2, \ldots, A_n$, respectively, $P_1, P_2, \ldots, P_n$ in any order.

Now let a compact one-dimensional space[8] be given. Let $\{\varepsilon_n\}$ and $\{\eta_n\}$ be two sequences of positive numbers that converge to zero. By Lemma 3, A is the sum of finitely many closed neighborhoods ($< \varepsilon_1$) $A_1, A_2, \ldots, A_n$, any two of which have intersection at most zero-dimensional, so that every point of A belongs to at most two sets A_i, and at the same time there exist polyhedra ($< \eta_1$) $P_1, P_2, \ldots, P_n$, so that P_i and P_j have points in common (and then the intersection is exactly a segment Q_{ik}) if and only if A_i and A_j have points in common.

Next Lemma 3 may be applied to each of the sets A_i and the corresponding polyhedron P_i, to specify sets $A_{i_1 i_2} < \varepsilon_2$ and polyhedra $P_{i_1 i_2} < \eta_2$ with corresponding relations to each other and to the boundaries of A_{i_1} and P_{i_1}. Continuing to apply Lemma 3, we may construct a finite neighborhood system $\{A_{i_1 i_2 \ldots i_n}\}$ and a corresponding system of polyhedra $\{P_{i_1 i_2 \ldots i_n}\}$, so that all neighborhoods of the nth level are $< \varepsilon_n$ and all polyhedra of the nth level are $< \eta_n$.

The two systems are homologous, since two polyhedra of the nth level have points in common if and only if the corresponding neighborhoods have points in common, while any three neighborhoods or any three polyhedra of the same level have no points in common. The kernel A of the neighborhood system is thus homeomorphic to the kernel of the system of polyhedra in R_3, which completes the proof of the theorem.

An analysis of the preceding proof gives occasion for a few re-marks. First, the connectedness relations of the sets $A_{i_1 i_2 \ldots i_n}$ play no role in the construction. Disconnected neighborhoods might corre-spond to (connected) polyhedra in Lemma 1 or Lemma 2. Second, compactness does not play an essential role, and presumably (by an easy modification of the construction) could be replaced by the as-sumption that A is a subset of a compact space. Third, one-dimension-ality of A is used essentially only for the following point in the proof: A one-dimensional set admits a finite neighborhood system, so that any two of the neighborhoods of the same level have intersection at most zero-dimensional. The construction is simplified by the fact that such systems exist, in which, in addition, every point belongs to at most two neighborhoods of the same level; but this is not essential.

---

[8] $A - Tr.$

Theorems that are connected with the preceding work, and which will be published soon, include the following: Necessary and sufficient conditions that a one-dimensional space be homeomorphic to a subset of the plane; a universal one-dimensional set, that is, a curve in R_3, which contains as subset a topological image of every compact one-dimensional space. Above all, the topics given here will be extended to higher-dimensional sets and number spaces, and lead to a fundamental topological theorem: *For a compact metric space to be homeomorphic to a set in a number space, it is necessary and sufficient that it possess finite dimension.*

Second Communication:
On universal n-dimensional sets

An n-dimensional set N will be called a *universal n-dimensional set* if it contains a subset homeomorphic to every n-dimensional compact metric space. We prove first:

There exist universal one-dimensional sets, among them are continuously parameterizable[9] *curves in R_3.* We can construct such a set as follows: Subdivide the unit cube in R_3 into 27 homothetic, mutually congruent subcubes with side $1/3$. Next we remove the innermost of these subcubes (that is, the subcube whose outer surface does not meet the outer surface of the unit cube) and the six subcubes that have a face in common with the innermost subcube. The remaining 20 subcubes should retain their entire boundaries. In each of these 20 subcubes, repeat the process, dividing into 27 smaller cubes, and removing the seven middle ones. Continuing in this manner, in the nth step one obtains 20^n cubes of side $1/3^n$; they will be called "cubes of the nth step of I", and their sum will be denoted I_n. The set $\prod_{n=1}^{\infty} I_n$ (which is clearly one-dimensional continuously parameterizable) is our universal one-dimensional set I.

One can also represent I in the following way: The nth scaffold of I is the sum G_n of the edges of all the cubes of the nth step of I. The space I is then the closed hull of its scaffold $G = \sum_{n=1}^{\infty} G_n$.

---

[9]A compact set $E \subseteq R_n$ is "continuously parameterizable" (*stetig durchlaufbar*) iff it is the image of the interval $[0, 1]$ under a continuous function. The Hahn-Mazurkiewicz Theorem states that a compact set $E \subseteq R_n$ is continuously parameterizable iff it is connected and locally connected [11], p. 230. –Tr.

The proof in the First Communication[10] that every one-dimensional compact metric space A is embeddable in R_3, was done by the introduction of a finite neighborhood system in A and a homologous system of polyhedra in R_3. The cubes of all steps of I will be called *cubes of I* for short, and a polyhedron that is a finite sum of cubes of I will be called an *interval polyhedron of I*. So in order to prove that A is embeddable in I, it suffices *to give a finite neighborhood system in A and a homologous system of interval polyhedra of I.*

We now suppose: Let A be a compact set, let $B_1, B_2, \ldots, B_m$ be m pairwise disjoint subsets of A. Let P be an interval polyhedron of I, and let $Q_1, Q_2, \ldots, Q_m$ be m pairwise disjoint square faces from the outer surface of P. Let A be written as a sum of n closed neighborhoods $A_1, A_2, \ldots, A_n$, no two of which have interior points in common, and so that every point of A belongs to at most two sets A_i, and every point of a set B_k belongs to exactly one set A_i. Now because of the fact that for every nth scaffold G_n of I, the set $G_n - G_{n-1}$ is connected, one may show by elementary geometric considerations that under the given conditions, the following is possible: If one chooses any n distinct points $p_1, p_2, \ldots, p_n$ in the interior of P, then one can give sufficiently small pairwise disjoint cubes $W_1, W_2, \ldots, W_n$ of I so that p_i is in the interior of W_i and so that pairwise disjoint polygons may be chosen from the scaffold of I, joining the outer surface of each W_i with the outer surfaces of all remaining W_k and with all Q_k. From this there follows immediately the existence of interval polyhedra $P_1, P_2, \ldots, P_n$ of I so that all $P_i < P$ and so that (for every i and k) P_i has points in common with P_k, respectively with Q_k, (and then the intersection is exactly a square face) if and only if A_i has points in common with A_k, respectively with B_k. Thus the special polyhedra in Lemma 1 of the First Communication may be replaced by interval polyhedra of I.

Lemma 2 of the First Communication requires no change. The refinement of Lemma 3 says that for a given compact one-dimensional space A one can find a decomposition into arbitrarily small neighborhoods, and at the same time for a given interval polyhedron of I one

---

[10]Cf. these Proceedings, vol. 29, 1926, p. 476.[11] The following lemma, which is simple but has many applications, is proved there: The kernels of homologous finite neighborhood systems are homeomorphic. The converse, that in homeomorphic compact spaces there exist homologous finite neighborhood systems, is trivial. *Thus for two compact spaces to be homeomorphic, it is necessary and sufficient that there exist in them homologous finite neighborhood systems.* If two spaces are given as kernels of finite neighborhood systems, the the question of whether the two spaces are homeomorphic is reduced to the question of whether the two given neighborhood systems can be derived from homologous neighborhood systems, and that is a question about the possibility of a countable sequence of finite operations.

[11]The original has the incorrect page number 4Ѕ6. –Ir.

can find a corresponding system of arbitrarily small interval polyhedra of I, fulfilling the two-sided boundary conditions. To see this, one must first find *some* decomposition of A into small enough parts with the desired properties, as in the proof of Lemma 3; then for this decomposition, one can use the refinement of Lemma 1 to find corresponding interval polyhedra of I, which also satisfy the corresponding boundary conditions. If these interval polyhedra are also sufficiently *small*, then one has reached the goal. Otherwise, one replaces each of the large interval polyhedra by a chain of interval polyhedra of I with the properties specified in Lemma 2, and by Lemma 2 determines for these polyhedra subneighborhoods of the neighborhood in A in corresponding position. This completes the proof of the refinement of Lemma 3. Based on that, one can construct a finite neighborhood system in a given one-dimensional space and a homologous system of interval polyhedra of I.

These considerations yield the two following corollaries: *Every one-dimensional compact metric space is homeomorphic to a set in R_3 for which every point admits arbitrarily small intervals with whose boundary the set has discontinuous intersection.* And: *Every compact one-dimensional space is homeomorphic to a subset of a surface of R_3.* Indeed, in the construction of the set I we can divide the cubes into 27 parts, but merely remove the center cube (except its boundary), and in the 26 remaining cubes repeat the process ad infinitum. The sum of the 26^n remaining cubes of side $1/3^n$ will be called II_n, and II will denote the intersection of all the sums II_n. The set so constructed is clearly a (locally connected) surface, which contains I as a subset.[12]

Investigation of these ideas in higher dimensions is interesting. The n-dimensional unit cube can be decomposed into 3^n mutually homothetic congruent subcubes with side length $1/3$. Each of these cubes corresponds to an n-tuple built from the numbers $0, 1, 2$: the kth coordinate of a cube, the distance from the $(n - 1)$-dimensional hyperplane through the origin perpendicular to the kth axis, falls in one of three intervals. Of these 3^n cubes, there are $\binom{n}{n-k}.2^{n-k}$ with exactly k

---

[12]One can characterize sets that are universal one-dimensional sets in various other ways: that they contain a subset homeomorphic to the set I; or that they contain a subset in which every open set can be passed through (its connectedness permitting) by the concatenation of finitely many sub-neighborhoods that were employed in the proof of the embeddability of all one-dimensional sets in the set I; or as closed hulls of a sufficiently ramified scaffold; and so on.

ones occuring among the coordinates. Now we remove each cube with at least $m + 1$ ones among its coordinates. We retain, then,

$$\mu(m) = \sum_{k=0}^{m} \binom{n}{n-k} \cdot 2^{n-k}$$

cubes (and return their outer boundaries if they were removed). Repeat with each of these remaining cubes the same process of decomposition and removal of the

$$\sum_{k=m+1}^{n} \binom{n}{n-k} \cdot 2^{n-k}$$

middle cubes. At the kth step we obtain $\mu(m)^k$ cubes[13] with sides of length $1/3^k$; their sum will be denoted $_kR_n^m$. The set $\prod_{k=1}^{\infty} {}_kR_n^m$, easily seen to be continuously parameterizable and m-dimensional, will be called R_n^m.[14]

Generalizing a conclusion of Sierpiński,[15] one easily shows that for every n, the set R_n^{n-1} contains a subset homeomorphic to every nowhere dense closed subset of R_n, or even every closed subset of R_n with dimension at most $n - 1$. But much more is true: it can be proved that *the set R_n^{n-1} contains a subset that is the topological image of every (closed or not) subset of R_n with dimension less than n.* By a theorem of dimension theory, every subset of R_n with dimension less than n is contained in a set, whose complement is countable and dense in R_n. By a theorem of Fréchet, all sets of this type are homeomorphic to each other. Thus it suffices to prove that a subset of the set R_n^{n-1} is homeomorphic to a set whose complement is countable and dense in R_n. But this happens, as one can easily see, for the G_δ-set that is obtained by starting with the open unit cube, and then removing the inner cube of the 27 subcubes *including its boundary*, and continuing this process in each of the 26 remaining cubes ad infinitem.

I believe it is very likely that in general the set R_n^m is a universal m-dimensional set with respect to R_n, that is, it contains a subset homeomorphic to every m-dimensional subset of R_n. The proof of this result should offer no substantial difficulties, although for the case $m < n - 1$ the proof may be quite long.

---

[13]The original has the obvious misprint $\mu(m$ in place of. $\mu(m)^k$. –Tr.

[14]Thus R_1^0 is the nowhere dense perfect Cantor set, R_2^1 is the universal plane curve of Sierpiński (Comptes Rendus, **162**, p. 629), R_3^1 is the universal one-dimensional set constructed above.

[15]*loc. cit.*

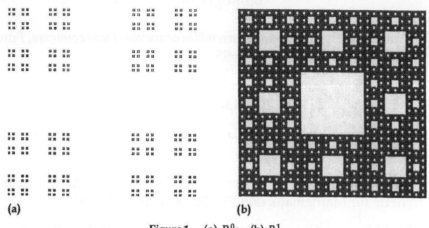

(a)   (b)

Figure 1. (a) R_2^0; (b) R_2^1.

In a subsequent communication it will be proved that every n-dimensional compact (and thus also every n-dimensional separable) metric space is homeomorphic to a subset of R_{2n+1}. Therefore, an example of a continuously parameterizable continuum that contains a subset homeomorphic to every compact finite dimensional space is obtained as the union $\sum_{n=1}^{\infty} I_n$, where I_n ($n = 1, 2, \ldots$) is an n-dimensional interval with side length $1/n$ containing a fixed point p.

The sponge I is universal for compact metric spaces with topological dimension 1. But, in fact, any separable metric space is homeomorphic to a subset of a compact metric space with the same topological dimension [11], Chap. IX, §1, [7], [9], [8], p. 65. So I is universal for separable metric spaces with topological dimension 1, compact or not.

Menger's sponge is shown in Plate 3. The two-dimensional sets R_2^n are shown in Figure 1. (a) is the Cartesian square of the Cantor set (Selection 2); (b) is known as the Sierpiński carpet. These two sets are combined in Plate 4.

Menger suggests—but does not prove—that the set R_{2n+1}^n is universal for (separable metric) spaces with topological dimension n. In [11], Chap. IX, Menger still provides only a "sketch" of the proof. The proof was carried out by S. Lefschetz [10]. A different universal n-dimensional space is due to G. Nöbeling [12] (a student of Menger). These universal sets prove, in particular, the assertion at the end of the "First Communication": *For a compact metric space to be homeomorphic to a set in Euclidean space, it is necessary and sufficient that it possess finite topological dimension.* —Ed.

Bibliography

[1] Waclaw Sierpiński, *Sur les ensembles connexes et non connexes*, Fundamenta Math. **2** (1921), 81–95.

[2] Stefan Mazurkiewicz, *Un théorème sur les lignes de Jordan*, Fundamenta Math. **2** (1921), 119–130.

[3] Karl Menger, *Über reguläre Baumkurven*, Math. Annalen **96** (1926), 572–582.

[4] Karl Menger, *Über die Dimension von Punktmengen, IITeil*, Monatshefte für Mathematik und Physik **34** (1926), 137–161.

[5] Karl Menger, *Allgemeine Räume und Cartesische Räume*, Proc. Kon. Ned. Akad. van Wetensch. Amsterdam **29** (1926), 476–482.

[6] W. Sierpiński, *Sur une courbe cantorienne que contient une image biunivoque et continue de toute courbe donnée*, Comptes Rendus Acad. Sci. Paris **162** (1916), 629–632.

[7] W. Hurewicz, *Über das Verhältnis separabler Räume zu kompakten Räumen*, Proc. Kon. Ned. Akad. van Wetensch. Amsterdam **30** (1927), 425–430.

[8] W. Hurewicz and H. Wallman, *Dimension Theory*, Princeton University Press, 1948.

[9] W. Kuratowski, *Sur les théorème de "plongement" dans la théorie de la dimension*, Fundamenta Math. **28** (1937), 336–342.

[10] S. Lefschetz, *On compact spaces*, Math. Annalen **32** (1931), 521–538.

[11] Karl Menger, *Dimensionstheorie*, B. G. Teubner, Leipzig, 1928.

[12] G. Nöbeling, *Über eine n-dimensionale Universalmenge in* R_{2n+1}, Math. Annalen **104** (1930), 71–80.

[13] S. Kass, *Karl Menger*, Notices of the American Mathematical Society **43** (1996), 558-561.

This selection introduces another fractional dimension. In many important cases, it coincides with the Hausdorff dimension, but in other cases it disagrees.

Georges Louis Bouligand (1889–1979) was born in Lorient, France, and was educated at the Ecole Normale Superieure in Paris. He held positions in Tours, Rennes, Poitiers, and Paris. [10]

Bouligand wrote a summary of the main results of this selection. Then apparently he gave it, in a sealed envelope, to the Paris Academy of Sciences. When the time was right, he asked that the envelope be opened; the summary was published in the *Comptes Rendus* of the Academy [2]. This method was used at that time to provide evidence in case of possible disputes over who first made a scientific discovery. Thus the evidence of your priority was on file, but it was sealed so that you would not have to reveal what you knew before you were ready.

SEVEN

Improper Sets and Dimension Numbers (Excerpt)

Georges Bouligand

This selection is only Part I of Bouligand's paper. Part II deals with the connection of the fractional dimension with the Dirichlet problem, potential theory, and sets of capacity zero. Part III deals with the calculus of variations. (I have not included the remainder of the paper for reasons of space.) These applications were Bouligand's motivation for investigation of the fractional dimension. Today we can see still more uses for it.

Note several variants to the definition of the Bouligand dimension. In section 5c Bouligand describes the "capacity" dimension (see Selection 17), where one counts the maximum number of disjoint spheres of the same radius ρ with centers in the set. The "box dimension" is described in section 5d. In section 7 we see that the Bouligand dimension of a Cartesian product of two sets is the sum of the dimensions of the factors.

Bouligand's dimension has been rediscovered several times in other equivalent forms: in this volume we will see it again in Selection 8 (Pontrjagin and Schnirelmann), Selection 15 (Besicovitch and Taylor), and Selection 17 (Kolmogorov and Tihomirov). Twelve equivalent formulations (for subsets of $\mathbb{R}$) are discussed in [11].–Ed.

---

1. During my research on the Dirichlet principle (Bull. Soc. Polon. Math. **5** (1925), p. 59–112) I proposed the notions of *improper sets* and *reduced boundary* (paragraph 14). I showed (paragraphs 13, 16, and 17) that improper sets (that is, those that can be removed from the region and added to the boundary without bounded and continuous conditions on these sets affecting the solution) are sets of *zero capacity*. As well, at the end of 1924, I proposed to distinguish sets of zero capacity and sets for which every (closed) subset intersecting an infinitesimal sphere is of positive capacity (that is, the reduced boundaries) by means of criteria based on consideration of the *dimension number*.[1]

This work will address the following topics:

1. Give a systematic exposition of considerations which, by different means than those of Hausdorff,[2] allowed me to independently rediscover elements of the study of closed sets from the dimensional point of view.

2. Supply details to the proof of results about the Dirichlet problem as relating to distinguishing improper sets, results whose justification I had only sketched.[3]

3. Show that the fundamental ideas of the preceding topic are not merely limited to the Dirichlet problem but, on the contrary, play a role in a large class of problems in the calculus of variations, the classical problem about the limits of potential theory being only a special case.

I. Dimensional properties of closed sets.

2. Maurice Fréchet was the first to consider the study of sets of points according to their dimensional properties: in this way he introduced *dimension type* which plays the role of a topological invariant, that is, with respect to the entire group of continuous and one to one transformations.[4] Actually, such an invariant is of no use for the questions that we have posed ourselves in this paper. It is quite possible to have two homeomorphic sets with one a reduced boundary and the other improper (see paragraph 16).

The area that we will investigate does not belong to pure topology but to the *restricted topology of the first order*, where one still considers continuous and bijective transformations but restricted to functions with continuous first derivative: for these transformations rectifiability of a curve or a surface are invariant properties. *The dimension order*

---

[1] *Dimension, étendue, densité* (C.R. Ac. Sc. **180** (1925), p. 245), pli cacheté deposé, November 17, 1924.

[2] *Dimension und äuszeres Masz*, Math. Ann. **79** (1918), p. 157–179.

[3] I gave such a sketch in l'Enseignement mathématique (1927), p. 240–250.

[4] *Les dimensions d'un ensemble abstrait*, Mathematische Annalen **68** (1909), p. 145–168. See also *Leçons sur les Espaces abstraits*, Gauthier–Villars, Paris 1928.

that we are about to define is also invariant under these transformations.

We will use the term *dimension order:*

1. In a different way than the concept of dimension type of Fréchet.

2. Noting that we are using an ordinal number on a scale that is infinitely finer than the real numbers. In fact, we will see that the dimension order is a growth order as in the classical concept due to Émile Borel.[5]

3.   Given the above this is how we will introduce dimension order. For example, let E be a closed set of Euclidean 3–space (generalizing to the n–dimensional case offers no difficulties). From each point Q of E draw a sphere of radius ρ. The set E_ρ of points in the interior of each of these spheres has a well determined volume $f(\rho)$, an increasing function of ρ. Such a volume was, according to a remark of Henri Lebesgue,[6] already proposed by Cantor. Minkowski also used it to give a definition of length of a curve or area of a surface. Thus we will denote by the name *construction of Cantor–Minkowski* the construction of the set E_ρ starting from the set E. This is also the set of points whose distance from E is less than ρ.

With this definition in hand we will obtain the dimensional properties of the set of infinitesimal characteristics of the volume $f(\rho)$ as ρ tends to zero in the Cantor–Minkowski construction. If we first choose the following: *A point, a line segment, a plane area,* then each of these $f(\rho)$ will have a well defined infinitesimal order whose value will be exactly equal to 3, the number of spatial dimensions, minus the number of dimensions (in the ordinary sense) of the set in question.

We are thus led to the following definition: *If $f(\rho)$ is of infinitesimal order α, then the dimension number of the set in question is $3 - \alpha$.*[7]

However, we must be more precise for two reasons:

1. Because the function $f(\rho)$ of the Cantor–Minkowski construction does not necessarily have a well defined infinitesimal order as a number.

---

[5] See *Leçons sur la théorie de la croissance.*
[6] *Leçons sur l'intégration,* 2nd edition, footnote 1 of page 37.
[7] The function $f(\rho)$ is not necessarily an infinitesimal: it can, as ρ tends to zero, tend to a finite limit—it is necessary and sufficient that the *outer measure* of E be non zero.

2. Because, in general, a closed set E will not have the degree of homogeneity required for the set to have at each point the same number of dimensions. For example, this negative property is realized for the set consisting of the union of a circular disk and a line segment.

One must thus give, instead of a dimensional number, a dimensional order, and also give to the notion of dimensional order a local character.

4. Before studying local dimensional properties it is important to note the following: we will always be able to work on the set E, *taken in its entirety,* when we will want to express the fact that *its dimensional order is bounded above* in a certain way: in fact, this amounts to saying that we know a function $f_1(\rho)$ such that the ratio

$$\frac{f(\rho)}{f_1(\rho)}$$

stays (in the neighborhood of $\rho = 0$) less than a fixed number or, in particular, tends to zero. If the function $f_1(\rho)$ has a well defined infinitesimal order α_1, one can say that the dimensional order of the set is $3 - \alpha_1$, and even in the special case when the above ratio tends to zero we will say that the dimensional order is $< 3 - \alpha_1$. But the set of comparison functions $f_1(\rho)$ could possibly be larger than that of the functions ρ^α, for example (this is once again only a special case), we could be led to compare $f(\rho)$ to a function of the form

$$\rho^\alpha (\log \rho)^{\alpha_1} (|\log \log \rho|)^{\alpha_2}$$

or of an analogous type with an arbitrary number of iterated logarithms.

On the other hand, if we want to make sure that the dimension order of the set E is at each point greater than a certain bound, we will no longer be able to work on the entire set E because the parts of E that contribute the most from the point of view of dimension furnish the principal part of $f(\rho)$. Thus, this function does not tell us anything about the subsets where the dimension order is less than that of the entire set. We will proceed as follows: let e be the subset of E not lying outside a closed sphere with center a point of E and radius ε arbitrarily small: applying the Cantor construction to e, we will write that the volume $f(\rho)$ that corresponds to e is such that its ratio to a function type (representing a given bound) stays greater than a given positive number, well defined for each value of ε.

5. We now present some remarks that will be useful for what follows:

a. Let E be a closed bounded set and for a given value of ρ let $f(\rho)$ be the volume of E_ρ. The set E_ρ, being an open set, is by definition the *sum* (in the usual set theoretic sense of this word) of spheres of radius ρ having as centers the points of E. The set of these spheres can be replaced by a countable subset, a trick often used by Poincaré. The set of centers of the spheres in this countable subset make up an everywhere dense subset of E.[8]

Denote by θ an arbitrary number between 0 and 1. We will establish the following theorem:

The volume of the set $E_{\theta\rho}$ obtained by applying the Cantor construction with radius $\theta\rho$ to E is greater or equal to the volume obtained by applying the Cantor construction with the same radius $\theta\rho$ to a homothetic image of E with ratio θ, i.e., θE. This construction gives rise to $(\theta E)_{\theta\rho}$.

In fact, let us consider the sequence of spheres whose sum is exactly the open set $E_{\theta\rho}$ and for which the first n covers a volume v_n that tends to $f(\theta\rho)$. Consider the set θE of images of the centers of these spheres and take these as centers of spheres equal to the previous ones. We will get a volume w_n that, as n increases indefinitely, tends to the volume $\theta^3 f(\rho)$ of the set $(\theta E)_{\theta\rho}$. Moreover, it is easy to see by induction that

$$w_n \leq v_n$$

and, in fact, the inequality is immediate for the case of two spheres. Assume that this holds up to n, then consider a system (v) of equal spheres with centers $A_1, \ldots, A_n, A_{n+1}$ and a system (w) consisting of the same spheres with centers $B_1, B_2, \ldots, B_{n+1}$ such that

$$\overrightarrow{B_{n+1}B_1} = \theta\,\overrightarrow{A_{n-1}A_1}, \quad \overrightarrow{B_{n+1}B_2} = \theta\,\overrightarrow{A_{n+1}A_2}, \quad \overrightarrow{B_{n+1}B_n} = \theta\,\overrightarrow{A_{n+1}A_n}.$$

The volume w_{n+1} is the sum of w_n and the part β of the sphere B_{n+1} that lies outside the sum of the sphere $B_1, B_2, \ldots, B_n$. Similarly, the volume v_{n+1} is the part α of the sphere A_{n+1} lying outside the sum of the spheres $A_1, A_2, \ldots, A_n$. It thus suffices to compare the part β of the sphere B_{n+1} to the part α of the sphere A_{n+1}. However, to clarify this point, one can always take it to A_{n+1} by modifying the center of the homothety and so B_{n+1} can be identified with it: it is then clear that the

---

[8]This follows from the fact that a sequence of closed sets f_n can be defined by the property that $f_n \subset f_{n+1}$ and $\lim_{n=\infty} f_n = E_\rho$ and that each f_n can be recovered by the Borel–Lebesgue lemma by means of a finite number of spheres of radius ρ with centers in E.

part removed from A_{n+1} (or B_{n+1}) by a sphere A_i will also be *a fortiori* removed by a sphere B_i which proves that one has $\beta \leq \alpha$. So

$$w_n - \beta \leq v_n - \alpha$$

and finally

$$w_{n+1} \leq v_{n+1}.$$

The claimed inequality thus holds for a finite number of spheres.

We now let n increase indefinitely: in the limit we will still have

$$w \leq v,$$

or, in more suggestive form,

$$1 > \frac{f(\theta\rho)}{f(\rho)} > \theta^3.$$

This inequality implies some interesting properties of the function $f(\rho)$. It follows by writing $\Delta\rho$ for a positive increment that

$$\frac{f(\rho)}{f(\rho - \Delta\rho)} > \left(\frac{\rho}{\rho - \Delta\rho}\right)^3,$$

from which

$$\frac{f(\rho + \Delta\rho) - f(\rho)}{f(\rho)} < \left(1 - \frac{\Delta\rho}{\rho}\right)^3 - 1,$$

and so

$$\lim_{\Delta\rho=0} \frac{f(\rho - \Delta\rho) - f(\rho)}{\Delta\rho} \leq 3\frac{f(\rho)}{\rho}.$$

Thus *our function has a right derivative which never exceeds the value* $3f(\rho)$: ρ. By similarly writing

$$\frac{f(\rho - \Delta\rho)}{f(\rho)} < \left(1 - \frac{\Delta}{\rho}\right)^3$$

we get

$$\lim_{\Delta\rho=0} \frac{f(\rho) - f(\rho - \Delta\rho)}{\Delta\rho} \leq \frac{3f(\rho)}{\rho}.$$

We thus get that *the function has a left derivative having the same property*. From this we deduce that *the function* $f(\rho)$ *is continuous and that the set* E_ρ *has the same inner and outer volume.*

Moreover, we know that the two derivatives of $f(\rho)$ will be equal except on a set of measure zero, but in this case it is easy to see that equality always holds. In fact, reexamining the function $v_n(\rho)$ which tends to $f(\rho)$ and has a continuous derivative, its derivative is the area of a polyhedron with spherical faces bounding the volume $v_n(\rho)$. By the above, we have that for any n

$$v_n'(\rho) \leq \frac{3v_n(\rho)}{\rho} < \frac{3f(\rho)}{\rho},$$

so the boundary of E_ρ consists of surfaces of finite total area. As well, by the above method (but substituting areas for the volumes) one gets

$$v_n'(\theta\rho) \geq \theta^2 v'(\rho),$$

and so the v_n are also continuous in an interval (ρ_1, ρ_2) (with $0 < \rho_1 < \rho_2$).[9] It follows that they have a continuous accumulation function whose primitive is an accumulation function of the $v_n(\rho)$, i.e., the primitive function is $f(\rho)$. We conclude that $f(\rho)$ has a continuous derivative representing the area of the surface $\rho = $ constant.

   b. The Cantor–Minkowski construction applied to a set E is a special case of the following: about a point O consider a region ω_1 of well defined volume (i.e., with equal inner and outer measure) and suppose that O is in the interior of this domain so that the absolute minimum distance δ of O to the boundary of ω_1 is $\neq 0$. Let Δ be the maximum of this distance. To each point Q of E we associate a region ω_ρ that corresponds to ω_1 by a homothety of ratio ρ taking the point Q to the point O. Let $F(\rho)$ be the volume of the set of interior points of the set ω_ρ. We then have

$$f(\rho\delta) < F(\rho) < f(\rho\Delta),$$

where

$$\left(\frac{\delta}{\Delta}\right)^3 < \frac{f(\rho\delta)}{f(\rho\Delta)} < \frac{F(\rho)}{f(\rho)} < \frac{f(\rho\Delta)}{f(\rho\delta)} < \left(\frac{\Delta}{\delta}\right)^3.$$

Thus the ratio $F(\rho) : f(\rho)$ and its inverse stay bounded and so, with regards to evaluating the dimension number, we can substitute the construction leading to the volume $f(\rho)$ (i.e., the Cantor–Minkowski construction) with the one leading to the volume $F(\rho)$.

---

[9]It goes without saying that one must exclude a small neighborhood of the value $\rho = 0$. The set E in question can be a non rectifiable *surface* even though the surfaces $\rho = $ positive const. can be rectifiable.

c. In some cases it will be convenient, given a certain radius ρ, to consider a system S of spheres of radius ρ that are either pairwise disjoint or tangent and that satisfy the following properties: (i) each has its center in E, (ii) each sphere of radius ρ with center in E that does not belong to S is neither disjoint nor tangent to one of the spheres of S. Let K be the cardinality of S. We have

$$\frac{4}{3} K\pi\rho^3 < f(\rho).$$

As well, if we draw spheres concentric to the ones in S but of radius 3ρ, it is clear that the volume of the set of interior points of at least one of these spheres will surpass $f(\rho)$. We finally get that

$$\frac{4}{3} K\pi\rho^3 < f(\rho) < 36K\pi\rho^3,$$

which implies that the ratio $f(\rho)/K\rho^3$ is bounded between two fixed positive numbers.

d. Let E be a bounded closed set and a (closed) cube containing it whose edge we will take as a unit of length. We then divided it into N equal parts obtaining N^2 equal cubes. We will call one of these cubes an *empty cell* if it does not contain a point of E in its interior or its boundary, otherwise, we will call it a *useful cell*. Given this definition, let us consider the total volume of the useful cells. This is a function of $\rho = 1/N$ bounded above by the volume $f(\rho\sqrt{3})$ arising from the Cantor construction on the set E with radius $\rho\sqrt{3}$, the diagonal of a subcube. As well, if we do the Cantor construction with radius ρ we get a volume lying in the interior of the one covered by homotethic cubes concentric to the useful cells and with edges three times larger. We get the inequalities

$$\frac{1}{27} f(\rho) < \varphi(\rho) < f(\rho\sqrt{3}),$$

and, as in *a*, the volume $f(\rho\sqrt{3})$ is less than $3\sqrt{3} f(\rho)$ and it is seen that one can determine the dimension order of E by substituting the volume $f(\rho)$ of the Cantor construction with the total volume of the useful cells.

6. Remarks *b, c, d* allow us to vary the evaluation process of the dimension order while still obtaining equivalent results. It follows that we can similarly benefit from a certain flexibility when, from the local point of view, we look for the *dimension order of the set at a point Q*.

In order to do this, we consider a sphere with Q as center and ε as radius and let e be the closed set consisting of points of E not lying outside this sphere. Let e/ε be a homotethic image of e with ratio $1/\varepsilon$ and effecting the Cantor construction on this set to obtain a function $f_\varepsilon^Q(\rho)$. If we know a function $f_1^Q(\rho)$ such that the ratio

$$\frac{f_1^Q(\rho)}{f_\varepsilon^Q(\rho)}$$

stays bounded (or tends to zero) independently of Q as ρ tends to zero, we say that at the point Q the dimension order of our set is not less than the infinitesimal order, in the large sense, of the function

$$\rho^2 : f_1^Q(\rho).$$

Similarly, if we know a function $f_2^Q(\rho)$ such that the ratio

$$\frac{f_\varepsilon^Q(\rho)}{f_2^Q(\rho)}$$

stays bounded (or tends to zero) independently of Q as ρ tends to zero, we will say that at the point Q the dimension order of the set does not surpass the infinitesimal order, in the large sense, of the function

$$\rho^2 : f_2^Q(\rho).$$

Finally, if we know a function $f^Q(\rho)$ such that the ratio

$$\frac{f_2^Q(\rho)}{f^Q(\rho)}$$

and its inverse are bounded independently of Q, we will say that the dimension order of E at Q is equal to the infinitesimal order, in the large sense, of the function $f^Q(\rho)$. In the case when this property holds with the same function $f(\rho)$ for all points of E, we will say that E is *isodimensional*.[10]

---

[10]This definition of isodimensionality is in some sense restrictive: the surface of revolution generated by the curve $y^2 = x^3$ rotating about Ox would not be isodimensional. Note also that at the exceptional points of a rectifiable line, the dimension order can be larger than one (see my article *Sur l'aire d'un domaine plan*, Bulletin des Sciences mathématiques, February 1928).

In these definitions one can simply consider a sequence of ε decreasing to zero. One can also substitute spheres with inscribed cubes containing the point Q in their interior and tending to this point, provided that the ratio of an edge of such a cube to the minimal distance from Q to one if its faces stays larger than a fixed θ_1. This precaution is necessary in order to insure that the results of the two above processes stay compatible. This is seen in the case when E contains a circular disk and a line segment perpendicular to the disk passing through its center and Q is taken to be the center of the disk. In this case the sphere covering will necessarily give an upper bound to the indeterminacy of the local dimension order at the point Q, that is 2. With inscribed cubes we can obtain any result between the limits 1 and 2: however, a condition like the preceding one will have the effect of opposing such an indeterminacy.

The inscribed cubes that we are led to consider in order to evaluate local dimension properties are the elements of an indefinite progressive packing that is determined by the following process: from a given initial cell we determine the cells of order 1 by dividing the edge of the cell into h_1 equal parts, where h_1 is an integer ≥ 2. A cell of order 2 is defined by dividing the edge of a cell of order 1 by an integer $h_2 \geq 2$, and so forth. Consider a point Q of E. One can always obtain it as the limit of a sequence of inscribed cubes whose general term is a cell of order i. By surrounding each c_i by the 26 contiguous cubes of the same order as c_i, we get a new sequence of inscribed cubes C_i such that the edge of each one is three times that of the corresponding c_i in the original sequence and such that the ratio between the edge of one of these cubes and the minimal distance from Q to its faces is $\geq 1/3$. Such a sequence of cubes can be substituted for a sequence of concentric spheres in the constructions relating to the dimension order (maximum at the point Q).

6bis. Actually, the problem that seems to arise most naturally in applications dealing with questions of dimensional properties consists of determining either that the dimension order of a set is less than or equal to the infinitesimal order of a certain function (which will most often be an ordinary number) or, on the contrary, that this dimension order is everywhere $\geq$ a known infinitesimal order. In the first case, as indicated in paragraph 4, we will give a condition relating to the set E taken as a whole. In the second case it will suffice to give a condition relating to the entirety of a closed spherical or cubic (that is, cut out by a sphere or a cube) subset e by adding the condition is satisfied for *every subset* obtained by a sphere having its center in e or by a cube containing at least one point in the interior of e.[11]

7.   We present another consequence of remark b.

Let $(x_1, x_2, \ldots, x_p)$ be a point of a set E in p dimensional Euclidean space and $(x_{p+1}, x_{p+2}, \ldots, x_{p+q})$ a point in another set E' of q dimensional Euclidean space. After Fréchet the *composition* of these two sets is the set (E, E') consisting of points $(x_1, \ldots, x_p; x_{p+1}, \ldots, x_{p+q})$ in $p + q$ dimensional Euclidean space, i.e., the set with first p coordinates a point in E and the last q coordinates a point in E'. So for two sets E, E' effect the Cantor–Minkowski construction on each of the sets and compose the two sets E_ρ and E'_ρ. We get a set that can also be constructed from (E, E') by doing not the Cantor construction proper but a variant that is the subject of remark b: the region ω_1 is obtained by composing a p–sphere of radius ρ in E space with a q–sphere of radius ρ of E' space, so it will be a kind of bicylinder. One is immediately led to a theorem about the addition of dimension order by composition. If E and E' are isodimensional then it is easy to see that (E, E') will also be. If we have two sets that are not isodimensional but there are two points Q and Q' for which these sets have well defined dimension order (in the sense of paragraph 6), then at the composed point (Q, Q') the dimension order of (E, E') is the sum of the two preceding ones. Otherwise, one can use inequalities to get a new form of the theorem about addition of orders by composition.

8.   When we have a line in the plane or a curve joining two points A and B, its dimension order is everywhere ≥ 1. This holds because the volume in the Cantor construction corresponding to the segment AB is the minimum volume that this construction gives for any continuous region containing A and B.

Similarly, *if a continuous region is the boundary of a domain*, it is easy to see that *its dimension order is everywhere* ≥ 2.[12] In fact, let Q be a point

---

[11]Using this point of view one can give a definition of isodimensionality more general than the one of paragraph 6. A set will have a given dimension order equal to the infinitesimal order of a known function $\varphi(\rho)$ if it satisfies both the following conditions:

1. The volume arising from the Cantor–Minkowski construction applied to the whole of E has bounded ratio $\rho^3 : \varphi(\rho)$.

2. The function $\rho^3 : \varphi(\rho)$ is bounded for the volume given by the Cantor construction for each closed spherical or cubic part of E when these contain at least one point of E in their interior.

[12]This result must be interpreted in the sense of paragraph 6b–it could happen that one could get less for the local dimension order.

of this continuous boundary. It is both the limit of interior and exterior points. Let I and E be two such points in the neighborhood of Q, so a path starting at the point I and ending Q must necessarily cross the continuous boundary in the neighborhood of Q. One can always arrange it that this point belongs to the line segment IE and a segment parallel to it as long as they are sufficiently close. The stated theorem corresponds to a new minimal property of the Cantor construction: given a cylinder of revolution, it gives a correspondence between a straight section of this cylinder and the minimum volume obtained by the same construction applied to a continuous region intersecting in at least one point each parallel to the generators lying in the interior of the cylinder.

It follows from this that, in three space, a set whose dimension number is < 2 can be the boundary of only a single set (its exterior). The same holds in n dimensions for a set whose dimension number is $< n - 1$.

Bibliography

[1] G. Bouligand, Bull. Soc. Polon. Math. **5** 1925, 59–112.

[2] G. Bouligand, *Dimension, étendue, densité*, Comptes Rendus Acad. Sci. Paris **180** (1925), 245–248.

[3] F. Hausdorff, *Dimension und äußeres Maß*, Math. Annalen **79** (1918), 157–179.

[4] G. Bouligand, *Nombre dimensionnel et ensembles impropres dans le problème de Dirichlet*, Enseignement Mathématique (1927), 240–250.

[5] M. Fréchet, *Les dimensions d'un ensemble abstrait*, Math. Annalen **68** (1909), 145–168.

[6] M. Fréchet, *Leçons sur les Espaces Abstraits*, Gautier-Villars, Paris, 1928.

[7] E. Borel, *Leçons sur la théorie de la croissance*, Paris, 1910.

[8] H. Lebesgue, *Leçons sur l'intégration et la recherche des fonctions primitive*, Paris, 1928.

[9] G. Bouligand, *Sur l'aire d'un domaine plan*, Bulletin des Sciences mathématiques **52** (1928), 55–63.

[10] A. Debus, editor, *World Who's Who in Science*, A. N. Marquis, Chicago, 1968.

[11] C. Tricot, *Douze définitions de la densité logarithmique*, Comptes Rendus Acad. Sci. Paris, Série I **293** (1981), 549–552.

This paper defines a fractional dimension in another way, called in the selection the **metric order**, nowadays sometimes known as the "capacity dimension". Note that the definition makes sense for an abstract compact metric space (F, ρ), not necessarily embedded in a Euclidean space.

Lev Semenovich Pontrjagin[1] was one of the leading Soviet mathematicians. He was born in Moscow in 1908, educated at Moscow State University, and later became professor at the same institution. His interests included topology and the theory of continuous groups. [7]

Lev Genrikhovich Schnirelmann[1] (1905–1938) was born in Gomel, Russia. He was educated and spent most of his career in Moscow. His mathematical work includes contributions to the calculus of variations and number theory. [10]

---

[1] These are the transliterations used in the *Annals of Mathematics* (1932). In Cyrillic: Понтрягин, Шнирельман.

EIGHT

On a Metric Property of Dimension

L. Pontrjagin & L. Schnirelmann

The metric order of F is compared to the topological dimension of F, defined in the way nowadays called the "covering dimension", due to Brouwer, Urysohn, and Menger. The definition is especially simple in the compact case: a compact metric space F has topological dimension $\leq r$ iff for every $\varepsilon > 0$, there is a cover of order $\leq r + 1$ of F by finitely many closed sets with diameter $< \varepsilon$. A cover has **order** $\leq r + 1$ iff every $r+2$ of the sets have empty intersection. (See [11], p. 55 or [8], Theorem 3.4.6; note that this use of "order" differs by 1 from that in the references.) The use of finite closed covers is the same as the use of open covers [8], Theorem 3.4.3.

Urysohn's **flattening coefficient** is used here. Let a non-negative integer r be given. The rth flattening coefficient of F is the largest number d such that every cover by closed sets of diameter $< d$ has order $> r$. This coefficient is positive if and only if the topological dimension of F is $\geq r$.

A **Lebesgue tiling** of Euclidean space $\mathbb{R}^n$ is a cover of $\mathbb{R}^n$ by small cubes (aligned with the coordinate axes), such that the intersection of any k of them is contained in an $(n-k+1)$-dimensional face $(2 \leq k \leq n+1)$. For example in the plane, we should tile by squares so that any two of them intersect at most in a line segment, three intersect at most in a point, and four or more do not intersect at all. (This is the way bricks are often arranged.)

Other facts that are used: A subset of $\mathbb{R}^n$ that has topological dimension n has an interior point [11], Theorem IV 3 or [12], p. 244. Approximation of an r-dimensional space by a sequence of polyhedra in $\mathbb{R}^{2r+1}$: [11], Chapter V Section 9 or [5] (also Selection 6 for $r = 1$).—Ed.

The aim of this note is to derive the dimension of a compact metric space using a covering method by "units of measure" whose number allows one to define dimension. This method is analogous to the method of measure used in elementary geometry.

The definitions of dimension always use purely topological notions that set aside metric properties. As will be seen, there nevertheless exists a quite striking relationship between metric properties of a compact set and its dimension.

We will use the following formulation for the definition of dimension (in the well known sense of Brouwer, Urysohn, and Menger):

A compact set F has dimension n if n is the smallest integer such that there exists (ε being an arbitrary small positive number) a finite system of closed sets covering F with diameters not exceeding ε and for which all $n + 2$-fold intersections are empty.

Let a compact metric space F be given. We will cover it by a finite system of closed sets whose diameter does not exceed ε. By the Heine-Borel theorem this is possible. We can then define the minimum number $N(\varepsilon)$ of such sets necessary to cover F. The number $N(\varepsilon)$ takes only positive values for all positive values of ε and increases indefinitely as ε tends to 0 (except when the set F only contains a finite number of points). This function clearly depends on the metric of the space F. We will call it the volume function of the space F relative to the given metric and write it as $N_F(\varepsilon)$.

Let r be an arbitrary real number, we will say that r belongs to the first class if there exists a positive number c such that

$$N_F(\varepsilon) \geq \frac{c}{\varepsilon^r} \quad \text{(for any } \varepsilon > 0),$$

and otherwise that r belongs to the second class. The Dedekind cut obtained by this determines a non negative k ($\leq +\infty$) that we will call *the metric order of the space* F. It is easily seen that

$$k = \liminf \left(-\frac{\log N_F(\varepsilon)}{\log \varepsilon} \right).$$

From the various metrics of F we will obtain the various volume functions and the various metric orders. For example, a simple Jordan arc metrized in an Euclidean way has as volume function (asymptotically) c/ε, where c is a constant. The metric order of such an arc is thus equal to *one*. On the other hand, we can define another metric on a simple arc in the following way: we take an everywhere discontinuous perfect set with positive measure in k-dimensional Euclidean space R^k. We can pass a simple arc through such a set. This arc automatically inherits the metric of R^k and its volume function is asymptotically c/ε^k and its metric order is equal to k.

So the metric order is not an invariant of F. However, one obviously obtains a topological invariant of the space F by considering the lower bound of the metric orders corresponding to the various metrics on F that do not change the topological properties of this space. The principal aim of this paper is to determine the topological invariant in question and *to show that it equals the Brouwer dimension of the space F.* One has

Fundamental Theorem. *The lower bound of the metric order*

$$k = \liminf \left(\frac{-\log N_F(\varepsilon)}{\log \varepsilon} \right)$$

for all the metrics of the compact space F equals its dimension.

Corollary. *If the lower bound in question is finite, it is necessarily a whole number.*

Proof.

1. Let K be a Euclidean cube in r dimensions. It is clear that, with its ordinary metric, one has (for $\varepsilon > 0$ sufficiently small and constants c and c' chosen suitably)

$$\frac{c'}{\varepsilon^r} > N_K(\varepsilon) > \frac{c}{\varepsilon^r}.$$

2. We will say that a transformation φ of a metric space F to a metric space F' is a transformation without dilation if the distance between each pair of points of F does not increase under this transformation, i.e., if

$$\varrho[\varphi(a), \varphi(b)] \leq \varrho(a, b).^1$$

Clearly every transformation without dilation is continuous.

It is easy to see that if it is possible to transform without dilation a metric space F into a metric space F', one has

$$N_{F'}(\varepsilon) \leq N_F(\varepsilon).$$

3. We show that for a compact space of finite or infinite dimension $\geq r$ one has

$$N_F(\varepsilon) > \frac{c}{\varepsilon^r}, \quad (\varepsilon \text{ sufficiently small, } c \text{ a positive constant that depends on } F).$$

---

[1] Transformations of this type were studied for the first time by M. Kolmogoroff in his work on measure theory.

We first prove the following proposition:

Auxiliary theorem: *A compact metric space F of finite or infinite dimension $\geq r$ can be transformed without dilation into an r–dimensional subset of r–dimensional Euclidean space.*

Proof of the auxiliary theorem. Let

$$(I) \qquad\qquad a_1, a_2, \ldots, a_n$$

be an arbitrary finite set of points of F. To each point x of F there corresponds a system of numbers

$$(II) \qquad\qquad \varrho_1, \varrho_2, \ldots, \varrho_n,$$

where $\varrho_1, \varrho_2, \ldots, \varrho_n$ are the distances $\varrho(a_i, x) = \varrho_i$ between x and the points in our finite set. We let x correspond to a point $\varphi'(x)$ in Euclidean n–space with Cartesian coordinates $x_1, x_2, \ldots, x_n$, where

$$(III) \qquad\qquad x_i = \frac{\varrho}{\sqrt{n}}.$$

We show that the transformation φ' defined in this way is without dilation. In fact, if p' and p'' are two arbitrary points of F and $r_1', r_2', \ldots, r_n'$, $r_1'', r_2'', \ldots, r_n''$, the corresponding coordinates of their images, then we have:

$$\varrho[\varphi'(p'), \varphi'(p'')] = \sqrt{(r_1'' - r_1')^2 + (r_2'' - r_2')^2 + \cdots + (r_n'' - r_n')^2}.$$

Let j be an index such that

$$|r_j'' - r_j'| \geq |r_i'' - r_i'| \quad (i = 1, 2, \ldots, n),$$

then we have

$$\varrho[\varphi'(p'), \varphi'(p'')] \leq (r_j'' - r_j') \sqrt{n} = \varrho_j'' - \varrho_j'.$$

But, by the triangle inequality, one has:

$$|\varrho_j'' - \varrho_j'| \leq \varrho(p', p'').$$

It follows that

$$\varrho[\varphi'(p'), \varphi'(p'')] \leq \varrho(p', p'').[2]$$

---

[2] This part of the proof is only a slight modification of an argument of Urysohn (Comptes Rendus Paris **178** (1924), p. 65).

We now show that, for a suitable choice of the system (I), the dimension of $F' = \varphi'(F)$ is not smaller than r. We first take (I) in such a way that to each point $x \subset F$ there corresponds a point a_j of the system (I) such that $\varrho(a_j, x) < d/2$, where d is the rth flattening coefficient[3] of the set F (this number is necessarily positive since the dimension of F is at least r).[4]

Let us first show that there exists a positive number δ satisfying the condition:

If $\varrho[\varphi'(p), \varphi(q)] < \delta$ then $\varrho(p, q) < d$.

In fact, suppose that the opposite is true and there exist on F two infinite convergent sequences of points

(IV) $$p_1, p_2, \ldots, p_m, \ldots$$

and

(V) $$q_1, q_2, \ldots, q_m, \ldots$$

such that

$$\varrho(p_m, q_m) \geq d \quad \text{with} \quad \varrho[\varphi'(p_m), \varphi'(q_m)] \to 0 \quad (m \to \infty).$$

Denoting by p and q the two limit points of p_m and q_m we have:

$$\varrho[\varphi'(p), \varphi(q)] = 0, \quad \varrho(p, q) \geq d,$$

i.e.,

(VI) $$\begin{cases} \varphi'(p) = \varphi'(q) \\ \varrho(p, q) \geq d. \end{cases}$$

By hypothesis, there exists a point a_j of the system (I) for which $\varrho(a_j, p) < d/2$. For this point $\varrho(a_j, q) > d/2$ (since otherwise we would have

$$\varrho(p, q) \leq \varrho(a_j, p) + \varrho(a_j, q) < d).$$

As a consequence the jth Cartesian coordinates of the points $\varphi'(p)$ and $\varphi'(q)$ are distinct, contradicting condition (VI).

Suppose that the dimension of F' were less than r. There would then exist a covering of F' by a system of closed sets $f_1', f_2', \ldots, f_k'$ having diameters less than δ with the order of this covering (that is,

---

[3] In the sense of Urysohn, Fund. Math. 8 (1926), p. 353.
[4] Such a choice of the system (I) is always possible, d being positive and F compact.

the maximum number of sets of the covering that contain the same points) not exceeding r.

Denoting by f_i the set of all points of F whose image belongs to f_i', we get a covering of the set F by sets f_i of diameter $< d$. The order of this covering does not exceed r. However, this last conclusion is impossible since the rth Uryshon number of F is equal to d.

Nevertheless, suppose that the system (I) is chosen such that the dimension of F' is not less than r and, with this hypothesis, proceed with the construction below: Let $R_1, R_2, \ldots, R_e$ be the set of all r–dimensional planes in R^n, denote by $\varphi_i(F')$ be the orthogonal projection of F' to R_i, and write F_i^* for the set $\varphi_i(F')$. We will show that at least one of the F_i^* has dimension r.

Let Q be the Lebesgue tiling[5] of R^n, where the faces are parallel to the coordinate planes. Let

(VII) $\qquad\qquad\qquad p_1, p_2, \ldots, p_t, \ldots$

be the set of all non empty intersections of $r + 1$ cubes of the tiling Q. The system (VII) only contains a finite number of elements intersecting F'. Let p_s be one of these elements. We show that if all the dimensions of the projections F_i^* are less than r then there exists an arbitrarily small parallel separation ψ of the space R^n such that $\psi(p_s)$ does not intersect F'.

In fact, let R^{n-r} be the $n - r$ dimensional space containing p_s. Because of the parallelism of R^{n-r} to one of the coordinate planes, R^{n-r} is perpendicular to one of the planes R_i, for example, to R_t, and intersects R_t at a point. By the assumption that the dimension of F_t^* is less than r, we conclude that F_t^* does not contain an interior point of R_t, that is, one can choose an arbitrarily small parallel separation ψ in such a way that the intersection of $\psi(R^{n-r})$ with F_t^*, and so $\psi(R^{n-r})$ with F' and $\psi(p_s)$ with F', are empty.

One can suppose that the separation ψ is so small that if the intersection of p_i with F' is empty then the intersection of $\psi(p_i)$ with F' is also empty. By repeating the operation of parallel separation a finite number of times, we get a parallel separation $\bar\psi$ such that all the intersections $\bar\psi(p_i)$ with F' are empty $(i = 1, 2, \ldots)$.

Denote by

(VIII) $\qquad\qquad\qquad q_1, q_2, \ldots, q_s$

the set of all cubes of the tiling $\bar\psi(Q)$ whose intersection with F' are not empty, and by $\bar f_i$ the intersection of q_i with F'. The intersections of all

[5]Fund. Math. 2 (1921), p. 266.

the products of the $r + 1$ cubes of the system (VIII) do not intersect F'. The system $\bar{f}_1, \bar{f}_2, \ldots, \bar{f}_s$ is thus a covering of the F' of order not exceeding r.

But since the tiling Q (and so the tiling $\bar{\psi}(Q)$) is arbitrarily small, the dimension of F' is necessarily smaller than r, which is impossible.

We can therefore conclude that at least one of the projections F_i^* of the set F', for example F_j^*, is of dimension r. As projection is a transformation without dilation so is the transformation $\varphi(F) = \varphi_j(\varphi'(F))$ and it takes F to an r–dimensional set located in r–dimensional Euclidean space. The auxiliary theorem is thus proved.

The set that we have just found necessarily contains an interior point of R^r and it easily follows that, for ε sufficiently small,

(1) $$N_F(\varepsilon) \geq N_{F_j^*}(\varepsilon) \geq \frac{c}{\varepsilon^r}.$$

4. Let us now show that r is exactly the lower bound of all exponents

$$\liminf \left[-\frac{\log N_F(\varepsilon)}{\log \varepsilon} \right].$$

To this end, it will suffice to find a metric on F for which $N_F(\varepsilon)$ becomes less than $c/\varepsilon^{r+\delta(\varepsilon)}$, where $\delta(\varepsilon) \to 0$ as $\varepsilon \to 0$.

We start with approximation of sets in Euclidean space.[6]

Given the r–dimensional compact metric space F it is possible to find in $2r+1$–dimensional Euclidean space R^{2r+1} a sequence $C_1, C_2, \ldots, C_n, \ldots$ of r dimensional complex polyhedra having the following properties

1. The sequence has a topological limit F' homeomorphic to F.

2. It can be constructed in such a way that if the first n terms $C_1, C_2, \ldots, C_n$ have already been constructed, then one can place the $(n + 1)$st term C_{n+1} in an arbitrary neighborhood of C_n.

We can assume that all the complexes of the sequence $C_1, C_2, \ldots$ are located in a finite part of the space R^{2r+1}. Since the complexes C_n are polyhedral and the combinatorial structure of C_i is determined beforehand, it is possible to choose for each index i and each positive number δ_i a positive number ε_i such that $N_{C_i}(\varepsilon) < (1/\varepsilon)^{r+\delta_i}$ (for $\varepsilon \leq \varepsilon_i$).

---

[6]For a proof see an article of L. Pontrjagin and Miss Tolstowa to appear in volume 105 of "Matematische Annalen." See also Lefschetz, Annals of Mathematics 32 (1931), number 3.

One can further suppose that $\varepsilon_i = \tau^{p_i}$, where $0 < \tau < 1$, the p_i are positive integers with $p_{i+1} > p_i$, and $\delta_1, \delta_2, \ldots$ is an arbitrary sequence of positive numbers tending to 0.

Suppose that the position of C_1 in G is determined. Denote by φ_j, $j = p_1, p_1 + 1, \ldots, p_2 - 1$, a covering of the complex C_1 by sets of diameter less than τ^j such that their number does not exceed $(1/\tau^j)^{r+\delta_1}$. By enlarging all the sets of the covering in such a way that the diameters of the enlarged sets are also less than τ^j we obtain a covering φ_j' in a neighborhood G_j' of the complex C_1. The number of elements in the covering φ_j' is equal to that of φ_j, that is, it does not exceed $(1/\tau^j)^{r+\delta_1}$.

Denoting by G_1 the intersection of the neighborhoods G_{p_1}', G_{p_1+1}', $\ldots, G_{p_2-1}'$ we get

$$N_{\bar{G}_1}(\varepsilon) \leq (1/\varepsilon)^{r+\delta_1} \quad (\text{for } \varepsilon = \tau^j, \ j = p_1, p_1 + 1, \ldots, p_2 - 1).$$

We construct the complex C_2 in the interior of the neighborhood G_1. By iterating the preceding constructions we obtain a sequence of complexes $C_1, C_2, \ldots, C_i, \ldots$ and their neighborhoods $G_1, G_2, \ldots, G_i, \ldots$ satisfying the following conditions:

1. The closure $\bar{G}_i$ of G_i lies inside G_{i-1}.

2. For each index j, $p_{i-1} \leq j \leq p_i - 1$, one has:

$$N_{\bar{G}_{i-1}}(\tau^j) \leq \left(\frac{1}{\tau^j}\right).$$

But the topological limit F' of the sequence $C_1, C_2, \ldots, C_n, \ldots$ lies in the intersection of all the $\bar{G}_1, \bar{G}_2, \ldots, \bar{G}_n, \ldots$.

It follows that:

$$N_{F'}(\tau^j) \leq \left(\frac{1}{\tau_j}\right)^{r+\delta'(j)} \quad \begin{array}{l}(j > 0, \ \delta'(j) > 0, \text{ and } \delta'(j) \text{ tends to } 0 \\ \text{as } j \text{ increases indefinitely}).\end{array}$$

Let $\varepsilon < \tau$ be an arbitrary positive number. We can write it in the form

$$\varepsilon = s\tau^j, \quad t < s \leq 1,$$

and we get

$$N_F(s\tau^j) \leq N_{F'}(\tau^{j+1}) < \left(\frac{1}{\tau^{j+1}}\right)^{r+\delta'(j+1)} \leq \left(\frac{1}{s\tau^j}\right)^{r+\delta'(j+1)+\frac{\log s - \log \tau}{-\log(s\tau^j)}}$$

that is, $N_F(\varepsilon) < (1/\varepsilon)^{r+\delta(\varepsilon)}$.

Our fundamental theorem on dimension is now completely proved.

The dimension defined here—the "metric order"—coincides with dimensions defined in other ways (at least for subsets of Euclidean space). For example, it agrees with the lower metric dimension in Selection 17 (Kolmogorov and Tihomirov). (See [6], p. 174, [8], (6.5.12), [9], §3.1)

What is the essential difference between this definition and the definition used for the Hausdorff dimension? To define the Hausdorff dimension, we cover our set F by sets of diameter $\leq \varepsilon$, and add up the sth power of the diameters; to define the Pontrjagin-Schnirelmann dimension, we cover our set F by sets of diameter $\leq \varepsilon$, and add up the sth power of ε (one term for each set); so in fact, it would make no difference if the sets have diameter exactly ε. The essential difference is in whether the sets are all the same size, or their different sizes are taken into account. It is therefore not difficult to see that the Hausdorff dimension is always $\leq$ the Pontrjagin-Schnirelmann dimension.

But also the Hausdorff dimension is $\geq$ the topological dimension. In fact, this is a consequence of the "Auxiliary Theorem": The compact space F of topological dimension r can be transformed "without dilation" onto a set $F_j^* = \varphi(F) \subseteq \mathbb{R}^r$ also of topological dimension r. Because φ is "without dilation", the Hausdorff dimension "dim" satisfies

$$\dim F \geq \dim F_j^*$$

([8], Exercise (6.1.9) or [9], Prop. 2.3). Since F_j^* has an interior point (as in the selection), of course $\dim F_j^* = r$. Thus $\dim F \geq r$.

So the "Fundamental Theorem" is also true for the Hausdorff dimension: *The lower bound of the Hausdorff dimension for all metrics on the compact space F is the topological dimension of F.*

The dimension defined here is

$$k = \liminf \left(-\frac{\log N_F(\varepsilon)}{\log \varepsilon} \right).$$

Another possibility is

$$\limsup \left(-\frac{\log N_F(\varepsilon)}{\log \varepsilon} \right).$$

Of course, in many cases the two coincide, but not in every case. —Ed.

Bibliography

[1] P. Urysohn, *Les classes (D) séparables et l'espace Hilbertien*, Comptes Rendus Acad. Sci. Paris **178** (1924), 65–67.

[2] P. Urysohn, *Mémoire sur les multiplicités Cantoriennes (suite)*, Fundamenta Math. **8** (1926), 223–359.

[3] H. Lebesgue, *Sur les correspondances entre les points de deux espaces*, Fundamenta Math. **2** (1921), 256–285.

[4] L. Pontrjagin and G. Tolstowa, *Beweis des Mengerschen Einbettungssatzes*, Math. Annalen **105** (1931), 734–745.

[5] S. Lefschetz, *On compact spaces*, Annals of Math. **32** (1931), 521–538.

[6] M. F. Barnsley, *Fractals Everywhere*, Academic Press, 1988.

[7] A. Debus, editor, *World Who's Who in Science*, A. N. Marquis, Chicago, 1968.

[8] G. A. Edgar, *Measure, Topology, and Fractal Geometry*, Springer-Verlag, New York, 1990.

[9] K. Falconer, *Fractal Geometry: Mathematical Foundations and Applications*, Wiley, 1990.

[10] C. C. Gillespie, editor, *Dictionary of Scientific Biography*, Scribners, New York, 1970.

[11] W. Hurewicz and H. Wallman, *Dimension Theory*, Princeton University Press, 1948.

[12] Karl Menger, *Dimensionstheorie*, B. G. Teubner, Leipzig, 1928.

In this selection the fractal dimension of certain subsets of the line (defined in terms of the frequencies of ones and zeros in the dyadic expansion) is computed. The "dimensional number" used is, of course, the Hausdorff dimension. The Hausdorff dimension of a set is denoted d(S).

Note the use of multiplication signs to mean intersection of sets. F × E means F ∩ E. The number of elements in a (finite) set A is denoted by N(A).

Abraham Samoilovitch Besicovitch (1891–1970) was born at Berdiansk, on the Sea of Azov, and graduated from the University of St. Petersburg in 1912. After teaching at the University of Perm and the Naval Engineering School, he left the Soviet Union in 1924. After brief periods in Stockholm, Oxford, and Liverpool, he settled at Cambridge University as Cayley Lecturer and then as Rouse Ball Professor of Mathematics. Many of the early papers on Hausdorff dimension were written by Besicovitch. Besicovitch made important contributions to the study of probability theory, almost periodic functions, complex variables, measure theory, differentiation in the general theory of functions of a real variable. He solved the "Kakeya problem" by showing that an interval can be moved in the plane and returned to its original position (by 180 degrees), while staying within a region

In this selection, the fractal dimension of certain subsets of the line (defined in terms of the frequencies of ones and zeros in the dyadic expansion) is computed. The "dimensional number" used is, of course, the Hausdorff dimension. The Hausdorff dimension of a set E is denoted dE.

Note the use of multiplication signs to mean intersection of sets: $F \times E$ means $F \cap E$. The number of elements in a (finite) set A is denoted by $\mathfrak{N}\{A\}$.

Abraham Samoilovitch Besicovitch (1891–1970) was born at Berdjansk on the Sea of Azov, and graduated from the University of St. Petersburg in 1912. After teaching at the University of Perm and the Naval Engineering School, he left the Soviet Union in 1924. After brief periods in Stockholm, Oxford, and Liverpool, he settled at Cambridge University, as Cayley Lecturer and then as Rouse Ball Professor of Mathematics. Many of the early papers on Hausdorff dimension were written by Besicovitch. Besicovitch made important contributions to the study of probability theory, almost periodic functions, complex variables, geometric measure theory, differentiation in the general theory of functions of a real variable. He solved the "Kakeya problem" by showing that an interval can be moved in the plane and returned to its original position rotated by 180 degrees, while staying within a region of arbitrarily small area; the Mathematical Association of America produced a motion picture illustrating the construction of such "Kakeya sets."[9]

NINE

On the Sum of Digits of Real Numbers Represented in the Dyadic System

(On sets of fractional dimensions II.)

A.S. Besicovitch in Cambridge (England)

§1.

The problem of this paper is connected with the well–known Hardy–Littlewood problem on the sum of digits of real numbers represented in the dyadic system.[1] Denote by $P(x, n)$ the sum of the first n figures of the number $x(0 < x < 1)$. Their result is

"The inequality

$$\left| P(x, n) - \frac{1}{2}n \right| < \sqrt{n \log n}$$

is ultimately satisfied almost everywhere in $(0, 1)$."

I consider the problem of the sets of points for which $P(x, n)$ lies outside the Hardy–Littlewood region of most probable values. The problem is

To characterise in terms of measure the set of points x for which

$$\overline{\lim} \frac{P(x, n)}{n} \leqq p < \frac{1}{2}.$$

---

[1]Hardy–Littlewood, Some problems on Diophantine approximations, *Acta Math.* **37** (1914), pp. 155–190.

The solution of this problem is given in terms of sets of fractional dimensions.[2] The r-dimensional measure of a linear set E is defined as follows:

$J(E, \delta)$ denoting any set of intervals l of lengths $\leq \delta$, containing the set E, the exterior r-measure of E is defined by the equations

$$m_r^* E = \varliminf_{\delta \to 0} \sum_{J(E,\delta)} l^r .$$

To any set E corresponds a number ϱ such that

$$m_r^* E = 0 \quad \text{for any } r > \varrho ,$$

$$m_r^* E = \infty \quad \text{for any } r < \varrho .$$

We say that ϱ is the *dimensional number* of the set E and we write $dE = \varrho$ or we express the same by saying that E is a ϱ-*dimensional* set or, more simply, a ϱ-set.

A set E is said to be r-measurable if for any linear set F the equation

$$m_r^* F = m_r^* (F \times E) + m_r^* (F - F \times E)$$

is satisfied.

The sets we consider in this paper are measurable but we do not dwell on proving that.

Two linear sets of Lebesgue measure zero can be compared with each other from the point of view of measure either by the values of their measures if they have the same dimensional number or by their dimensional numbers, whereas the one with a larger dimensional number is infinitely larger than the one with the smaller, in the same sense as an area is infinitely larger than a rectifiable curve.

We shall use the following theorem, of which the proof is obvious.

Theorem. *The sum of an enumerable set of r-dimensional sets is an r-dimensional set.*

§2.

The solution of the problem of this paper is given by the following theorem:

[2]Hausdorff Dimension und äußeres Maß, *Math. Annalen* **79** (1918), pp. 157–179.–A.S. Besicovitch, On linear sets of fractional dimensions, *Math. Annalen* **101** (1929), pp. 161–193.

Theorem. *The real numbers of the interval $0 < x < 1$ for which*

$$\overline{\lim} \frac{P(x, n)}{n} \leq p < \frac{1}{2}$$

form a ϱ–dimensional set, where ϱ is defined by the equation

$$2^\varrho = \frac{1}{p^p q^q} \qquad (p + q = 1) .$$

The case of $p \geq \frac{1}{2}$ has its solution in the Hardy–Littlewood theorem.

We shall first prove three lemmas.

We shall write o for $o(1)$.

§3.

Lemma 1. *Let p be a constant satisfying the inequality $0 < p < \frac{1}{2}$ and ϱ the constant defined by the equation*

$$2^\varrho = \frac{1}{p^p q^q} \qquad (p + q = 1) ,$$

further let $s(n)$ satisfy the inequality

$$0 \leq s(n) \leq p\,n - 1 ,$$

then

$$\sum_{s(n) < i \leq p\,n} C_n^i = 2^{(\varrho + o)n}$$

where $C_n^i = \frac{n!}{i!(n-i)!}$ is the number of combinations of i elements out of n elements.

Denoting the integral part of $p\,n$ by $p'n$ so that

$$p' \leq p < p' + \frac{1}{n}$$

and observing

$$\frac{C_n^{i-1}}{C_n^i} = \frac{i}{n - i + 1} < \frac{i}{n - i}$$

we see that for all $i \leq p\,n$

$$\frac{C_n^{i-1}}{C_n^i} < \frac{p'n}{n - p'n} = \frac{p'}{q'}, \quad \text{where } p' + q' = 1,$$

so that

$$C_n^i \leq \left(\frac{p'}{q'}\right)^{p'n-i} C_n^{p'n}$$

and

(1)
$$\sum_{s(n) \leq i \leq p\,n} C_n^i < C_n^{p'n} \sum_{i=0}^{p'n} \left(\frac{p'}{q'}\right)^{p'n-i} < \frac{q'}{q'-p'} C_n^{p'n}.$$

We have also

(2)
$$\sum_{s(n) \leq i \leq p\,n} C_n^i \geq C_n^{p'n}.$$

Writing $A \sim B$ for the statement

$$\lim \frac{A}{B} = 1$$

we have by the Stirling formula

$$C_n^{p'n} = \frac{n!}{(p'n)!(q'n)!} \sim \sqrt{\frac{1}{2\pi pqn}} \left(\frac{1}{p'^{p'} q'^{q'}}\right)^n = 2^{(\varrho+o)n}.$$

Obviously we have also

$$\frac{q'}{q'-p'} C_n^{p'n} = 2^{(\varrho+o)n}$$

and thus from (1) and (2) the lemma follows.

$$\S 4.$$

Lemma 2. *Given constants* $\varepsilon > 0$, $\eta > 0$, $0 < p < \frac{1}{2}$ *there exists a constant* N *such that for every* $n > N$

(1)
$$\sum_{n_1 > N} \sum_{m_1 > (p+\varepsilon)n_1} C_n^{m_1} C_n^{m_2} < \eta C_n^{p'n}$$

and

(2)
$$\sum_{n_1 \geq N} \sum_{m_1 < (p-\varepsilon)n_1} C_{n_1}^{m_1} C_{n_2}^{m_2} < \eta C_n^{p'n}$$

where $n_1 + n_2 = n$, $m_1 + m_2 = p'n$, n_1, n_2, m_1, m_2 being non negative integers and p' having the same meaning as in Lemma 1.

Take $m_1 > n_1 p$ and write $m_1 = n_1 p_1$, $m_2 = n_2 p_2$ so that

$$p_1 > p > p_2, \quad p_1 n_1 + p_2 n_2 = p'n.$$

Consider now the ratio

(3)
$$R(m_1) = \frac{C_{n_1}^{m_1+1} C_{n_2}^{m_2-1}}{C_{n_1}^{m_1} C_{n_2}^{m_2}} = \frac{(n_1 - m_1)m_2}{(m_1 + 1)(n_2 - m_2 + 1)}$$

$$< \frac{(n_1 - m_1)m_2}{m_1(n_2 - m_2)}$$

$$< \frac{n_1 - m_1}{m_1} \frac{p}{q}.$$

Assuming ε small enough we see that for $m_1 > (p + \varepsilon)n_1$

(4)
$$R(m_1) < 1 - \frac{\varepsilon}{pq} + O(\varepsilon^2)$$

$$< 1 - \frac{\varepsilon}{2pq}.$$

Then in the same way as in Lemma 1 we have

(5)
$$\sum_{m_1 > (p+\varepsilon)n_1} C_{n_1}^{m_1} C_{n_2}^{m_2} < \frac{2pq}{\varepsilon} C_{n_1}^{m_1'} C_{n_2}^{m_2'}$$

where $C_{n_1}^{m_1'} C_{n_2}^{m_2'}$ is the largest term of the left hand side, i.e. m_1' is the least integer $> (p + \varepsilon)n_1$.

Consider now values of $m_1 \leq (p + \varepsilon)n_1$. We have

$$\frac{n_1 - m_1}{m_1} \frac{p}{q} = 1 - \frac{m_1 - p n_1}{pqn_1} + O\left\{ \frac{(m_1 - p n_1)^2}{n_1^2} \right\} < 1 - \frac{m_1 - p n_1}{2pqn_1}.$$

Denoting $[a]$ the integral part of the positive number a we have

$$C_{n_1}^{m_1'} C_{n_2}^{m_2'} < C_{n_1}^{[n_1 p+1]} C_{n_2}^{np-[n_1 p+1]} \prod_{m_1=[n_1 p_1+1]}^{m_1'-1} \left(1 - \frac{m_1 - p n_1}{2pqn_1} \right)$$

$$< C_n^{p'n} \prod_{m_1=[n_1 p_1+1]}^{m_1'-1} \left(1 - \frac{m_1 - p n_1}{2pqn_1} \right).$$

Since

$$\log \prod_{m_1=[n_1 p_1+1]}^{m_1'-1} \left(1 - \frac{m_1 - p\, n_1}{2pqn_1}\right) < - \sum_{m_1=[n_1 p+1]}^{m_1'-1} \frac{m_1 - p\, n_1}{2pqn_1}$$

$$< - \frac{n_1 \varepsilon^2}{5pq}$$

we have

(6) $$C_{n_1}^{m_1'} C_{n_2}^{m_2'} < e^{-\frac{n_1 \varepsilon^2}{5pq}} C_n^{p'n}.$$

By (5) and (6)

$$\sum_{n_1 > N} \sum_{m_1 > (p+\varepsilon)n_1} C_{n_1}^{m_1} C_{n_2}^{m_2} < C_n^{p'n} \frac{2pq}{\varepsilon} \sum_{n_1 \geq N} e^{-\frac{n_1 \varepsilon^2}{5pq}}.$$

The series

$$\sum_1^\infty e^{-\frac{n_1 \varepsilon^2}{5pq}}$$

being convergent we obviously can choose N large enough to satisfy the inequality

$$\frac{2pq}{\varepsilon} \sum_{n_1 > N} e^{-\frac{n_1 \varepsilon^2}{5pq}} < \eta$$

and thus to satisfy the inequality (1). Similarly the inequality (2) can be proved.

Remark. The above lemma has the following interpretation. Take all the different combinations of np' figures 1 and np' figures 0. Their numbers is $C_n^{np'}$. Then there exists a number N such that the number of combinations, for which the sum of the first n_1 figures is $\leq (p+\varepsilon)n_1$ for all $n_1 \geq N$, is $> (1 - \eta)C_n^{p'n}$. Obviously we have a similar result for the sum of last n_1 figures.

§5.

Denote by $F(n, p)$ $(0 < p < \frac{1}{2})$ the set of all n–figured proper fractions in dyadic system, the sum of whose digits is $np'(np' \leq np < np' + 1)$, and by $J(N, \varrho_1)$ any set of intervals l of the form

$$k\, 2^{-m} \leq x < (k+1)\, 2^{-m}$$

(k and $m \geq N$ integers), such that

(1) $$\sum_{J(N,\varrho_1)} l^{\varrho_1} < 1.$$

E being a set of a finite number of elements we denote by $\Re(E)$ the number of elements of E.

Lemma 3. *Let $p(0 < p < \frac{1}{2})$ be a constant and ϱ the constant defined by the equation*

(2) $$2^{\varrho} = \frac{1}{p^p q^q}$$

and $\varrho_1 < \varrho$ another constant. Then for sufficiently large N and $n > N$ and for an arbitrary set $J(N, \varrho_1)$ we have

(3) $$\Re\{F(n, p) \times J(N, \varrho_1)\} < \frac{1}{2}\Re\{F(n, p)\} < \frac{1}{2} C_n^{np'}.$$

We first define an upper bound for the number of terms of $F(n, p)$ included in s intervals of the form $\{k\,2^{-n_1}, (k + 1)\,2^{-n_1}\}$ (closed on the left, but open on the right), n_1 satisfying the inequality

(4) $$p\,n < \frac{1}{2}(n - n_1).$$

The first n_1 figures of all numbers of the set

$$F(n, p) \times \{k\,2^{-n_1}, (k + 1)\,2^{-n_1}\}$$

are figures of the fraction $k\,2^{-n_1}$. Denoting by m_1 the sum of the figures of $k\,2^{-n_1}$ we see that

(5) $$\Re\{F(n, p) \times (k\,2^{-n_1}, \overline{k + 1}\,2^{-n_1})\} = C_{n-n_1}^{p'n-m_1}.$$

>From (4) we see that this number decreases, as m_1 increases, so that for $m_1 = 0$ it has its maximum value, then follow values for $m_1 = 1$, then for $m_1 = 2$ and so on. The number of the fractions $k\,2^{-n_1}$ for which the sum of the figures is m_1 is $C_{n_1}^{m_1}$. Denoting by m_1^0 the least integer satisfying the inequality

(6) $$\sum_{m_1=0}^{m_1^0} C_{n_1}^{m_1} \geq s$$

we see that the required upper bound can be given by the sum

(7)
$$\sum_{m_1=0}^{m_1^0} C_{n_1}^{m_1} C_{n-n_1}^{p'n-m_1} .$$

We now represent the number $\Re\{F(n,p) \times J(N,\varrho_1)\}$ as the sum of four numbers

$$\Re\{F(n,p) \times J(N,\varrho_1)\} = \Re_1 + \Re_2 + \Re_3 + \Re_4$$

where

$\Re_1$ is the number of terms of $F(n,p)$ belonging to the intervals of $J(N,\varrho_1)$ of length 2^{-n_1} for $N \le n_1 \le (1-2p)n$;

$\Re_2$ is the number of terms of $F(n,p)$ belonging to the intervals of $J(N,\varrho_1)$ of length 2^{-n_1} for $(1-2p)n < n_1 \le n-N$, and for which the sum of the last $n-n_1$ figures is included between $(p-\varepsilon)(n-n_1)$ and $(p+\varepsilon)(n-n_1)$; the value of ε will be given in the course of the proof;

$\Re_3$ is the number of terms of $F(n,p)$ belonging to the same intervals as the preceding group but for which the sum of the last $n-n_1$ figures does not satisfy the above condition;

$\Re_4$ is the number of terms of $F(n,p)$ belonging to the intervals of $J(N,\varrho_1)$ of length 2^{-n_1} for $n_1 \ge n-N$.

We shall prove the lemma by showing that each of the four numbers $\Re_1, \Re_2, \Re_3, \Re_4$ is $< \eta C_n^{np'}$, $(n < \frac{1}{8})$, for sufficiently large N and n.

(I) To define an upper bound of $\Re_1$ we first define a constant $p_1 < \frac{1}{2}$ by the equation

$$2^{\varrho_1} = \frac{1}{p_1^{p_1} q_1^{q_1}}, \quad p_1 + q_1 = 1$$

and we take another constant p_2 satisfying the inequalities

$$p_1 < p_2 < \frac{1}{2}.$$

We conclude at once from Lemma 1 that for sufficiently large N

$$\sum_{0 \le m_1 < p_2 n_1} C_{n_1}^{m_1} > 2^{\varrho_1 n_1} .$$

Then by (6) and (7) the number of the terms of $F(n,p)$ included in the intervals of $J(N,\varrho_1)$ of the form $\{k\,2^{-n_1}, (k+1)\,2^{-n_1}\}$ is less than

$$\sum_{0 \le m_1 < p_2 n_1} C_{n_1}^{m_1} C_{n-n_1}^{np'-m_1}$$

and consequently

$$\mathfrak{R}_1 < \sum_{N \leq n_1 \leq (1-2p)n} \sum_{0 \leq m_1 < p_2 n} C_{n_1}^{m_1} C_{n-n_1}^{np'-m_1}.$$

Now by Lemma 2

$$\mathfrak{R}_1 < \eta C_n^{np'}$$

for sufficiently large N.

(II) The first n_1 figures are the same for all numbers x of $F(n, p)$ belonging to an interval $\{k\,2^{-n_1}, (k+1)\,2^{-n_1}\}$ i.e. satisfying the inequalities

$$k\,2^{-n_1} \leq x < (k+1)\,2^{-n_1}$$

and therefore for all of them the sum of the last $n - n_1$ figures is also the same, say m_2. Thus either every number of $F(n, p)$ belonging to this interval is counted in $\mathfrak{R}_2$ or none of them. As now for all numbers counted in $\mathfrak{R}_2$

(8) $$(p - \varepsilon)(n - n_1) < m_2 < (p + \varepsilon)(n - n_1)$$

and as the number of intervals $J(N, \varrho_1)$ of length 2^{-n_1} is $< 2^{\varrho_1 n_1}$, we have

$$\mathfrak{R}_2 < \sum_{(1-2p)n < n_1 \leq n-N} 2^{\varrho_1 n_1} C_{n-n_1}^{[(p+\varepsilon)(n-n_1)]}.$$

Take now a constant ϱ_2 satisfying the inequalities

$$(1 - 2p)\varrho_1 + 2p\varrho < \varrho_2 < \varrho$$

and take ε so small that the small number δ defined by the equation

$$2^{\varrho+\delta} = \frac{1}{(p + \varepsilon)^{p+\varepsilon}(q - \varepsilon)^{q-\varepsilon}}$$

satisfies the inequality

$$(1 - 2p)\varrho_1 + 2p(\varrho + \delta) < \varrho_2,$$

then by Lemma 1

$$C_{n-n_1}^{[(p+\varepsilon)(n-n_1)]} = 2^{\varrho_1 n_1 + (\varrho+\delta+o)(n-n_1)}.$$

and thus

$$\mathfrak{R}_2 < \sum_{(1-2p)n < n_1 \leq n-N} 2^{\varrho_1 n_1 + (\varrho+\delta+o)(n-n_1)}.$$

Consequently

$$\mathfrak{R}_2 < n \, 2^{\{\varrho_1(1-2p)+(\varrho+\delta+o)2p\}n} \, .$$

Hence for large n

$$\mathfrak{R}_2 < 2^{\varrho_2 n} < \eta C_n^{np'} \, .$$

(III) For the numbers counted in $\mathfrak{R}_3$ the sum of the last $n - n_1$ figures does not satisfy the inequalities (8) and therefore by Lemma 2

$$\mathfrak{R}_3 < \eta C_n^{p'n} \, .$$

(IV) The number of all n–figured fractions x belonging to an interval

$$k \, 2^{-n_1} \leqq x < (k+1) \, 2^{-n_1}$$

being 2^{n-n_1} we conclude that

$$\mathfrak{R}_4 < \sum_{n_1 > n-N} 2^{n_1 \varrho_1 + (n-n_1)}$$

$$< N \, 2^{(n-N)\varrho_1 + N} = 2^{n(\varrho_1 + o)}$$

where o is small if $\frac{n}{N}$ is large, and thus for sufficiently large n

$$\mathfrak{R}_4 < \eta C_n^{p'n} \, .$$

The results of (I), (II), (III), (IV) prove the lemma.

§6.

We shall now pass to the proof of the theorem of §2. Denote by $E(p)$ the set of numbers x of the interval $(0, 1)$ for which

$$\overline{\lim} \frac{P(x, n)}{n} \leqq p$$

and by $E(p, N)$ the set of numbers for which

$$P(x, n) \leqq p \, n \qquad \text{for all } n > N \, .$$

We split the set $E(p, N)$ into a finite number of subsets of G_i,

$$E(p, N) = \sum G_i \, ,$$

each of which consists of all the numbers of $E(p, N)$ having the same first N figures. Replace the first N figures of all the numbers of G_i by zeroes and denote by H_i the set into which G_i is transformed. Obviously H_i is congruent with G_i and belongs to $E(p, 0)$. Thus

$$dG_i = dH_i \leqq d\, E(p, 0)$$

and consequently by the theorem of §1

$$d\, E(p, N) = d \sum G_i \leqq E(p, 0).$$

Applying this theorem again we have

$$d \sum_{N=0}^{\infty} E(p, N) = d\, E(p, 0).$$

We now observe that

$$E(p) \supset E(p, 0).$$

On the other hand taking any $p_2 > p$ we have

$$\sum_{N=0}^{\infty} E(p_2, N) \supset E(p).$$

Hence

(1) $$d\, E(p, 0) \leqq d\, E(p) \leqq d\, E(p_2, 0)$$

for any $p_2 \geqq p$.

We shall now evaluate $d\, E(p, 0)$. Let $\Phi(n, p)$ be the set of all n–figured fractions, the sum of whose figures is $\leqq np$. Writing as before $[np] = np'(p' \leqq p < p + \frac{1}{n})$ the number of members of $\Phi(n, p)$ is

$$\sum_{0 \leqq i \leqq np'} C_n^i$$

which, by Lemma 1, is equal to $2^{(\varrho+o)n}$ where ϱ is defined by the equation

$$2^\varrho = \frac{1}{p^p q^q}.$$

Denote by $J_{n,p}$ the set of intervals of length 2^{-n} drawn to the right from each point of $\Phi(n, p)$. Obviously $J_{n,p}$ contains all the points of $E(p, 0)$. Let ϱ_2 be any number $> \varrho$. We have for large n

$$\sum_{J_{n,p}} l^{\varrho_2} = 2^{-n\varrho_2} \sum_{0 \leqq i \leqq p\, n} C_n^i = 2^{-n\varrho_2 + n(\varrho+o)}$$

i.e.

$$\lim_{n \to \infty} \sum_{J_{n,p}} l^{\varrho_2} = 0$$

and thus

$$\lim_{\delta \to 0} \sum_{J\{E(p,0),\delta\}} l^{\varrho_2} = 0 \qquad \text{(notation of §1).}$$

which means

$$d\, E(p,0) \leqq \varrho_2$$

for any $\varrho_2 > \varrho$ and consequently

(2) $$d\, E(p,0) \leqq \varrho.$$

We shall now prove that for all $\varrho_1 < \varrho$

(3) $$\varliminf_{\delta \to 0} \sum_{J\{E(p,0),\delta\}} l^{\varrho_1} \geqq \frac{1}{4}.$$

>From (2) and (3) it follows that

(4) $$d\, E(p,0) = \varrho.$$

Assume then that (3) is false: there exist sets $J\{E(p,0),\delta\}$ with arbitrarily small δ, such that

(5) $$\sum_{J\{E(p,0),\delta\}} l^{\varrho_1} < \frac{1}{4}.$$

The number $p_1 (< p)$ is defined by the equations

$$2^{\varrho_1} = \frac{1}{p_1^{p_1} q_1^{q_1}} \qquad p_1 + q_1 = 1$$

and p_1' is any number such that

$$p_1 < p_1' < p.$$

By Lemma 2 (p. 323) there exist numbers M, n_0 corresponding to p_1', p such that for $n > n_0$

$$\mathfrak{R}\{F(n, p_1') \times E(p, M)\} > \frac{1}{2}\mathfrak{R}\{F(n, p_1')\}.$$

Now with each interval l

$$2^{-n-1} \leqq l < 2^{-n}$$

of the set $J\{E(p,0),\delta\}$ of (5) we associate two intervals

$$\{(j-1)2^{-n}, j2^{-n}\}, \qquad \{j2^{-n},(j+1)2^{-n}\}$$

where j is defined by the condition that the interval l is included in the interval $\{(j-1)2^{-n},(j+1)2^{-n}\}$.
>From (5) it follows that

$$\sum l^{\varrho_1} < 1$$

where the summation is extended over all the associated intervals. Among these intervals there are only a finite number containing points of the set $E(p,M)$, since this set is closed. Thus we have arrived at the set $J(N_1,\varrho_1)$ as defined in Lemma 3, which contains the whole of the set $E(p,M)$ and more than half of the points of $F(n,p_1')$ which for large values of n is impossible by Lemma 3.

Thus equation (4) is proved and then from (1) it follows that

$$d\,E(p) = \varrho$$

which proves the theorem.

The computation in the paper is done in a purely combinational manner. The result may be proved in other ways, as well—for example using the "strong law of large numbers" in probability theory ([6], [4], § 14, [5], pp. 206–210, [7], Proposition 10.1). But of course, the usual proofs of the strong law involve combinatorial estimates as well.

This computation relates to a "multifractal decomposition" of the interval. Although almost all points x satisfy $P(x,n) \sim (1/2)n$, not all points x have this typical behavior.

Here is mathematician's view of the multifractal decomposition in general. When a set K has fractal dimension d and supports a "natural" finite measure μ, we may expect "typically", for $x \in K$ and $\varepsilon > 0$, that the measure $\mu(B_\varepsilon(x))$ of the ball of radius ε centered at x is roughly equal to $(2\varepsilon)^d$, the dth power of the diameter of the ball. This might mean that

$$0 < \limsup_{\varepsilon \downarrow 0} \frac{\mu(B_\varepsilon(x))}{(\operatorname{diam} B_\varepsilon(x))^d} < \infty$$

or, more generally, that

$$\lim_{\varepsilon \downarrow 0} \frac{\log \mu(B_\varepsilon(x))}{\log \operatorname{diam} B_\varepsilon(x)} = d$$

for all $x \in K$. Multifractal decomposition will be interesting exactly when this does not happen—for many different values of the parameter α, the set

$$K^{(\alpha)} = \left\{ x \in K : \lim_{\varepsilon \downarrow 0} \frac{\log \mu(B_\varepsilon(x))}{\log \operatorname{diam} B_\varepsilon(x)} = \alpha \right\}$$

is non-trivial. The sets $K^{(\alpha)}$ may be thought of as the **multifractal components** of K. Write $f(\alpha)$ for the Hausdorff dimension of $K^{(\alpha)}$. The function $f(\alpha)$ is called the **dimension spectrum** of the set K or, more precisely, of the measure μ.

It is important to note that we are using the *Hausdorff dimension* of $K^{(\alpha)}$, and not the box dimension. Typically, all the sets $K^{(\alpha)}$ are dense in K, so they all have box dimension equal to the box dimension of K itself. But the Hausdorff dimension $f(\alpha)$ varies with α.

Mandelbrot [8] has extended the interpretation of the function $f(\alpha)$. When α is too large or too small, then the set $K^{(\alpha)}$ is empty. There is still a useful value for $f(\alpha)$, but it is negative. But this "negative dimension" follows many of the classical formulas of geometric measure theory. —Ed.

Bibliography

[1] G. Hardy and J. Littlewood, *Some problems on Diophantine approximations*, Acta Mathematica **37** (1914), 155–190.

[2] F. Hausdorff, *Dimension und äußeres Maß*, Math. Annalen /bf 79 (1918), 157–179.

[3] A. S. Besicovitch, *On linear sets of fractional dimensions*; Math. Annalen **101** (1929), 161–193

[4] P. Billingsley, *Ergodic Theory and Information*, Wiley, 1965.

[5] G. A. Edgar, *Measure, Topology, and Fractal Geometry*, Springer-Verlag, New York, 1990.

[6] H. G. Eggleston, *The fractional dimension of a set defined by decimal properties*, Quarterly J. Math. **20** (1949), 31–36.

[7] K. Falconer, *Fractal Geometry: Mathematical Foundations and Applications*, Wiley, 1990.

[8] B. Mandelbrot, *Negative fractal dimensions and multifractals*, Physica A **163** (1990), 306–315.

[9] S. J. Taylor, *Abram Samilovitch Besicovitch*, Bull. London Math. Soc. **7** (1975), 191–210.

This paper was received 16 January, 1934;
read 18 January, 1934

If $q > 2$, let E_q be the set of real numbers r that can be "well-approximated" by rational numbers in the sense that there are infinitely many rational numbers m/n with

$$\left| r - \frac{m}{n} \right| < \frac{1}{n^q}.$$

Then E_q has Lebesgue measure zero (Khintchine 1926 [10]). A much stronger determination of the size of E_q is in this paper: it has Hausdorff dimension exactly $2/q$.

The "dimensional number" used is the Hausdorff dimension. The Hausdorff dimension of a set E is denoted dE. Note the use of addition and multiplication signs to mean union and intersection of sets: $F + E$ means $F \cup E$ and $F \times E$ means $F \cap E$. The number of elements in a (finite) set F is denoted $\mathfrak{N}\{F\}$.

TEN

Sets of Fractional Dimensions (IV): On Rational Approximation to Real Numbers

A. S. Besicovitch

1. It is a well–known result that there exist infinitely many rational approximations m/n to any real number r with an error less than n^{-2}. The problem of this article is the study of the sets of real numbers with stronger rational approximations. The solution of this problem will naturally be given in terms of sets of fractional dimensions*.

Let E_q, $q > 2$, be the set of real numbers r of the interval $(0, 1)$ for which the inequality

$$\left| r - \frac{m}{n} \right| < \frac{1}{n^q}$$

is satisfied by infinitely many rational numbers m/n. We shall prove the

Theorem. *The dimensional number of the set E_q is $2/q$.*

---

*Hausdorff, "On linear sets of fractional dimensions", *Math. Annalen,* 79 (1918), 157–179; A.S. Besicovitch, "On linear sets of fractional dimensions (II): On the sum of digits of real numbers represented in the dyadic system"; "Sets of fractional dimensions (III): Set of points of non–differentiability of absolutely continuous functions and of divergence of Fejér sums", *Math. Annalen* (in the press).

2. Denote by $I(\lambda, \rho)$ any set of intervals $l < \lambda$ for which

(1)
$$\sum_{I(\lambda,\rho)} l^\rho < 1.$$

The theorem will be proved by proving the two statements:

(i) *For any $\lambda > 0$ and $\rho > 2/q$ there exists a set $I(\lambda, \rho)$ including the set E_q.*

(ii) *No set $I(\lambda, \rho)$ with $\rho < 2/q$ and λ small includes the whole of the set E_q.*

3. (i) Given $\lambda > 0$ and $\rho = (2 + \delta)/q$, $\delta > 0$, define an integer n_0 to satisfy the inequalities

(2)
$$\frac{2}{n_0^q} < \lambda, \quad 2^{(2+\delta)/q} \sum_{n>n_0} \frac{1}{n^{1+\delta}} < 1,$$

and take for a set $I\{\lambda, (2 + \delta)/q\}$ the set of the intervals

$$\left(\frac{m}{n} - \frac{1}{n^q}, \frac{m}{n} + \frac{1}{n^q} \right)$$

corresponding to all the rational numbers m/n, $n > n_0$, of the interval $(0, 1)$. This set obviously includes the set E_q, and, by (2), the condition (1) is satisfied. Thus the statement (i) is proved.

4. (ii) We shall now prove that no set $I(\lambda, \rho)$ with $\rho < 2/q$ and λ small includes the whole of the set E_q.

We first define an increasing sequence of integers

$$n_0, n_1, n_2, \ldots,$$

and we assume n_0 to be sufficiently large, and the sequence to increase sufficiently rapidly, to satisfy certain inequalities based on the order of magnitude of their terms.

Take now a set $I(\lambda, \rho)$ with

$$\lambda = n_0^{-q}, \quad \rho = \frac{2 - \delta}{q}, \quad \delta > 0.$$

Denote by F_1 the set of intervals $\left(\frac{m}{n} - \frac{1}{n^q}, \frac{m}{n} + \frac{1}{n^q} \right)$ corresponding to all proper fraction m/n with a prime denominator n varying in the interval

$$n_1 \leq n < 2n_1.$$

We write
$$F_1 = F_1' + F_1'',$$
where F_1' consists of those intervals of F_1 which are covered for a quarter or more of their length by a single interval of $I(\lambda, \rho)$, and F_1'' consists of all the other intervals of F_1. Introducing the notation $\Re(H)$ for the number of elements of the set H, we see that

$$\Re(F_1) > \frac{n_1^2}{\log n_1}.$$

We shall now define $\Re(F_1')$.

Denote by H_1 the set of those intervals of $I(\lambda, \rho)$ each of which covers at least a quarter of an interval of F_1', and we write

$$H_1 = H_1' + H_1'' + H_1''',$$

where the lengths l of the intervals of H_1', H_1'', H_1''' satisfy respectively the inequalities

$$\lambda = \frac{1}{n_0^q} \geq l > \frac{1}{n_1^\beta} \quad \left(\frac{1}{2} < \beta < 1\right),$$

$$\frac{1}{n_1^\beta} \geq l > \frac{1}{8n_1^2},$$

$$\frac{1}{8n_1^2} \geq l$$

(n_1 is supposed to be large enough to satisfy the inequality $n_1^\beta > n_0^q$). We have
$$\sum_{H_1'} l^{(2-\delta)/q} < 1,$$

and consequently

(1)
$$\sum_{H_1'} l < \lambda^{1-\{(2-\delta)/q\}} = n_0^{2-\delta-q}.$$

Since the intervals of H_1' are large in comparison with $1/n_1$, the number $\mathfrak{G}'$ of the intervals of F_1 covered completely or partly by intervals of H_1' is approximately equal to

$$\Re(F_1) \sum_{H_1'} l$$

and, *a fortiori*,

$$\mathfrak{G}' < 2\mathfrak{R}(F_1) \sum_{H_1'} l.$$

By (1),

(2) $$\mathfrak{G}' < 2n_0^{2-\delta-q}\mathfrak{R}(F_1).$$

The largest number $\mathfrak{G}''$ of intervals of F_1, covered by at least a quarter of their length by intervals of H_1'', is obtained if all the intervals of H_1'' have the greatest length, *i.e.*, $n_1^{-\beta}$. Denoting by s the number of intervals of H_1'', in this case we have

$$sn_1^{-\beta\{(2-\delta)/q\}} < 1,$$

i.e.,

$$s < n^{\beta\{(2-\delta)/q\}}.$$

Since the distance between the centres of any two intervals of F_1 is greater than $1/(4n_1^2)$, it follows easily that

(3) $$\mathfrak{G}'' < 8n_1^{2-\beta\{(q-2+\delta)/q\}}.$$

The greatest number $\mathfrak{G}'''$ of intervals of F_1, covered at least for one quarter by intervals H_1''', is obtained if each interval of H_1''' is exactly equal to a quarter of an interval of F_1. Thus

$$\mathfrak{G}''' \left\{ \frac{1}{2(2n_1)^q} \right\}^{(2-\delta)/q} < 1,$$

i.e.,

(4) $$\mathfrak{G}''' < 2^{(1+q)\{(2-\delta)/q\}} n_1^{2-\delta}.$$

By (2),(3),(4)

(5) $$\mathfrak{R}(F_1') < 2n_0^{2-\delta-q}\mathfrak{R}(F_1) + 8n_1^{2-\beta\{(q-2+\delta)/q\}} + 2^{(1+q)\{(2-\delta)/q\}} n_1^{2-\delta}$$

$$< \mathfrak{R}(F_1)4n_0^{2-\delta-q}$$

($n_1^{\delta/2q}$ is supposed to be large in comparison with $n_0^{q+\delta-2}$), and

(6) $$\mathfrak{R}(F_1'') > (1 - 4n_0^{2-\delta-q})\mathfrak{R}(F_1) > \frac{1}{2}\frac{n_1^2}{\log n_1}.$$

Cut a quarter from each end of each interval of F_1'' and denote the set of remaining central halves (with the ends of these halves) by G_1. Obviously G_1 has no point in common with any interval of $I(l, \rho)$ of length greater than or equal to n_1^{-q}. We have

(7)
$$\sum_{G_1} l > \frac{1}{(2n_1)^q} \Re(F_1'') > \frac{1}{(2n_1)^q} \frac{n_1^2}{\log n_1}.$$

Denote now by F_2 the set similar to F_1 corresponding to prime values of n belonging to the interval

$$n_2 \leq n < 2n_2;$$

denote further by I_1 the subset of $I(\lambda, \rho)$ consisting of all the intervals of $I(\lambda, \rho)$ of length less than n_1^{-q}, and write

$$F_2 = F_2' + F_2'',$$

where F_2' consists of those intervals of F_2 which are covered for a quarter or more of their length by a single interval of I_1, and F_2'' consists of all the other intervals of F_2. In the same way as before, we see that

(8)
$$\Re(F_2') < \Re(F_2) 4 n_1^{2-\delta-q}.$$

Since each interval of G_1 is large in comparison with n_2^{-q}, the number of intervals of F_2 included in G_1 is approximately equal to $\Re(F_2) \sum_{G_1} l$ and, a fortiori, is greater than $\frac{1}{2}\Re(F_2) \sum_{G_1} l$. By (8) the number of intervals of F_2'' included in G_1 is greater than $\frac{1}{2}\Re(F_2) \sum_{G_1} l - \Re(F_1) 4 n_1^{2-\delta-q}$. Cut a quarter from each end of each of these intervals and denote the set of remaining halves by G_2. We have

$$\Re(G_2) > \Re(F_2) \left\{ \frac{1}{2} \frac{1}{(2n_1)^q} \frac{n_1^2}{\log n_1} - 4 n_1^{2-\delta-q} \right\}$$

$$> \frac{1}{4} \frac{(n_2 n_1)^2}{(2n_1)^q \log n_2 \log n_1},$$

and

$$\sum_{G_2} l > \frac{1}{4} \frac{(n_2 n_1)^2}{(2n_2 2n_1)^q \log n_2 \log n_1}.$$

In the same way we introduce the set F_3 and then define the set G_3 which is entirely included in $G_2 \times F_3$ and for which the inequality

$$\sum_{G_3} l > \frac{1}{4^2} \frac{(n_3 n_2 n_1)^2}{(2n_3 2n_2 2n_1)^q \log n_3 \log n_2 \log n_1}$$

is satisfied.

Similarly define a sequence of sets

$$G_4, G_5, \ldots$$

Each of these sets is included in the preceding set. The set G_1 has no point in common with any interval of $I(\lambda, \rho)$ of length greater than or equal to n_1^{-q}, the set G_2 no point in common with any interval of $I(\lambda, \rho)$ of length greater than or equal to n_2^{-q}, and so on. Thus the set

$$G = G_1 \times G_2 \times G_3 \times \cdots$$

has no point in common with any interval of $I(\lambda, \rho)$. On the other hand, G is not an empty set (as the limit of a decreasing sequence of closed sets, each different from an empty set) and is included in the set

$$F = F_1 \times F_2 \times F_3 \times \cdots$$

which in its turn belongs obviously to E_q. Thus we have arrived at the conclusion that

there exists a set G of points E_q which do not belong to the set $I(\lambda, \rho)$,

which proves the statement (ii) and consequently our theorem.

The Hausdorff dimension computed in this selection is done immediately from the definition. The combinatorial details are intricate, but elementary. A more modern proof is found in [7, Theorem 10.3]. This dimension was, in fact, previously computed by V. Jarnik [9], so the assertion dim $E_q = 2/q$ is often known as **Jarnik's theorem.**

In fact, Jarnik proved some more general results as well. He considered Hausdorff functions $f(x)$ other than power functions x^r, and approximation measures other than power functions $1/n^q$. Jarnik also considered simultaneous rational approximation of several real numbers: Let s be a positive integer, and let q be a real number with $q > 1 + 1/s$. Let E_q^s be the set of all s-tuples $(r_1, r_2, \ldots, r_s)$ for which

$$\left| r_i - \frac{m_i}{n} \right| < \frac{1}{n^q}, \qquad i = 1, 2, \ldots, s$$

for infinitely many choices of integers $m_1, m_2, \ldots, m_s, n$. Then [9, Satz 1] the set E_q^s has Hausdorff dimension $(s + 1)/q$.

The sets E_q decrease as q increases:

$$E_2 \supseteq E_p \supseteq E_q \quad \text{for } 2 \leq p \leq q.$$

What can be said about the set $E_\infty = \cap_{q=2}^\infty E_q$ of numbers r that can be approximated

$$\left| r - \frac{m}{n} \right| < \frac{1}{n^q}$$

for every q? Such numbers r are called **Liouville numbers**. A theorem of Liouville (1844) (for example [11, §11.7] or [5, p. 1–2]) shows that an irrational number that is algebraic of degree q does not belong to $E_{q+\varepsilon}$ for any $\varepsilon > 0$. So irrational Liouville numbers are transcendental. By the theorem of this selection, E_∞ has Hausdorff dimension 0. But it is not empty; it contains the set $\mathbf{Q}$ of rational numbers. In fact, E_∞ contains more than just the rationals; it is uncountable. Each E_q is a G_δ set: that is, a countable intersection of open sets. And each E_q is dense in $\mathbf{R}$, for example, because it contains $\mathbf{Q}$. The "Baire Category Theorem" [13, p. 110] tells us (in a precise sense) that a countable intersection of dense G_δ sets is a very "big" set (called **comeager** or **residual**), and its complement is a very "small" set (called **meager** or **first category**). This example illustrates that a big set in the sense of category is not necessarily big in the sense of Hausdorff dimension, and a small set in the sense of category is not necessarily small in the sense of Hausdorff dimension.

Liouville's theorem was improved over the years. The best result is by Roth [12]; see [5, Chapter 7] for an exposition. An irrational algebraic number belongs to no set $E_{2+\varepsilon}$ for $\varepsilon > 0$.

Although all irrational Liouville numbers are transcendental, there are many non–Liouville transcendentals as well. For each q, simply by the Hausdorff dimension results, we know that there are uncountably many numbers in E_q but not in any E_p for $p > q$; so certainly there are transcendentals among these numbers. (By Roth's theorem, all such numbers are transcendental.) The most popular transcendental number is π–it is not a Liouville number. In fact, $\pi \notin E_{42}$: *If m and $n \geq 2$ are positive integers, then*

$$\left| \pi - \frac{m}{n} \right| > \frac{1}{n^{42}}.$$

This surprising result is due to Kurt Mahler [11]. Mahler's theorem has also been improved over the years; the best result known today seems to be $\pi \notin E_{8.0161}$, due to M. Hata [14]. See [6, §11.3] for an exposition of some of these improvements. –Ed.

Bibliography

[1] F. Hausdorff, *Dimension und äußeres Maß*, Math. Annalen **79** (1918), 157–179.

[2] A. S. Besicovitch, *On linear sets of fractional dimensions*, Math. Annalen **101** (1929), 161–193.

[3] A. S. Besicovitch, *On the sum of digits of real numbers represented in the dyadic system*, Math. Annalen **110** (1934), 321–330.

[4] A. S. Besicovitch, *Sets of points of non–differentiability of absolutely continuous functions and of divergence of Fejér sums*, Math. Annalen **110** (1934), 331–335.

[5] A. Baker, *Transcendental Number Theory*, Cambridge University Press, 1975.

[6] J. Borwein and P. Borwein, *Pi and the AGM*, John Wiley & Sons, 1987.

[7] K. Falconer, *Fractal Geometry: Mathematical Foundations and Applications*, Wiley, 1990.

[8] G. H. Hardy and E. M. Wright, *An Introduction to the Theory of Numbers*, Oxford University Press, Fifth Edition, 1979.

[9] Vojtěch Jarník, *Über die sumultanen diophantischen Approximationen*, Math. Zeitschrift **33** (1931), 505–543.

[10] A. Khintchine, *Zur metrischen Theorie der diophantischen Approximationen*, Math. Zeitschrift **24** (1926), 706–714.

[11] K. Mahler, *On the approximations of π*, Indag. Math. **15** (1953), 30–42.

[12] K. F. Roth, *Rational approximations to algebraic numbers*, Mathematika **2** (1955), 1–20, corrigendum, p. 168.

[13] K. R. Stromberg, *An Introduction to Classical Real Analysis*, Wadsworth, 1981.

[14] M. Hata, *Rational approximations to π and some other numbers*, Acta Arithmetica **63** (1993), 335-349.

Received 3 July, 1936, read 14 January, 1937.

This selection investigates the Hausdorff dimension of the graph of a continuous function f $\mathbb{R} \to \mathbb{R}$. If the function is differentiable, then that graph has dimension 1. So a computation of the Hausdorff dimension $d > 1$ is a more precise assertion than non-differentiability. The dimension is related to the Lipschitz condition satisfied by f.

Harold Douglas Ursell (1907–1969) was born in Warwickshire. He held fellowships at Cambridge and Harvard, then taught at Leeds and Calgary. His work shows the influence both of Besicovitch and of Wittgenstein. Ursell's writings include papers on geometry, point-set topology, function theory, mathematical physics, almost periodic functions, and measure theory. [6]

ELEVEN

Sets of Fractional Dimensions (V): On Dimensional Numbers of Some Continuous Curves

A. S. Besicovitch & H. D. Ursell

The d-dimensional **Hausdorff measure** of a set E is denoted here $d\text{-}mE$. In the original it looks misleadingly like a subtraction: $d - mE$.

The "dimensional number" of a set is the Hausdorff dimension. The **oscillation** of a function f on an interval I is $\sup \left\{ |f(x) - f(y)| : x, y \in I \right\}$. If $\alpha(h)$ and $\beta(h)$ are positive functions defined for $h > 0$, then

$$\alpha(h) \prec \beta(h) \qquad \text{means} \qquad \lim_{h \to 0} \frac{\alpha(h)}{\beta(h)} = 0.$$

The function f satisfies a **Lipschitz condition** of order δ iff there is a constant C such that

$$|f(x) - f(y)| \le C\, |x - y|^{\delta}$$

for all $x, y \in \mathbb{R}$. The collection of all such functions is called the **Lipschitz δ-class**, Lip^{δ}. A Lipschitz condition is also sometimes called a **Hölder condition**. [In 1864, Rudolph Otto Sigismund Lipschitz singled out these functions. In 1889, Ludwig Otto Hölder generalized the definition to functions of several variables.] The exponent δ must satisfy $0 < \delta \le 1$.

Question. Why is the Lipschitz condition uninteresting when $\delta > 1$? —Ed.

1. We first recall the numerical definition of d–dimensional measure of a "measurable set".*

Definition. Given a plane set of points E, denote by C_ρ any set of circles of radii less than or equal to ρ, covering all the points of E. Given $0 < d < 2$, d–measure is defined by the equation

$$d - mE = \varliminf_{\rho \to 0} \sum_{C_\rho} (2r)^d ,$$

where r is the radius of the general circle of C_ρ and $\sum_{C_\rho}$ denotes the summation extended over all circles of C_ρ.

To every plane set E corresponds one of the three possibilities:

(i) There exists a number $0 < d < 2$ such that, for any $d' > d$, $d' - mE = 0$, and for any $d' < d$, $d' - mE = \infty$. Then we say that E is a d–dimensional set and we call d the dimensional number of E.

(ii) $d - mE = \infty$ for any $d < 2$. Then we say that E is a 2–dimensional set.

(iii) $d - mE = 0$ for any $d > 0$. Then we say that E is a 0–dimensional set.

2. Theorem. *The dimensional number d of the curve $y = f(x)$, where $f(x)$ belongs to the Lipschitz δ–class (Lip^δ); satisfies the inequalities*

$$1 \le d \le 2 - \delta .$$

Consider the curve for $0 \le x \le 1$. Suppose first that the coefficient in the Lipschitz inequality can be taken equal to 1, so that to any x corresponds an interval $(x - k, x + k)$ such that, for any $x + h$ of this interval,

(1) $$|f(x + h) - f(x)| < |h|^\delta .$$

By the Heine–Borel theorem, there exists a finite set,

$$(0, k_0), \ (x_1 - k_1, x_1 + k_1), \ \ldots, \ (x_{n-1} - k_{n-1}, x_{n-1} + k_{n-1}), \ (1 - k_n, 1)$$

*F. Hausdorff, *Dimension und äusseres Mass*, *Math. Annalen,* 79 (1918), 157–179; A.S. Besicovitch, "On linear sets of fractional dimensions", *Math. Annalen,* 101 (1929),161–193; "Sets of fractional dimensions (II)," *Math. Annalen,* 110 (1934), 321–329; "Sets of fractional dimensions (III)", *Math. Annalen,* 110 (1934), 331–335; "Sets of fractional dimensions (IV)", *Journal London Math. Soc.,* 9 (1934), 126–131.

of overlapping intervals of the above kind covering the whole of $(0, 1)$. Denoting by c_i an arbitrary point between x_{i-1} and x_i belonging to both of the intervals

$$(x_{i-1} - k_{i-1}, x_{i-1} + k_{i-1}), \; (x_i - k_i, x_i + k_i),$$

we have

$$0 < c_1 < x_1 < c_2 < x_2 < \ldots < x_{n-1} < c_n < 1.$$

The oscillation of $f(x)$ in the interval (c_{i-1}, c_i) is less than $2|c_i - c_{i-1}|^\delta$, and thus the part of the curve corresponding to the interval (c_{i-1}, c_i) can be enclosed in a rectangle of height $2|c_i - c_{i-1}|^\delta$ and of base $c_i - c_{i-1}$, and consequently in $[2(c_i - c_{i-1})^{\delta-1}] + 1$ squares of side $c_i - c_{i-1}$ or in the same number of circles of radius $(c_i - c_{i-1})/\sqrt{2}$ circumscribed about each of these squares.

Given an arbitrary $\rho > 0$, we can always assume all $c_i - c_{i-1} < \rho$. Denote by C_ρ the set of all the above circles and consider

$$\sum_{C_\rho} (2r)^{2-\delta}.$$

The sum of the terms corresponding to the interval (c_{i-1}, c_i) is

$$\{[2(c_i - c_{i-1})^{\delta-1}] + 1\}\{(c_i - c_{i-1})\sqrt{2}\}^{2-\delta} < 6(c_i - c_{i-1}),$$

and thus

$$\sum_{C_\rho} (2r)^{2-\delta} < 6 \sum (c_i - c_{i-1}) < 6,$$

which shows that the $(2-\delta)$–dimensional measure of the curve is finite and hence that the dimensional number of the curve is less than or equal to $2 - \delta$.

If now $f(x)$ satisfies the Lipschitz condition with a variable coefficient

$$|f(x + h) - f(x)| < C \, |h|^\delta,$$

and C is not bounded, then, for any $\varepsilon > 0$, it satisfies the condition

$$|f(x + h) - f(x)| < |h|^{\delta-\varepsilon}$$

for sufficiently small h, and thus the dimensional number of the curve is less than or equal to $2 - \delta + \varepsilon$, i.e., less than or equal to $2 - \delta$. This completes the proof.

Corollary. *The dimensional number of the curve* $y = f(x)$, *where* $f(x)$ *has a finite derivative at all points, is 1.*

This follows at once from the fact that $f(x)$ belongs to Lip$^\delta$ for $\delta = 1$.

3. A curve of class Lip$^\delta$ may have any dimension number in the range $1 \leq d \leq 2 - \delta$, as we now show by examples. We write $\varphi(x)$ for the function equal to $2x$ in $0 \leq x \leq \frac{1}{2}$ and defined elsewhere by the relations

$$\varphi(x) = \varphi(-x) = \varphi(x + 1).$$

We consider curves

$$y = f(x) = \sum a_n \varphi(b_n x),$$

where

$$a_n = b_n^{-\delta} \quad (0 < \delta < 1),$$

and we write

$$s_\nu(x) = \sum_{n \leq \nu} a_n \varphi(b_n x).$$

I. *If* $b_{n+1} \geq B b_n$, *where* $B > 1$, *then* $f(x)$ *is of class* Lip$^\delta$.

We have

$$0 \leq \varphi \leq 1, \quad |\varphi'| = 2,$$

and hence

$$|\varphi(b_n x + b_n h) - \varphi(b_n x)| \leq 1,$$

$$|\varphi(b_n x + b_n h) - \varphi(b_n x)| \leq 2b_n h.$$

Hence

$$|f(x + h) - f(x)| \leq \sum a_n |\varphi(b_n x + b_n h) - \varphi(b_n x)|$$

$$\leq \sum_{n \leq \nu} 2a_n b_n h + \sum_{n > \nu} a_n$$

$$\leq 2h b_\nu^{1-\delta}[1 + B^{\delta-1} + B^{2(\delta-1)} + \ldots + B^{\nu(\delta-1)}]$$

$$+ b_{\nu+1}^{-\delta}[1 + B^{-\delta} + B^{-2\delta} + \ldots]$$

$$= K_1 b_\nu^{1-\delta} h + K_2 b_{\nu+1}^{-\delta},$$

the numbers K_1, K_2 depending only on B and δ. Now choose $\nu = \nu(h)$ so that

$$\frac{1}{b_\nu} > h \geq \frac{1}{b_{\nu+1}}.$$

Then we get at once
$$|\Delta f| \leq (K_1 + K_2)h^\delta.$$

II. $f(x)$ is not of any higher Lipschitz class.

For, taking $x = 0$ and $h = 1/2b_\nu$, we get

$$\Delta f = f(h) > b_\nu^{-\delta}\varphi(1/2) = K_3 h^\delta,$$

and h here is arbitrarily small.

III. If $1 < d < 2 - \delta$, $b_1 > 1$, $b_{n+1} = b_n^{\mu_n}$, where $\mu_n \geq \frac{1-\delta}{\delta}\frac{2-d}{d-1}$, then the d-measure of the part of the curve arising from a finite range of x is finite or zero: and the curve is of dimension less than or equal to d.

As in I we get

$$|\Delta y| \leq K_1 b_\nu^{1-\delta}h + K_2 b_{\nu+1}^{-\delta} = H, \quad \text{say.}$$

By dividing the range of x into intervals of length h we are able to cover the curve with rectangles of width h and height H, and these in turn can be covered with squares of side h. Choose $h = h_\nu$ so that

$$b_\nu^{1-\delta}h_\nu = b_{\nu+1}^{-\delta},$$

whence

$$h_\nu = b_\nu^{-\delta\mu_\nu - 1 + \delta} \leq b_\nu^{-(1-\delta)/(d-1)} \to 0 \quad \text{as } \nu \to \infty.$$

The number of squares of side h_ν required is less than

$$\left(\frac{1}{h_{\nu_1}} + 1\right)\left(\frac{H_\nu}{h_\nu} + 1\right),$$

l being the length of the range of x. For $h_\nu < l$ this is less than

$$\frac{2l}{h_\nu}[(K_1 + K_2)b_\nu^{1-\delta} + 1] < K_4 b_\nu^{1-\delta}h_\nu^{-1},$$

and the corresponding approximation to the d–measure is less than

$$K_5 b_\nu^{1-\delta}h_\nu^{d-1} \leq K_5.$$

Since h_ν is arbitrarily small, this proves the result stated.

Note that it is sufficient to have

$$\mu_n \geq \frac{1-\delta}{\delta}\frac{2-d}{d-1} \quad \text{for } n \geq n_0,$$

instead of for all n, provided that $b_{n_0} > 1$. If we now construct an example in which $b_n \to \infty$, $\mu_n \to \infty$, then we can take d arbitrarily near to 1 in the above argument and hence the curve is of *dimension precisely* 1. This will be true, for instance, if

$$b_n = 2^{2^{2^n}}.$$

IV. *If* $1 < d < 2 - \delta$, $b_1 > 1$, $b_{n+1} = b_n^\mu$, *where* $\mu = \frac{1-\delta}{\delta} \frac{2-d}{d-1}$, *then the* d*–measure of the curve* $y = f(x)$ *is greater than zero.*

Let φ be a square of side h with its sides parallel to the axes of coordinates. The points of the curve $y = f(x)$ which lie in φ give by orthogonal projection on the x–axis a set which we denote by E_φ. We show that the linear measure of E_φ is less than Kh^d. The desired result then follows immediately.

The gradient $s'_\nu(x)$ is dominated by its final term $\pm 2b_\nu^{1-\delta}$, and the remainder $f - s_\nu$ is dominated by its first term when ν is large. We suppose that

$$|s'_{\nu-1}(x)| < b_\nu^{1-\delta}, \quad \text{so that } |s'_\nu(x)| > b_\nu^{1-\delta},$$

and

$$0 \le f - s_\nu < 2a_{\nu+1}$$

for all the relevant values of ν.

We choose $\nu = \nu(h)$, as in I. We have, of course, $a_\nu > b_\nu^{-1} > h$: choose κ so that

$$a_{\nu+\kappa-1} > h \ge a_{\nu+\kappa}.$$

Then

$$b_{\nu+\kappa-1}^{-\delta} > h \ge b_{\nu+1}^{-1},$$

$$(b_1^{\mu\nu+\kappa-2})^{-\delta} > b_1^{-\mu^\nu},$$

$$\delta\mu^{\nu+\kappa-2} < \mu^\nu,$$

$$\delta_\mu^{\kappa-2} < 1,$$

$$\kappa < \kappa_0 = 2\log_\mu \delta.$$

We distinguish two cases: (i) $h \ge h_\nu$, (ii) $h < h_\nu$.

(i) Suppose first that $\kappa = 1$ or $h \ge a_{\nu+1}$. If (x, f) lies in φ, then (x, s_ν) lies in a rectangle φ' obtained by prolonging φ downwards a distance $2a_{\nu+1} \le 2h$. The range h of x we divide now into subintervals in which s'_ν is of constant sign: since $h < b_\nu^{-1}$, there are at most three of them. In any one of them the curve $y = s_\nu(x)$ can lie in φ' only in an interval of x of length at most $3hb_\nu^{\delta-1}$. For the height of φ' is at most $3h$ and the gradient of the curve at least $b_\nu^{1-\delta}$. Hence

$$\text{measure of } E_\varphi < 9hb_\nu^{\delta-1} < 9h^d h_\nu^{\delta-1} = 9h^d .$$

If $\kappa > 1$ we first divide the range h of x into at most three parts in which $s'_\nu(x)$ is of constant sign. The rectangle φ' is of height $h + 2a_{\nu+1} \leq 3a_{\nu+1}$, and hence in each of these parts we have only to consider a subinterval of length at most $3a_{\nu+1}b_\nu^{\delta-1}$. These we now further subdivide into intervals in which $s'_{\nu+1}(x)$ is also of constant sign. The number of new intervals obtained from each of the old is less than

$$3a_{\nu+1}b_\nu^{\delta-1}2b_{\nu+1} + 1 < 7\left(\frac{b_{\nu+1}}{b_\nu}\right)^{1-\delta} .$$

If $\kappa > 2$, we repeat the construction, obtained from each of the intervals last constructed a set of intervals in which $s'_{\nu+2}(x)$ is also of constant sign, in number less than

$$7\left(\frac{b_{\nu+2}}{b_{\nu+1}}\right)^{1-\delta} .$$

Finally we get E_φ covered by a set of intervals in which $s'_{\nu+\kappa-1}$ is of constant sign. The number of these intervals is less than

$$3.7^{s-1}\left(\frac{b_{\nu+\kappa-1}}{b_\nu}\right)^{1-\delta} .$$

In each of them (x, f) lies in φ only if $(x, s_{\nu+\kappa-1})$ lies in a rectangle $\varphi^{(\kappa)}$ of height $h + 2a_{\nu+\kappa} \leq 3h$, and hence only in an interval of length less than $3hb_{\nu+\kappa-1}^{\delta-1}$. Thus

$$\text{measure of } E_\varphi < 9.7^{\kappa-1}hb_\nu^{\delta-1} < K_6 h^d .$$

(ii) In this case $a_{\nu+1} > h_\nu > h, \kappa > 1$. If $\kappa = 2$, we divide the range h of x into intervals in which $s'_{\nu+1}$ is of constant sign. The number of these is less than

$$2b_{\nu+1}h + 1 < 3b_{\nu+1}h .$$

In each such interval (x, f) lies in φ only $(x, s_{\nu+1})$ lies in a rectangle φ'' of height $h + 2a_{\nu+2} \leq 3h$, and hence only in an interval of x of length less than $3hb_{\nu+1}^{\delta-1}$. Thus

$$\text{measure of } E_\varphi < 9h^2 b_{\nu+1}^\delta < 9h^d h_\nu^{2-d} b_{\nu+1}^\delta = 9h^d .$$

If $\kappa > 2$, we construct a covering of E_φ as in (i) by intervals in which $s'_{\nu+\kappa-1}$ is of constant sign: only the first step in the construction is changed as above. Thus

$$\text{measure of } E_\varphi < 9 \, . \, 7^{\kappa-2} h^2 b^\delta_{\nu+1} < K_6 h^d \, .$$

The curve considered in IV, in which[1]

$$b_n = b_1^{\mu^{n-1}} \, ,$$

is of dimension precisely d. We cannot take $d = 2 - \delta$ in the above work, since this gives $\mu = 1$ and every term will be the same. However, the same argument can be used to establish an example of dimension $2 - \delta$, as follows.

V. *If the ratio b_{n+1}/b_n increases to infinity, and if also $\mu_n \to 1$, then the curve $y = f(x)$ is of dimension precisely $2 - \delta$.*

We show that, for any fixed positive α, the linear measure of E_φ is less than $K_\alpha h^{2-\delta-\alpha}$, indeed is $o(h^{2-\delta-\alpha})$ as $h \to 0$, and hence that the dimension of the curve is greater than or equal to $2 - \delta - \alpha$. Write

$$\frac{b_{n+1}}{b_n} = \beta_{n+1}$$

and define κ as before. It is no longer bounded, but

$$b^{-\delta}_{\nu+\kappa-1} > b^{-1}_{\nu+1}$$

gives

$$(\beta_{\nu+2}\beta_{\nu+3} \cdots \beta_{\nu+\kappa-1})^{-\delta} > b^{-1+\delta}_{\nu+1} \, ,$$

$$\beta^{\delta(\kappa-2)}_{\nu+1} < b^{1-\delta}_{\nu+1} < (b_1 \beta^\nu_{\nu+1})^{1-\delta} \, ,$$

$$\delta(\kappa - 2) < (1 - \delta)(\nu + \log_{\beta_{\nu+1}} b_1) \, ,$$

$$\kappa < \frac{1-\delta}{\delta} \nu + 3 < A\nu \quad \text{when } \nu \text{ is large,}$$

where A depends only on δ. Hence, as in IV, we get

$$\text{measure of } E_\varphi \text{ (i)} < 9 \, . \, 7^{A\nu} h b^{\delta-1}_\nu \quad \text{if } h \geq h_\nu \, ,$$

$$(ii) < 9 \, . \, 7^{A\nu} h^2 b^\delta_{\nu+1} \quad \text{if } h < h_\nu \, .$$

---

[1] The exponent is not clear in the original. The formula should be b_1 raised to the power μ^{n-1}.
—Ed.

Now

$$b_\nu > \beta_{[\frac{1}{2}\nu]}^{\frac{1}{2}\nu} \succ K^\nu \quad \text{for any fixed } K ,$$

$$b_\nu^{\frac{1}{2}\alpha} \succ 9.7^{A\nu} .$$

So also

$$\left(\frac{b_{\nu+1}}{b_\nu}\right)^{\delta(1-\delta-\alpha)} = b_\nu^{\delta(1-\delta-\alpha)(\mu_\nu-1)} = b_\nu^{o(1)} \prec b_\nu^{\frac{1}{2}\alpha} .$$

Hence we get

$$\text{measure of } E_\varphi \prec h^{2-\delta-\alpha} ,$$

and it follows that the $(2 - \delta - \alpha)$–measure of the curve is infinite. We get an example of this type if we take

$$b_1 = 1, \quad \mu_n = 1 + n^{-\frac{1}{2}} .$$

Trinity College, Cambridge
The University, Leeds.

Bibliography

[1] F. Hausdorff, *Dimension und äußeres Maß*, Math. Annalen **79** (1918), 157–179.

[2] A.S. Besicovitch, *On linear sets of fractional dimensions*, Math. Annalen **101** (1929), 161–193.

[3] A.S. Besicovitch, *On the sum of digits of real numbers represented in the dyadic system*, Math. Annalen **110** (1934), 321–330.

[4] A.S. Besicovitch, *Sets of points of non–differentiability of absolutely continuous functions and of divergence of Fejér sums*, Math. Annalen **110** (1934), 331–335.

[5] A.S. Besicovitch, *On rational approximations to real numbers*, J. London Math. Soc. **9** (1934), 126–131.

[6] L.C. Young, *Harold Douglas Ursell*, Bull. London Math. Soc. **2** (1970), 344–346.

This selection is a treatment of **self-similarity**. The idea originated implicitly in a paper of von Koch (Selection 3), and was formulated explicitly by Cesàro [3]. But Lévy shows that there are many diverse examples of curves and surfaces that may be constructed using the same ideas.

Paul Lévy (1886–1971) was born and educated in Paris, where he held positions at the Ecole Polytechnique and the Ecole Nationale Supérieure des Mines. Although he did considerable work on other branches of mathematics, such as functional analysis, his main contributions were to all aspects of modern probability theory [15]. Lévy's autobiography and philosophy of mathematics is [11].

TWELVE

Plane or Space Curves and Surfaces Consisting of Parts Similar to the Whole

Paul Lévy

The curves that we will study here have a particular case known for a long time: the curve of von Koch. The first memoir of von Koch[1] was followed by an interesting study by Cesàro[2] which attracted attention to a remarkable property of this curve, that it consists of parts similar to the entire curve. "It is," he wrote, "this similarity between the whole and its parts, even infinitesimal ones, that makes us consider this curve of von Koch as a line truly marvelous among all. If it were gifted with life, it would not be possible to destroy it without anihilating it whole, for it would be continually reborn from the depths of its triangles, just as life in the universe is."

However, the curve of von Koch is not the only one to have this marvelous property; it shares it with other plane curve, with space curves, and there even exist surfaces having this property, and among the plane curves, there exists at least another one which is possibly at least as curious as the one of von Koch. There is no doubt that the general study of these curves and surfaces will not be judged useless.

Let us recall that von Koch himself indicated in his 1906 memoir generalizations of his initial curve; but our present objective is quite different from the problem that he posed, to study in a geometric way

---

[1] *Arkiv för Matematik, Astronomi och Fysik*, 1914, p. 681–702. See also *Acta Mathematica*, vol. 30, 1906, p. 145. These two memoirs are written in French.
[2] *Atti d. R. Accademia d. Scienze d. Napoli*, 2, XII, number 15.

functions without derivatives. One can even ask oneself whether he noticed the property alluded to by Cesàro. In any case, it is clear that he did not appreciate its importance and that the extensions indicated by von Koch are unrelated to the ones studied here.

In Chapter I we make a general study of curves consisting of p parts similar to the entire curve. Chapter II is devoted to the case of plane curves and, in particular, to the special type of curve consisting of two symmetric parts similar to the entire curve. These depend on a parameter α running between zero and $\pi/3$ and, for each value of α, there exist two curves answering the question: one is the curve of von Koch, a Jordan curve without double points if $\alpha < \pi/4$, and filling a triangle if $\alpha \geq \pi/4$. The other has, in all cases, an infinite number of double points and if $\alpha < \pi/4$, it makes up a set of points of zero plane measure. For $\alpha = \pi/4$, it covers an area consisting of a countable infinity of distinct parts the largest of which has area about three thousandths of the entire area. Moreover, it has the very curious property that, if it were materialized, one could tile the plane with it, i.e., one can cover the plane with identical copies of this curve such that no two copies intersect, and such that no space is left uncovered (more precisely, there are points that would remain empty, or that would be covered more than once, depending on whether the copies represent only the interior of the areas that they cover on the entire curve, but these points constitute a set of measure zero).

Chapter III gives some examples relating to space, and extend the results obtained for curves to surfaces. This subject we have studied the least and would be grateful if this would instigate further research in this area.

The results of this memoir were presented to the Société Mathématique de France on February 22, 1938; while those relating to the curve of von Koch had in large part already been presented to the Société on March 11, 1908.[3]

---

[3] I should be excused for recalling my first exposure to the curve of von Koch which had not been at the time discovered. It was in November 1902 that, having set myself the task of proving the existence of curves without derivatives by studying a curve consisting entirely of infinitely small detours, that I was to define exactly this curve, and to show that, for α lying between 0 and $\pi/4$, one obtains a continuous Jordan curve without double points, having no tangent at any point and that does not fill an area. I also observed that for $\alpha = \pi/4$, the curve would fill an area. About a year later I obtained the main result of Cesàro relating to the behavior of the curve near rational heights. Unfortunately, I was not encouraged at the time to publish my results and the resulting delay lost me priority.

It was only on March 11, 1908 that I decided to speak about my curve to the Société Mathématique de France. The members in attendance were most encouraging, but I soon learned of the existence of the memoirs of von Koch and Cesàro and I published nothing. Perhaps I

Chapter I. General properties of curves C of order p.

1. Definition. We will call a *curve* a set of points whose coordinates x, y, z are continuous functions of a parameter which will be called the height. We will confuse the point $[t]$ or the value of the height t; we denote by $[t_0, t_1]$ the arc consisting of points on the curve whose height lies between t_0 and t_1, and the segment $[t_0, t_1]$ will mean the line segment joining these two points. Other expressions such as "the triangle $[t_0, t_1, t_2]$" will be self evident. We assume that t varies from zero to one without restriction.

We will say that two arcs $[t_0, t_1]$ and $[t'_0, t'_1]$ are *directly similar* if one can define a continuous *increasing* function $t' = \varphi(t)$ varying from t'_0 to t'_1 as t varies from t_0 to t_1 and such that the points $[t]$ and $[t']$ correspond by similitude with *positive functional determinant*. Three other *types* of similitudes can also be defined by decreasing functions and transformations with negative determinant.

We will call a curve C *of order p* (p an integer > 1) any curve that can be decomposed into p arcs $[\tau_{i-1}, \tau_i]$ ($\tau_0 = 0 < \tau_1 < \cdots < \tau_{p-1} < \tau_p = 1$) each of which is similar to the entire curve. If the similitudes under consideration are direct, we will denote the curve by C_0. If these are not direct, but are all of the same type, we will denote the curve by C' (C_1 denoting the curve C' for which direction on the curve is conserved by the similitudes, but not that of triples[4]). All other curves C will be denoted by C''.

The arcs $[\tau_{i-1}, \tau_i]$ are similar to the entire curve so each has the same property of being in turn decomposable into p partial arcs similar to themselves, and therefore to the original curve. Replacing in this way each of the initial arcs with p partial arcs, and repeating this for the new arcs, we get an nth order operation which allows us to consider C as being the union of p^n arcs each similar to the entire curve. These arcs correspond to disjoint intervals in the t parameter whose union is the interval $(0, 1)$: a curve C of order p can

---

was mistaken since I had already passed (not in 1902 but in 1908) the point reached by Cesàro concerning the behavior of the curve at an arbitrary point. I had communicated to the Société Mathématique the theorem presented in paragraph 7 below (see footnote 13). It was by recalling this unpublished result that I was led to formulate more generally the problem of the definition of a curve similar to its parts.

[4]A similitude is "direct for triples" if it has positive determinant. Three vectors emanating from a point at right angles to each other that constitute a right-handed system (in a certain order) are transformed to three vectors that again constitute a right-handed system (in the same order). –*Ed.*

thus be considered, for any positive integer n to be a curve C of order p^n.[5]

The square of a similitude is always direct (both for triples and direction on the corresponding arcs of C), so it follows from the preceding remark that a curve C' of order p can be considered as a curve C_0 of order p^2. This is false for the curves C''. We merely observe that after each operation keeping arcs directly similar to C and subdividing the rest we can, in the limit, consider a curve C'' as a curve C_0 of finite order.

In what follows the order p will be considered to be finite unless stated otherwise. In any case, the study of curves of infinite order is an easy extension of those of finite order; so we will only give indications relating to this extension.

2. Parametric representation.   Up till now we have not made precise the choice of parameter t. We can choose it by a most convenient method as follows: the p arcs constructed by the first operation correspond to equal intervals in t. The same is true for further applications of this operation. If we now represent t in base p by writing

(1) $$t = \frac{a_1}{p} + \frac{a_2}{p^2} + \cdots + \frac{a_n}{p^n} + \cdots$$

(each of the a_n takes on one of the values $0, 1, \ldots, p-1$ which we will call nth *decimal* of t).[6] The numbers $a_1, a_2, \ldots, a_n$ determine the p^n arcs constructed by the nth operation on the point $[t]$.[7]

---

[5] We could also consider a method of subdivision into more and more numerous arcs in which only one arc is divided at a time. It follows from this remark that a curve C of order p can always be considered to be of order

$$P = 1 + \nu(p-1),$$

for any positive integer ν.

[6] Due to a lack of appropriate nomenclature in the study of numbers in base p, we will continue to use this terminology: it is understood, of course, that this does not imply that $p = 10$. We will use this terminology even if $p = 2$, so that a multiple of 2^{-n} will be called a decimal number and not a dyadic number.

[7] The choice of the parameter t is related to the idea that, if one wants to randomly choose a point on the curve, the p^n arcs are considered to be equally probable. One can thus suppose that the a_ν are determined by random draw giving each of the p possible values the probability $1/p$.

We can also consider other probabilistic models relating to other parametric representations. For example, when the p lengths $|\tau_{i-1} - \tau_i|$ are all different, we can suppose that for each value of n the probabilities associated to the p values a_n are proportional to their lengths. The probabilities associated to the p^n arcs constructed by the nth operation will then be proportional to the lengths of the corresponding chords.

The relation between the parameter t' corresponding to such a choice and the parameter t originally considered is quite curious. Each is an increasing function of the other. Each of these functions has in any interval its extremal derived values equal to zero and $+\infty$. To a set of t values of measure 1 there can correspond a set of t' values of measure zero and conversely. We will return to this point at the end of paragraph 7.

To each value of t there corresponds a nested sequence of arcs and in the limit, these reduce to a point or to an arc. We will see in paragraph 4 that the latter never occurs: we might be concerned that the parameter thus chosen is not well defined, and that x, y, z might be continuous functions of a parameter other than t, but not of t itself. We will see that this is not possible.

3. Construction of C. A constructive definition of one of our curves C can be obtained by considering it as a limit, for n finite, of the polygonal line Γ_n derived from the nth operation, i.e., the line inscribed in C having as vertices the points that are multiplies of p^{-n}. To give this construction explicitly, Γ_1 must first be specified. For each side $[(h-1)/p, h/p]$ of Γ_1 one fixes the type of similitude that gives the correspondence with the whole curve and fixes a half plane bounded to the right by the side in question containing the first vertex of Γ_2 not lying on the side among those that follow the point $[(h-1)/p]$.[8] The p similitudes thus defined induce a correspondence between each line Γ_n and the part $\gamma_{n+1}^{(h)}$ of Γ_{n+1} contained between the points $[(h-1)/p]$ and $[h/p]$ and this allows us to successively construct the lines Γ_n.

Taking the distance between the points $[0]$ and $[1]$ to be the unit of length, we denote by q the length of the largest side of the broken line Γ_1, and by ℓ the largest of the diagonals of this line having an endpoint at one of the extremal vertices.[9] We will suppose that $q < 1$, and we will show that this does not exclude any solution to our problem—it is a necessary and sufficient condition for the existence of a curve C defined as a limit of the lines Γ_n.

If the expansion of t in formula (1) is finite then the point $[t]$ is a vertex of one of the lines Γ_n. Otherwise, we will define it to be the endpoint of the *approach line* L having vertices the points of height

$$t_n = \frac{a_1}{p} + \frac{a_2}{p} + \cdots + \frac{a_n}{p^n} \qquad (n = 1, 2, \ldots).$$

----

[8] We assume that all the vertices of Γ_1 do not lie on a line (since the curve would then reduce to a line). Therefore, the part of Γ_2 lying between $[(h-1)/p]$ and $[h/p]$ (similar to Γ_1) contains at least one vertex not lying on this line. By fixing the orientation of the half plane containing this line and this vertex we can unambiguously define the similitude in question.

For each line Γ_1 we obviously get 4^p families of curve C each depending on p parameters. If the line Γ_1 is symmetric either with respect to the center of the line segment $|0, 1|$ or with respect to its mediating plane, the number of families reduces to 3^p. This number is further reduced if one only asks for curves C having these symmetries themselves.

[9] It would be simpler to introduce the notion of the length of this line. The method used in the text has the advantage of allowing the case in which p is infinite, even if the length of Γ_1 is infinite. Of course, if p is infinite, the conditions imposed by our definition of the curve must still hold for Γ_1 (first line of paragraph 1).

The length of this line is majorized by the finite sum

$$\ell\,(1 + q + q^2 + \cdots + q^n + \cdots) = \frac{\ell}{1 - q},$$

and so the point $[t]$ is always well defined.

To see that this point varies in a continuous way, let us observe that the distance to the point $[t_n]$ is less than

$$(2) \qquad\qquad \ell\,q^n\,(1 + q + \cdots) = \frac{\ell q^n}{1 - q}.$$

This is also true for its distance to the point with height $t'_n = t_n + p^{-n}$, so what was shown for the approach line to the point $[t]$ also holds for the *retreat line* (broken line having the points t'_n as vertices). So when n is large enough to make the expression (2) less than ε (where ε is an arbitrarily small positive number) and

$$|t - \tau| < 2^{-n},$$

then the points $[t]$ and $[\tau]$ either belong to the same arc $[t_n, t'_n]$ or to two contiguous arcs having a common vertex. Since their distance to this vertex is less than $\varepsilon/2$, the distance between them is less than ε and this establishes the continuity of the curve.

We now indicate two obvious properties of this curve:

1. Apart from the trivial case when the curve is a straight line, the length L of Γ_1 is greater than one, so the length L^n of Γ_n increases indefinitely with n. The length of the curve C is thus infinite. Since any arc of this curve contains parts similar to the entire curve, all of these have infinite length.

2. There is no tangent at any point, not even a half tangent to the approach arc $[0, t]$ or to the retreat arc $[t, 1]$, each considered separately (apart from the case when the curve reduces to a line segment).

To fix our ideas let us now consider the approach arc. If t is decimal, the arc $[t - p^{-n}, t]$ is, for large enough n, similar to the entire curve and the variation in direction of the half line $[t, t']$ as the point $[t']$ describes this arc is the same as those of the half line $[1, T]$ or of $[0, T]$ (assuming that for the similitude under consideration the point $[t]$ corresponds to one of the endpoints of C) as the point $[T]$

describes the entire curve. These variations reproduce themselves infinitely and there is no tangent–the limit set[10] of these arcs is equal to all these points and does not reduce to a line.

If the expansion of t is unbounded then $a_n < p - 1$ infinitely often in this case. The figure formed by the arc $[t_n - p^{-n}, t_n]$ and the point t is then similar to the one formed by the entire curve C_0 and a point A located a distance r from one of the endpoints of C (the one that corresponds to $[t_n]$ under the similitude in question), and it is easy to bound it above. If $n = 1$, let q' denote the length of the smallest side of Γ_1 and R the maximum distance between a point on the curve C and one of its endpoints so that the distance between $[t_1]$ and $[t]$ cannot be greater than qR. One gets r by multiplying by a similarity factor equal at most to $1/q'$. Thus $r < R' = qR/q'$ and this conclusion holds for all n.

Let us consider the apparent width $\varphi(A)$ of the curve as seen from A, that is, the supremum of the angle MAM' (lying between 0 and π) as M and M' describe the curve. It has a positive minimum ω in the circle $r \le R'$. In any case, we only need to know three points that lie on this curve, not all on a line, in order to determine a positive lower bound for ω.

Taking into account the similitude between the arc $[t_n - p^{-n}, t]$ and the entire curve, we can then, for arbitrarily large n, find two numbers τ and τ' between $t_n - 2^{-n}$ and t_n such that the angle between the two half lines $[t, \tau]$ and $[t, \tau']$ is at least equal to ω. We thus see that not only does the approach arc have no tangent at $[t]$ but that its limit set contains two half lines forming an angle of at least ω.

4. Discussion. We now show that there are no other curves consisting of p arcs similar to the entire curves other than those that we have just defined. However, we will see that there are continuous objects which are not curves but that still have this property.

First, let C be a curve having this property. It is the union of p arcs $[0, \tau_1], [\tau_1, \tau_2], \ldots, [\tau_{p-1}, \tau_p]$ similar to the entire curve. Knowledge of their endpoints and the elements that define the similitudes under consideration suffices to determine all the lines Γ_n of the previous construction.

---

[10]The author uses the term *contingent* throughout the paper to denote the limit set of *lines* joining a given point on a curve to a point approaching it on the curve. This describes a nontrivial sector in the plane if the curve has no tangent at the point in question. This concept does not seem to have a name in English and I translate it in the text as "the limit set of the approach arc" or simply "the limit set." –Tr.

It is impossible that $q > 1$. If, in fact, R is the maximal distance between two points of C, then on the largest of the arcs $[\tau_{h-1}, \tau_h]$ there exist two points a distance qR apart. However, this distance cannot be greater than R since these points lie on C, so $q \leq 1$.

Let us consider the possibility that $q = 1$. There would then be an arc $[t_1, t_1']$ equal to C and on this arc there would exist an arc $[t_2, t_2']$ equal to C and so on. By this equality the distances $[0, t_1], [t_1, t_2], \ldots,$ would be equal, which is impossible unless they were all zero. In fact, the increasing sequence t_n must have a limit t and, and since C is a continuous curve, the points $[t_n]$ tend to $[t]$.

All the points $[t_n]$ therefore coincide with zero and the points $[t_n']$ with [1]. In fact, $p > 1$, i.e., Γ_1 does not reduce to a single side $[t_1, t_1']$ coinciding with $[0, 1]$. On at least one of the arcs $[0, t_1]$ and $[t_1', 1]$ we must find a point τ_1 which is not an endpoint of the arc. Let us denote by τ_n its homologue under the transformation (with or without symmetry) that takes the entire curve onto the arc $[t_n, t_n']$. If, to fix our ideas, τ_1 lies between 0 and t_1 and the points $[0], [\tau_1], [t_1], [\tau_2], \ldots$ are the vertices of a broken line inscribed in C with decreasing values of the parameters and all sides equal. Thus we cannot escape the impossibility noted above.

We cannot escape it because we are considering a curve. It is perhaps not entirely useless to indicate that removing this restriction introduces new solutions.

Consider points $A_1, A_2, \ldots$ as vertices of an open broken regular line inscribed in a circumference. We choose two points O and O' on the axis of this circumference and take as Γ_1 the broken line

$$O\Lambda_1 O\Lambda_2 \ldots O\Lambda_n O \ldots OO',$$

with corresponding heights of the vertices given by

$$0, \frac{1}{2}\left(1 - \frac{1}{2}\right), \frac{1}{2}\left(1 - \frac{1}{3}\right), \frac{1}{2}\left(1 - \frac{1}{4}\right), \ldots, \frac{1}{2}, 1. \bullet$$

There will be a discontinuity at $t = 1/2$. This does not preclude that the sides of Γ_1 other than the last can be used as the starting point for a construction analogous to the one in paragraph 3 and, providing that the common distance of $A_1, A_2, \ldots$ to O is less than OO', there is convergence. We arrive at a well defined continuous object C which contains all the intermediate Γ_n.

It has the following properties: it can be traversed by a point whose coordinates are functions of a parameter t. These functions have a countable number of points of discontinuity, but are continuous in intervals whose total length is equal to one. The figure obtained in this way is the union of three curves similar to C that correspond to the intervals of variation of t separated by the values $1/4$ and $1/3$; for the third of these there is an equality.[11] There exist arbitrarily small parts of C that are similar to C, but these cannot be located arbitrarily since the interval of variation of t decomposes into intervals each of which corresponds to a line segment.

We have not yet finished the proof that there are no other curves answering our problem other than those already obtained. In fact, in the case where $q < 1$, the only one left to consider, we constructed a curve C which certainly belongs to the one we supposed was given at the start, since this curve C is a continuous object that contains all the vertices of Γ_n. But there remains to show that C does not contain points other than those of C. In other words, one must answer the question already posed at the end of paragraph 2: can the parameter t, chosen for the convenient construction of C, be improper, i.e., a non decreasing function of another parameter T which could also define C, but that is not always increasing? There would then be on C arcs γ (necessarily closed, each corresponding to an interval of variation of T) not located on C.

However, C is the union of arcs $C_1, C_2, \ldots, C_p$ similar to C and having endpoints belonging to C. Each of the arcs γ must therefore belong to one of the C_h (nothing prevents there being two contiguous arcs γ' and γ'' belonging to two arcs C_h and C_{h+1}). Since C_h is similar to the entire curve γ, considered as a subset of C_h, is similar to a larger arc γ_1 itself similar to a larger arc γ_2 and so on. Similarly, due to the similitude that defines them, all these arcs do not belong to $\bar{C}$ and are never divided by the vertices of the lines Γ_n, so nothing prevents one from continuing the construction. In any case, the ratio between the homothety from γ_n to γ is at least q^{-h} and one would then get arbitrarily large arcs, which is impossible. The proof is now complete.

5. Properties of points of rational height.   The similitude between the curve and its parts implies an important consequence already noted by Cesàro in the case of the von Koch curve.

---

[11] We have assumed that we run over each $O\Lambda_n$ of Γ_1 successively in both directions. By allowing a parameter to run discontinuously over Γ_1, each of its sides being run only from O to Λ_n, we obtain a continuous object made up of two parts similar to the whole, the one from $O\Lambda_1$ and the rest.

Let us denote by γ_0 an arc similar to the entire curve and by γ an arc of γ_0 with similarity ratio $Q < 1$. Under the similitude that gives the correspondence between γ and γ_0, the arc γ has as image a smaller arc γ_2 and so on. After n iterations γ_0 is transformed into an arc γ_n similar to γ_0 with a ratio of similarity Q^n tending to zero as n goes to infinity. Each of these arcs belongs to the previous one, so they have a limit point $[t]$ on the curve, and this point is invariant under the similitude.

Moreover, we can suppose that the similitude is direct, even if the curve is a C' or a C'' (either from the point of view of orientation of triples or from direction of travel on the curve). One simply considers the square of the similitude and only considers every second arc of the γ_n.

The height t of the limit point that we have just defined is a rational number. In fact, the operations that lead from γ_0 to γ_1 repeat themselves at a smaller scale to get from γ_1 to γ_2, then from γ_2 to γ_3, and so on. In each case, however, these operations are defined by a certain number m of decimals that indicate, among the p^m similar arcs whose union is γ_n, which one is γ_{n+1}. To define t one must, after having written down the common decimals for the heights of all the points of the arc γ_0, periodically repeat the same sequence of m decimals that define the operations to get from γ_n to γ_{n+1}. It is known that such periodicity characterizes rational numbers.

Conversely, given a rational number t defined by its decimal expansion, the digits preceding the period define a certain arc γ_0 and in the interior of γ_0 the period, necessarily written twice, defines a certain arc γ directly similar to γ_0. The point $[t]$ is then identified with the fixed point of the similitude that takes γ_0 to γ_1.

The point P is therefore easy to determine effectively once the construction of the broken lines Γ_n allows one to determine the endpoints of the arcs γ_0 and γ_1 and the elements of the similitude which gives their correspondence. Let us note that what we are asserting about points of rational height in particular applies to the vertices of the lines segments of Γ_n, that is, to points whose heights are finite decimals.[12]

Let us now study the behavior of the curve in the neighborhood of the point P with rational height t. The similitude in question, being

---

[12]To study the approach arc of such a point $[t]$, a vertex of Γ_n, one must of course consider it as the common endpoint of arcs γ_ν, each of which has as origin the point $[t - p^{-(n+\nu)}]$. Otherwise, the retreat arc is considered as the common origin of the arcs $[t, t + p^{-(n+\nu)}]$. The similitude that reproduces the retreat arc in the neighborhood of such a point will then be (except in exceptional cases like the von Koch curve) distinct from the one that reproduces the retreat arc. Otherwise, for points with rational non decimal height, these two similitudes will be identified.

direct, is the product of a homothety (of ratio $Q < 1$) and a rotation (of angle Ω, possibly zero) with one of these leaving the point P fixed. We then consider the origins $M_0, M_1, \ldots$ of the arcs $\gamma_0, \gamma_1, \ldots$ so that these points, being vertices taken from m to m of the approach line to the point P, are the successive images of the first point under the same similitude. If the similitude reduces to a homothety, they lie on the line PM_0. Otherwise, they are on the same cone of revolution with axis of rotation D and the plane DM_h turning by the angle Ω, and the distance PM_h being multiplied by Q when h increases by one. They are regularly spaced on a helix (a curve cutting all the generators by the same angle and whose tangents make a constant angle with the parallels of the axis) lying on this cone of revolution. The projection of this helix to a plane perpendicular to the axis is a logarithmic spiral. It can happen that this cone reduces to a plane so that the helix itself becomes a logarithmic spiral. This is important to note in the study of plane curves.

We have just talked about points $M_0, M_1, \ldots$, but the arcs $M_h M_{h+1}$, being successive images of the arc $M_0 M_1$ by the same transformation, these remarks give an idea of the way the curve approaches the point P. The same remarks hold for the retreat arc. If one observes that, for the approach arc as well as for the retreat arc, points of rational height are everywhere dense and that what was just said for the point P also holds at any one of these points, one sees that the way the curve winds is more than our imagination can conceive: "It will tire of conceiving faster than nature will provide," and one understands the admiration of Cesàro for the marvelous properties of the curve of von Koch. There is no question that our intuition foresaw that the absence of a tangent and the infinite length of the curve were related to these infinitely small detours that one could not dream of drawing;[13] but one is left helpless by the inability of our imagination to even go through the first few steps of the construction of these infinitely small detours.

Let us mention that points of rational height are characterized by a certain regularity in the structure of the approach arc and the retreat arc, and one must expect to find even more complicated phenomena for points of irrational height—the curve would have as many infinitely small detours in the neighborhood of these points, but they would not be as evenly spaced. The study of plane curves will allow us to make

---

[13]I emphasize the role of intuition since I have always found it surprising to have it said that geometrical intuition led ineluctably to the consideration of continuous differentiable functions. From my first exposure to the notion of derivate, my experience has proved the contrary.

this remark precise and also, for points of rational height, to indicate
when the similitude by which the curve reproduces itself in the neighborhood of these points reduces to a homothety.

Chapter II. The case of plane curves. Some particularly remarkable curves.

6. Preliminary remarks. The area measure of the curves C.   To define
a plane curve C of order p one must first choose a plane broken line
Γ_1 with p sides satisfying the condition $q < 1$,[14] and for each of its
sides specify whether the similitude that defines $\gamma_2^{(h)}$ preserves signs
of angles or direction along the curve. For each line Γ_1 not having any
symmetries, there correspond 4^p curves C, among which one C_0 curve
and three C' curves (we denote by C_1 the one for which the similitudes
preserve direction along the curve). If Γ_1 is symmetric with respect to
the median of the segment $[0, 1]$ or to the center of the segment, there
are only 2^p distinct curves C and a single C' (two being identified and
the third identified with C_0).

We denote by $\ell_1, \ell_2, \ldots, \ell_p$ the length of the sides of Γ_1 (the distance between the points $[0]$ and $[1]$ still being taken as unity) and set

$$(3) \qquad\qquad k = \ell_1^2 + \ell_2^2 + \cdots + \ell_p^2 .$$

Let σ be the area of a region containing C, for example, the smallest convex region containing this curve. The p arcs $[(h - 1)/p, h/p]$,
$h = 1, 2, \ldots, p$, are similar to C and have ratio of similitude equal
to $\ell_1, \ell_2, \ldots, \ell_p$, respectively, so they are contained in regions of area
$\ell_1^2 \sigma, \ell_2^2 \sigma, \ldots, \ell_p^2 \sigma$. The curve C is the union of these arcs and so is contained in a region of area at most equal to $k\sigma$. A similar argument
applied to the p^n arcs separated by the vertices of Γ_n shows that C is
contained in a region of area at most $k^n \sigma$. Therefore, *if $k < 1$, the curve
is a set of points of area measure zero.*

Suppose now that $k > 1$ and instead of σ, consider the area s of
the smallest convex polygon (that we denote by S) containing all the
vertices of Γ_1. Denote by S_n the area of the union of the set of p^n areas
similar to S by the similitudes that gives the correspondence between
C and the arcs separated by the vertices of Γ_n. The sum of the areas of
these polygons is $k^n s$. The area s_n of S_n is thus less than or equal to $k^n s$

---

[14]The original has $q > 1$. –*Ed.*

depending on whether these polygons are disjoint or not. However, the domain S_n is (due to the hypothesis $q < 1$) contained in a finite region S' independent of n, so if $k > 1$ the parts of S' covered by S_n are, on average, covered an increasing number of times with n with order of growth k^n.

Finally, we consider the case when $k = 1$. The numbers s_n are all equal to s when the p^n parts of S_n are disjoint, no matter how large n is. Otherwise, they cannot decrease. If, in fact, one considers the p arcs $[(h - 1)/p, h/p]$ separately, the region similar to S_n corresponding to each of these has area $\ell_h^2 s_n$ and the union of these p regions constituting the region S_{n+1} has area $s_{n+1} \leq k s_n = s_n$. In any case, the numbers s_n form a non increasing sequence having a limit s'.[15]

The two possibilities $s' = s$ and $s' < s$ can actually both occur. The first is clearly realized for the curve C when $k = 1$, since S_1 is then a right triangle and S_2 is the union of two triangles interior to S_1 and separated by its height. All the areas S_n and the limit of the curve C_1 fill the triangle exactly in this case. It is thus a Peano curve filling a triangle. The Hilbert curve filling a square can also be easily obtained by our method by taking p to be the square of an odd integer. It is a C_0 curve and one can even define a C'' curve filling the rectangle.

Returning to the case of a right triangle ($p = 2$, $k = 1$), we consider the curve C_0. We will study in a more complete way the case of an isosceles right triangle and show that in this case $s' = s$. If, on the other hand, the right triangle is not isosceles, it suffices to draw a figure to see that the eight triangles of S_3 are disjoint, but that two of the sixteen triangles of S_4 have a common part (the eight and tenth if $k_1 < k_2$, the seventh and ninth if $k_1 > k_2$) and so $s' < s$.

In any case, let us show that the area measure m of the curve C is at least equal to s'. On the one hand this follows from the fact that the points of C form a closed set (which is thus measurable) and, on the other hand, that the largest of the regions that make up the area S_n, being similar to S with a similarity ratio of at most q^n, gets small as n gets large. However, all these regions are bounded by polygonal lines having their vertices on C. The maximal distance between a point of S_n to C thus tends to zero for infinite n, and it is thus less than an arbitrarily small number η, for n quite large ($n > N(\eta)$).

Let Σ then be a convex region of area σ containing C and so all the S_n. The points interior to Σ and not lying on C, if there are any,

---

[15] A problem that I have yet to resolve is the following: is the statement "if $s' < s$ then $\sqrt[n]{s_n}$ has limit less than 1" correct? If it is not, can one assert that s' can have no other values other than zero or s?

make up an open set Σ' of area $\sigma - m$. No matter how small ε is, one can define a finite number of circles interior to Σ' with total area greater than $\sigma - m - \varepsilon/3$, and a number η small enough so that when η is decreased, the area covered by these circles stays greater than $\sigma - m - \varepsilon$. For $n > N(\eta)$, this area, having distance to C greater than η, has no common point with S_n. Its measure is thus at most equal to $\sigma - s_n$ and so $s_n \leq m + \varepsilon$ and in the limit $s' \leq m + \varepsilon$. Since ε was arbitrary, it follows that $s' \leq m$. Q.E.D.

7. The behavior of the curve at an arbitrary point. For this study we will make a restrictive hypothesis: the region S_1 lies inside S and the p polygons similar to S whose union is S are pairwise disjoint apart from the vertices of Γ_1, so each vertex necessarily belongs to two consecutive polygons. These hypotheses imply that $q < 1$ and that the line Γ_1 is without double points and its endpoints are vertices of the polygonal area S. They imply that each region S_n lies in interior of the previous one and has area $q^n s$ tending to zero for infinite n (S_n still represents the same fraction of S_{n-1}) and that it is composed of p^n polygons similar to S that form a chain for which two consecutive polygons only have a common vertex and non consecutive polygons are disjoint. The curve C, the limit of S_n for infinite n, is then interior to all the S_n, has area measure zero, and has no double points.[16]

We will suppose that the curve C in question is of type C_0 (recall that a curve C' of order p can be considered as a C_0 of order p^2, so that the von Koch curve is not excluded from what follows).

We consider a point $[t]$ on the curve and propose, in order to fix our ideas, to study the approach arc, and in particular to see whether it winds infinitely around this point before reaching it. Letting ρ and ω denote the polar coordinates of the point $[\tau]$ on the curve (the origin being $[t]$) this reduces to deciding whether ω stays finite or not as τ tends towards t by small values.[17]

To study this problem we can, by virtue of the hypotheses that have just been made, substitute the approach curve with a polygonal approach line $\mathcal{L}$ defined as follows: as for the line L in paragraph 5, it passes through all the points $[t_n]$ but, instead of following a straight

---

[16]The case where the p parts of the area S_1 exactly cover the polygon S except that two consecutive parts have a common line and $q = 1$ (a limiting case of the one considered in the text) can be dealt with by an analogous method to the one in the text thanks to the fact that when C has double point, the curve does not cross over itself.

[17]The curve C not having double points, the retreat arc does not cross over the approach arc, so that the behavior of one of these arcs informs us about the other.

line from $[t_{n-1}]$ to $[t_n]$, it follows the line Γ_n by its a_n sides leading from the first point to the second. Let us clarify that if t is decimal we will take t_n (for n large enough) to be the value $t - p^{-n}$ approached by default in such a way as to obtain in all cases a line $\mathcal{L}$ inscribed in the approach arc and having an infinite number of sides. This line $\mathcal{L}$ clearly has no double points.

To successively determine the values of the polar angle ω at the vertices of $\mathcal{L}$ it does not matter whether one chooses at a vertex this line or the curve C. This clearly follows from the fact that each of the sides of $\mathcal{L}$ belongs to a line Γ_n and the arc of C that it subtends is located entirely within one of the convex polygons that make up S_n and that $[t]$ lies inside this polygon. It is therefore sufficient to study the shape of the line $\mathcal{L}$.

We now show that to study this line we can, with finite error, substitute for ω the angle θ with the x axis that is made at each instant by an observer describing this line. At each vertex of this polygonal line, the abrupt change of θ is assumed to lie between $-\pi$ and $+\pi$,[18] and at each point $[t_n]$ we will take as angle θ the angle that $[t_n, t'_n]$ makes with the x axis. The angle θ is thus well defined at each point of $\mathcal{L}$.

Consider two vertices $[t_n]$ and $[t_v]$ of $\mathcal{L}$ ($v > n$) and denote by $\mathcal{L}_n^v$ the segment of this line lying between these vertices, then this segment does not penetrate the region S_v that contains the segment $[t_v, t]$ and it has no double points. We will study the variation of $\theta - \omega$ between two endpoints of $\mathcal{L}_n^v$ and show that if the initial value $\theta - (\omega + \pi)$ (the angle between $[t_n, t'_n]$ and $[t_n, t]$) lies between $-\pi$ and $+\pi$, then the same holds for its terminal value.

This follows from the fact that the variation of θ and ω, and so of $\theta - \omega$, between the points $[t_n]$ and $[t_v]$ are not changed by a continuous deformation of this line verifying the following conditions: its endpoints are fixed and the line is still without a common point with the segment $[t_v, t]$ except for its endpoint $[t_v]$ (which implies the consequence noted for ω). The line does not have any double points within the interior of the region of S_n containing $[t]$ where it has $[t_n]$ as a vertex, and in the exterior of the region of the S_v containing $[t]$ where it has $[t_v]$ as a vertex. A deformation of $\mathcal{L}_n^v$ varying these conditions allows one to eventually eliminate the detours of this line and to reduce

---

[18]The hypotheses made about Γ_1 exclude the possibility that one would have to choose between $+\pi$ and $-\pi$. The two polygonal regions similar to S and adjacent to two sides of $\mathcal{L}$ between which there might be a retracing of the curve, would in fact have the smaller of these sides in common.

it to its essentials: a line winding a certain number of times around a convex region containing $[t_\nu, t]$, with endpoints at $[t_n]$, $[t_\nu]$, and with tangents that one can assume coincide with $[t_n, t'_n]$ and $[t_\nu, t'_\nu]$.

As well, one changes the terminal values of θ and ω, but not their difference, modifying on the one hand the endpoint $[t_\nu]$ of the line $\mathcal{L}^\nu_n$ and its tangent at this point in such a way that the figure constructed at this point, this tangent, and the point $[t]$ remains similar to itself. A new continuous deformation applied under these conditions allows one to bring the point $[t_\nu]$ to a point A_ν lying on the segment $[t_n, t]$ and the line obtained in this way no longer rotates around the point $[t]$, but consists of a line segment very close to $[t_n]A_\nu$, whose endpoints lie on tangents making angles between $-\pi$ and $+\pi$ with this segment. The extreme values of $\theta - \omega - \pi$ are then both between $-\pi$ and $+\pi$.

One can even observe that the figure formed by the points $[t_n]$, $[t'_n]$, and $[t]$ is similar to the one formed by the points $[0]$, $[1]$, and a point $[\tau_n]$ suitably chosen on C, thus in the interior of the convex area S having O as a vertex. The angle between $[t_n, t'_n]$ and $[t_n, t]$ is the value of $\theta - \omega - \pi$ at the point $[t_n]$ and is thus equal to the angle that the fixed direction $[0, 1]$ makes with the direction $[0, \tau_n]$ which lies in the interior of S and so is inside a fixed angle $\alpha < \pi$. It thus varies, as n increases, between two bounds whose difference is α.

Moreover, the maximal oscillation of ω between the points $[t_n]$ and $[t_{n+1}]$ is less than $p\pi$ for all n. Therefore, there is no need to preoccupy ourselves with the variation of ω between these two points and knowing whether the line $\mathcal{L}$ or the curve C winds indefinitely around the point $[t]$ before reaching it reduces to the study of the sequence of numbers θ_n (the values of θ at the points $[t_n]$).

However, this is a very easy problem due to the similitude between the part of the line $\mathcal{L}$ lying between the points $[t_{n-1}]$ and $[t_n]$ with the set of the first a_n sides of Γ_n. If we denote by $\varphi(1), \varphi(2), \ldots,$ $\varphi(p)$ the values of θ relative to the p sides of Γ_1 we have

(4) $$\theta_n - \theta_{n-1} = \varphi(a_n + 1).$$

It is quite important to note that $\varphi(0) = 0$ (as well as $\varphi(p) = 0$). In fact, if this were false then, since the direction $[0, p^{-n}]$ makes an angle of $n\varphi(0)$ with $[0, 1]$, the points $[p^{-n}]$ would be a spiral surrounding the point $[0]$ and the curve would leave this point only by turning indefinitely around it. Because of the similarity of the curve and its parts, the same would be true at any vertex $[t_n]$ of the line Γ_n and this would contradict the hypothesis at the beginning of this paragraph by which the approach arc at such a point and retreat arc at such a point respectively lie in the interior of two convex polygonal regions with no common point other than $[t_n]$.[19]

This remark is essential since, if $\varphi(0)$ were not zero and if v decimals of t following a positive decimal a_n were zero, formula (4) would give for the variation of θ on the line $\mathcal{L}$ between the side terminating at the point $[t_n]$ and the side leaving this point (common to $\mathcal{L}$ and Γ_{n+v+1}) the value

$$\varphi(a_n + 1) - \varphi(a_n) + (v + 1)\varphi(1),$$

not necessarily lying between $-\pi$ and $+\pi$. The angle θ computed using formula (4) would not be the one which we wanted to compute. On the contrary, since $\varphi(1)$ would be zero, we could find a value which, by the definition of $\varphi(h)$, would lie between $-\pi$ and $+\pi$. It follows that

$$(5) \qquad \theta_n = \varphi(a_1 + 1) + \varphi(a_2 + 1) + \cdots + \varphi(a_n + 1).$$

The function $\varphi(h)$ necessarily takes on values of either sign (the contrary hypothesis would imply a spiral form inconsistent with the fact that $[0]$ is a vertex of the convex polygon S containing Γ_1). It can then happen that the sequence of values θ_n is bounded. This occurs for a set of values t which is everywhere dense and has the cardinality of the continuum. For points $[t]$ corresponding to such values the variation in the polar angle ω is finite both for the approach arc and the retreat arc (for a curve without double points the two arcs are the same from this point of view). In particular, these points contain the points of C lying on the contour of S or on any of the convex polygons similar to S but which constitute different regions S_n.

But the fact that θ_n, and therefore ω, are finite is exceptional. Known theorems in probability allow us to assert this. Let us suppose that, in order to choose t, we successively choose its decimals by a random process that would give at each step p possible values a_n with positive probability independent of n. If these p probabilities are equal, the probability is equal to the measure on the t axis,[20] and if

---

[19] The study of the variation of θ on the line $\mathcal{L}$ is clearly of interest independent of this hypothesis. But in the case of lines with double points one cannot assert that its changes in orientation imply that this line turns around the point $[t]$. One cannot even say whether this has any significance concerning the curve C. If, for example, $\varphi(1) = 0$, $\varphi(p) > 0$, the curve C gets to each point $[t_n]$ by turning indefinitely around it and leaves it inside a fixed angle. The increase in θ seems infinite and the fact that the line $\mathcal{L}$ avoids the detours in C and approaches this point in a straight line does not constitute, in the case of C, a reason to assume that the value of θ is finite at this point or on the arc of the curve leaving this point.

[20] The Normal Number Theorem, for example [6], p. 1. –Ed.

they are not all equal, one can define a parameter t', a continuous increasing function of t (see footnote 7) giving another linear representation of the curve, and such that the probability in question would be equal to the measure on this other linear representation.

Denote by μ the probable value[21] of $\varphi(h)$ and by μ'^2 that of $[\varphi(h) - \mu]^2$, where h is the variable chosen at random according the law in question. By known theorems alluded to above,[22] the following properties of the sequence θ_n are almost sure (that is, they are realized except as an event of probability zero): if μ is non zero, θ_n is infinitely large and equivalent to μn, and $\theta_n - \mu n$ is sometimes of one sign and sometimes of another and has absolute value of order $\mu' \sqrt{n}$ except for a rather rare set of values of n. These values actually occur and the indicated order of growth is surpassed infinitely often. The true upper bound for $\theta_n - \mu n$ is $\mu' \sqrt{2n \log \log n}$ in the sense that the accumulation points of the quotient of these two numbers are (again almost surely) numbers in the closed interval $(-1, +1)$ and only these.

Of the above results we note that it is almost sure that θ_n and so ω (on the approach arc as well as on the retreat arc) do not remain finite. They increase indefinitely with values having a well determined sign when μ is non zero and alternating sign when μ is zero. In each of these cases one has a fairly good idea of the general behavior on both the approach arc and retreat arc which must leave the point $[t]$ without crossing the approach arc.[23]

Let us remark now that the value of μ depends on the probability law in question: since $\varphi(h)$ has possible values of either sign, the sign of μ depends on this choice. The only result independent of an arbitrary choice is that the different possibilities indicated above are all realized on a set of points which is everywhere dense and has the cardinality of the continuum. But each of these sets has measure zero or one depending on the linear representation used to represent the curve and define the measure.[24]

In the case where the line Γ_1 is symmetric with respect to the median of the segment $[0, 1]$, the curve C itself is symmetric and it is clearly indicated that one should choose either the parameter t of

---

[21] The "mathematical expectation" of the random variable. –Ed.

[22] The Strong Law of Large Numbers [6], p. 7, the Central Limit Theorem [6], p. 308, the Law of the Iterated Logarithm [6], p. 127. –Ed.

[23] It was the result relating to the case $\mu = 0$ that I communicated on March 11, 1908, to the Société Mathématique de France in the special case of the von Koch curve. It is hardly necessary to add that in this case the verification is quite simple–the precautions necessary for a general theory are pointless when considering a particular curve.

[24] See footnote 7.

paragraph 3 or another parameter t' without introducing any element of asymmetry (that is, that the values p and $p - h - 1$ be equiprobable and that two symmetrical points on the curve C would correspond on the t' axis to two symmetrical points with respect to $t' = 1/2$). So $\mu = 0$ and for a value t (or t') chosen at random all real numbers are accumulation points of ω on the approach arc to $[t]$ as well as for the retreat arc.

8. Points of rational height. In the case of plane curves it is easy to make precise the results of paragraph 3 relating to such points. The first results to be presented are independent of the hypotheses made at the beginning of paragraph 7 and we suppose that we are dealing with a curve C_0.

First of all, concerning points with decimal heights, it is quite clear that if $\varphi(1) = 0$ then each of these is the center of a homothety (with homothety ratio ℓ_1) leaving invariant the retreat arc in a neighborhood of the point. If, on the other hand, $\varphi(1)$ is non zero, then one has a similitude which is a composition of the above homothety with a rotation of angle $\varphi(1)$ and the retreat arc winds indefinitely around our point. The behavior of the approach arc also depends on the value of $\varphi(p)$. The two arcs can be qualitatively different and it is only in the case of a curve without double points that they behave in the same way.

We now consider a non decimal rational number t. The similitude which leaves invariant the part of the curve in a neighborhood of this point is well defined by m decimals that constitute the period of the decimal expansion of t. If the digits $0, 1, \ldots, p - 1$ occur in this period $m_1, m_2, \ldots, m_p$ times, respectively, the ratio of similitude Q and the angle of rotation Ω have values

(6) $$Q = \ell_1^{m_1} \ell_2^{m_2} \cdots \ell_p^{m_p},$$

(7) $$\Omega = m_1 \varphi(1) + m_2 \varphi(2) + \cdots + m_p \varphi(p),$$

and one can, with finite error, take as $\frac{n}{m} \Omega$ the value approached by the angle θ_n defined by formula (5).

If Ω is not a multiple of π then for a point A_0 on the approach arc which quite close to the point $[t]$, its successive images $A_1, \ldots, A_\nu, \ldots$ under the similitude in question are evenly spaced on a logarithmic spiral terminating at the point $[t]$ (the same holds for the retreat arc, for rational non decimal values of t, the similitude defined by formulas

(6) and (7) that leaves it invariant is the same for both arcs). If, in this case, the point $[t]$ on C is not a multiple point of infinite order, the different arcs $A_\nu A_{\nu-1}$ do not pass through this point, and are related to it in the same way. It follows that the curve turns infinitely around this point.

The same conclusion holds if Ω is an odd multiple of π, the points A_ν being on the same line, but sometimes on one side of the point $[t]$ and sometimes on the other. The similitude that takes $A_{\nu-1}A_\nu$ to $A_\nu A_{\nu+1}$ is an inverse homothety and the union of both these arcs necessarily corresponds to an odd number of rotations around the point $[t]$.

Suppose now that the conditions stated at the beginning of paragraph 7 hold. Then the rotation around the point $[t]$ is given, with finite error, by θ_n and so by $\frac{n}{m}\Omega$, and it does not matter whether Ω is a multiple of π or of 2π. If Ω is non zero the spiral pattern is realized and, since the arc has no double points, the retreat arc runs between the arms of the approach arc. If, on the other hand, $\Omega = 0$, then ω varies periodically between two fixed limits and if these limits differ by less than 2π, then the limit set of the approach arc does not include all the directions coming out of the point $[t]$. The same remark applies to the retreat arc.

Let us discuss the different possibilities depending on the values of $\varphi(1), \varphi(2), \ldots, \varphi(p)$.

First of all, it can happen that the formula (7), where $m_1, m_2, \ldots, m_p$ are non negative and non zero, never gives a multiple of π. Then, even without considering the hypotheses of paragraph 7, we can show that the spiral shape of the approach arc and retreat arc is realized at all points of rational height. The points where C intersects the contour of the smallest convex region surrounding the curve are therefore never points of rational height.

It can also happen that this impossibility only occurs for certain values of m. For example, if $p = 4$ the values of $\varphi(h)$ are $\alpha, \beta, \alpha - \beta, -\alpha$, so the numbers α, β, and π are linearly independent and $\Omega = 0$ implies that $m_1 = m_4, m_2 = m_3$ which cannot happen if m is odd.

The previous cases are incompatible with the hypotheses of paragraph 7 since these imply that $\varphi(1) = \varphi(p) = 0$. On the contrary, the following remarks apply in the particular case when these hypotheses hold and as a consequence $\Omega \neq 0$ implies the spiral pattern. These remarks serve to show that from a certain point of view the spiral pattern is the general case.

This point of view consists of thinking of choosing a rational number by first choosing m, the number of digits in its period, and then considering these m digits as being chosen at random, i.e., that the p^m combinations are considered equiprobable. Naturally, it remains to define the digits preceding the period, but, by varying these digits and even their number, we obtain different points that correspond to each other under the similitude between the curve and its parts. The choice of these digits does not therefore influence the shape of the curve in the neighborhood of the curve $[t]$. We thus consider that a circumstance reflects the general case if its probability, well defined for each value of m, tends to one for infinite m.

However, we must address whether Ω is zero or not. The question comes up only if, independent of any probabilistic computation, this value of Ω is possible, either for arbitrary m or for m increasing indefinitely. In any case, Ω arises as a sum of m independent terms not all equal and each satisfying the same probability law. Known principles of probability teach us that, in these conditions, the probability that Ω belongs to any fixed interval tends to zero for infinite m. The case when $\Omega = 0$ thus appears to be exceptional,[25] thus for C_0 curves without double points, the spiral formation for the approach curve and retreat curve appears to be the general case at points of rational height.

We remark that, even though a curve C_1 of order p is a curve C_0 of order p^2, this result does not apply to the curves C_1. For such a curve and odd m, one must group two periods to obtain a sequence of $2m$ decimals defining a direct similitude leaving invariant the arc of the curve near the point $[t]$, and the angle Ω would always have to be zero under these conditions (the similitude under consideration being the square of an inverse similitude). This does not prevent the fact that if one considers the curve C_1 as a curve C_0 of order p^2, the p^m periods corresponding to a zero value of Ω, once they are paired off, will only be a small fraction (when m is large) of the p^{2m} periods grouped together to study points of rational height of a curve C_0 of order p^2. There is no paradox here: our "general case" is dependent on a suitable grouping of the rational points. This grouping depends on p and for a curve C_1, the grouping differs depending on whether it is considered as a curve C_1 of order p or as a curve C_0 of order p^2.

We end our study of points of rational height with some remarks of a different nature. Such a point, being the center of a well defined similitude defined by a finite number of operations, can be considered as well defined even if the approach arc does not converge to it. This

---

[25]This conclusion does not apply to the study of the case $\Omega = 0$ (mod 2π). Though it remains true for most of the curves C, it is false if all the ratios $\varphi(h)/\pi$ are rational. In that case, for each value of m, Ω has a finite number N of possible values and the probabilities of these values tend towards each other as m increases indefinitely. In fact, the value zero (mod 2π) becomes a possible value, either for m large enough or for those in a certain arithmetic progression, and its probability does not go to zero. In any case, it tends to zero in all other cases.

comes down to saying that $x + iy$ for a point of rational height is the sum of a geometric progression with factor $Q\,e^{i\Omega}$ (preceded by a finite number of terms) and that this sum may have meaning even if $Q \geq 1$, one merely substitutes the function that it represents for $Q < 1$.

The following question poses itself quite naturally: if $q \geq 1$, is it not possible to define as in the construction of paragraph 3 in conjunction with suitable methods of analytic continuation or of divergent series, a set of points that would not be a curve but would have the property of being made up of parts similar to the whole?

The answer is affirmative. It suffices to have $q = 1$ and a single side of Γ_1, whose rank we denote by $h + 1$, to have its length equal to one. Then, for a number t whose decimal expansion contains an infinite number of decimals other than h, the x and y coordinates of a point $[t]$ are well defined as sums of convergent series. If, on the other hand, t has from some point on all its decimals equal to h, then it is rational and the point $[t]$ is again well defined–the successive positions of the segment $[t_n, t'_n]$ being derived from the previous one (from a certain value of n on) by the same translation repeated indefinitely. If, on the other hand, $\varphi(h + 1)$ is not zero, then the point $[t]$ is at a finite distance and the successive positions of the segment $[t_n, t'_n]$ are computed recursively for sufficiently large n by a rotation of angle Ω around this point. In this last case the curve lies entirely in a finite region of the plane and there exist a certain degree of continuity in the sense that the distances $[t_n, t]$ and $[t'_n, t]$ do not tend to zero but have, for n sufficiently large and the values of t in question, well defined positive values r and r' and there are only a finite number of values of t for which the maximum of r and r' exceeds a positive number ε. If one were to observe this curve with a microscope unable to distinguish points less than ε distance apart, for any $\varepsilon > 0$, we would only see a finite number of discontinuities.

Moreover, there is no other extension possible *using summability of divergent series* other than the one alluded to above (which is applicable to space curves as well as to plane curves). This result follows from a known theorem about series with terms which are independent stochastic variables: if they are essentially divergent, i.e., it is not possible to make them almost surely convergent by adding non stochastic terms, then they are almost surely non summable, no matter what summability method is used.[26] It might be possible to extend this theorem to the case of series with non independent terms that are almost

---

[26] A modern reference is [10], Chapter II, Theorem 1. –*Ed.*

independent, that is, that two distant terms are nearly independent. This condition holds here.[27]

If $q > 1$ then one can define the probability law for choosing the successive decimals of t in such a way that the probable value of λ is positive. It is then true, almost surely, that the distance $[t_n, t_{n+1}]$ increases indefinitely with n, its logarithm having order of growth $n\lambda$. The series that defines $[t]$ as the limit of $[t_n]$ is then essentially divergent and so not summable almost surely and therefore not summable independent of the underlying probability distribution for an everywhere dense uncountable set of points.

The same is true when $q = 1$ in which case this maximum is achieved for at least two sides of Γ_1 of ranks $h + 1$ and $h' + 1$. The preceding argument applies if one supposes that, after having chosen in any way an arbitrary large number of decimals, one randomly draws the following decimals between h and h'. The values of t thus defined again comprise an uncountable everywhere dense set and in this set the series defining $[t]$ is once more everywhere not summable.

One should remark that if one foregoes the condition that the point $[t]$ be well defined for all values of t between 0 and 1, then one is led to the study of sets of points that we have just thrown away. It is quite evident that if one is not interested in defining the point $[t]$ for any t, then the construction of paragraph 3 leads, without further restriction, to a well defined set of points corresponding to decimal values of t or even rational values consisting of parts similar to the whole.

9. The curve of von Koch.   The von Koch curve is a curve C_1 of order 2 that can also be considered to be a C_0 curve of order 4. Adopting the former point of view, one defines the curve from a given isosceles triangle with unit base. The line Γ_1 is made up of the two equal sides and the area S is that of the triangle. If α is defined to be the common value of two equal angles of this triangle, we have

$$\varphi(1) = -\varphi(2) = \alpha, \qquad t_1 = t_2 = q = \frac{1}{2\cos\alpha}.$$

---

[27]Paul Levy, *Sur la sommabilité des séries aléatoires divergentes*, [Bull. Soc. Math. France 63 (1935), p. 1–35,[28] Theorem I, p. 11]. The theorem relating to series with independent terms is Theorem I, p. 11. The last paragraph is devoted to an extension somewhat analogous to the one indicated here. Further exposition of this result is beyond the scope of this work but would allow us to prove the result stated here.

[28]The original has the misprint: p. 1395. –*Ed.*

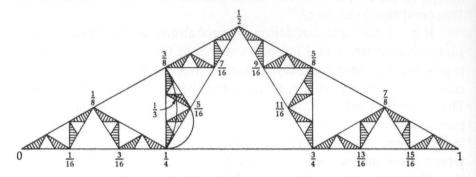

Figure 1.

The necessary condition $q < 1$ gives $\alpha < \pi/3$. For $\alpha = \pi/4$, the triangle S is a right isosceles triangle and in this case the curve fills the triangle. It is the simplest curve filling an area. For $\alpha > \pi/4$, the curve fills the triangle an infinite number of times. We will be mostly interested in the case $\alpha < \pi/4$ when we get a continuous Jordan curve without double points.

The area S is composed of two triangles interior to S having no common point other than their common vertex [1/2]. Without knowing anything about the preceding general study, we immediately see that: each of the areas S_n lies in the interior of the preceding one and is made up of a chain of 2^n triangles with no pair having common points except for consecutive ones having a common vertex. The linear size of these triangles tends to zero like q^n and they tend, in the limit, to a continuous line C without double points. The areas s_n of these regions form a geometric progression since $2q^2 < 1$. On the other hand, the lengths of the lines Γ_n are terms in an increasing geometric progression since $2q > 1$.

In Figure 1 we have drawn the lines $\Gamma_1, \Gamma_2, \ldots, \Gamma_n$, shading in the 32 triangles that comprise S_5, and have marked the heights which are multiplies of $1/16$. We have highlighted the point [1/3] from which we see the segments $[0, 1/4]$ and $[1/4, 5/16]$ under the angle 2α, so this point is the intersection of two easy to draw arcs. The figure was done for $\alpha = \pi/6$ and the center of the circles are the points [1/8] and [9/32], a circumstance which fails to hold in general.

One can find in the memoirs of von Koch and Cesàro cited above some more accurate renderings of the von Koch curve.

The points of the curve located on the segment $[0, 1]$ are evidently those whose heights can be written in base 4 without using the digits 1 and 2 (it should be noted that $1/4$ is written as $0.0333\ldots$ and not 0.1).

It follows that the points of the curve lying on Γ_n are those for which fractional parts of $2^n t$ can be written in base 4 using only the digits 0 and 3. For such a point, whether t is rational or not, the spiral formation is excluded–in a neighborhood of this point the curve lies entirely on one side of Γ_n adjacent to the point.

We can be even more precise. At such a point the limit set of the approach arc consists of all half lines lying inside an angle. This angle is α when $[t]$ is a vertex of one of the lines Γ_n and we shall compute the boundary values in the other case. We will not restrict ourselves by supposing that $t > 3/4$ and for the point $[t]$ lying on the segment $[0,1]$, the segment $[0,t]$ is one of the sides of the angle in question. The angle subtended by the limit set is then the supremum of angles from which certain homothetic quadrilaterals are visible from t. The first of these has vertices marked by $0, 8, 10,$ and 12 in Figure 1, and in general they contain in their interior the different arcs of the approach arcs (some of these quadrilaterals can reduce to points). A homothety magnifying the figure q^{4n} times takes the $(n+1)$st quadrilateral, if it is not reduced to a point, to the first quadrilateral by taking the points $[t]$ to a point $[\tau_n]$ located between the points $[3/8]$ and $[13/32]$ or between the points $[15/32]$ and $[1/2]$ depending on its base 4 expansion, i.e., whether the $(n+1)$st decimal of t is 0 or 3. However, since $[t]$ is not a vertex of a line Γ_n, these two possibilities alternate indefinitely and it is the former that gives a larger value to the angle from which the first of these quadrilaterals is seen from the point of view of $[\tau_n]$. The angle of the limit set is thus the angle that this quadrilateral shows from the point of view of the segment of the curve between the points $[3/8]$ and $[13/32]$. Thus it is at most equal to 3α and at least equal to an angle β (lying between α and 2α) representing the angle of $[5/32, 3/16]$ as seen from the point $[13/32]$, which is the same as the angle of the side $[0,1/4]$ as seen from the point $[5/8]$. We remark that t, and therefore τ_n, are not decimals (in base 4), so the extremal cases indicated above for the point $[\tau_n]$ when $a_{n+1} = 0$ are never attained. The values 3α and β are nevertheless possible values for the angle of the limit set of the approach arc: the first is realized if in the base 4 representation of t one finds the digit 3 followed by an arbitrarily large number of consecutive zeros, and the second if the opposite is true, i.e., arbitrarily long sequences of 3's separated by a single zero. There is an evident asymmetry in this result. The maximum 3α is attained only by smaller values at a point where the angle of the limit set is 3α and the approach arc lies entirely within it. On the other hand, if this angle is β

then there will be an infinite number of points lying outside the limit set, but surrounded by a curve tangent at the boundary of this angle. The asymmetry manifests itself as well in the simultaneous study of the approach arc and retreat arc. The minimum β of the angle of the limit set can be attained for one of these arcs only if the maximum 3α is attained for the other. But the maximum 3α can be realized simultaneously for both arcs if in the expansion of t one finds zeros separated by a large number of 3's as well as the opposite. In this case, if $\alpha > \pi/6$ the retreat arc penetrates the limit set infinitely often (without crossing the approach arc).

If we now consider the limit set of the approach arc at an arbitrary point $[t]$ then we already know that in general it contains all directions. The maximum of its angle, 3α as found above, is thus exceeded. We will show that, on the contrary, excepting the single case of the vertices of the lines Γ_n when the angle of the limit set is α, this angle is always greater than or equal to β. This result clearly follows from the fact that t, not being decimal in base 2, has a decimal equal to one followed by a zero infinitely often, i.e., the triangle $[t_n, t_{n+1}, \tau]$ is similar to the triangle $[0, 1/2, \tau]$, where τ is between $1/2$ and $3/4$.

The minimum angle subtended by the segment $[0, 1/2]$, as seen from a point inside the triangle $[1/2, 3/4, 5/8]$, has value β attained only at the point $[5/8]$ and the angle of the vertex $[\tau]$ is larger. It thus happens infinitely often that the angle $[t_n, t, t_{n+1}]$ is larger than β. The angle of the limit set of the approach arc is thus at least equal to β. This minimum cannot be attained if the angle in question does not turn (or turns infinitely little) from one time to the next, and it is easy to see that this is possible only if the point $[t]$ lies on one of the lines Γ_n, that is, if one of the numbers t or $2t$ has only a finite number of 1's and 2's in base 4.

We now deal with points of rational height other than the vertices of the lines Γ_n. If one considers the curve as a curve C_0 of order four, one has

$$\varphi(1) = \varphi(4) = 0, \qquad \varphi(2) = -\varphi(3) = 2\alpha,$$

so that a necessary and sufficient condition for the angle Ω defined by formula (7) to be zero is that the period of the base 4 expansion of t contains exactly as many 1's as 2's. For all these numbers, the part of the curve neighboring this point reproduces itself under a homothety leaving this point fixed. The spiral pattern is not realized, and for each of these numbers t the approach arc (as well as the retreat arc) lies inside a sector whose angle tends to zero with α so that for α quite

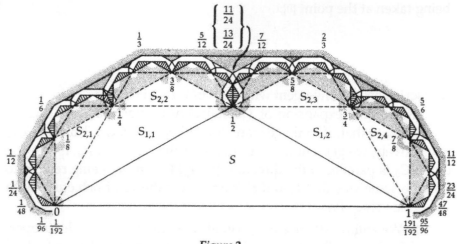

Figure 2.

small the limit set does not include all directions. We see that this case is possible even in the case when the point $[t]$ is not on a line Γ_n. If, on the other hand, the period does not have the same number of 1's and 2's, the spiral pattern is realized.

10. The curve C_0, symmetric of order two. **1.** The initial triangle whose vertices are the points $[0]$, $[1/2]$, and $[1]$ is an isosceles triangle S as in the von Koch curve. We are still assuming that $\alpha < \pi/3$ so $q < 1$. But the two triangles similar to S making up the area S_1 lie outside S and are situated as in Figure 2: the sides $[1/4, 1/2]$ and $[1/2, 3/4]$ of these triangles which are the second and third sides of Γ_2 are parallel to the side $[0, 1]$ of the initial triangle, while for the von Koch curve the first and fourth sides of Γ_2 are the ones having this property. The triangles comprising the areas S_n can be recursively constructed without difficulty.

It is particularly easy here to express $x + iy$ as a function of t. By setting

$$(8) \qquad t = \frac{a_1}{2} + \frac{a_2}{4} + \cdots + \frac{a_n}{2^n} + \cdots, \qquad a'_n = 1 - 2a_n,$$

the angle of the direction $[t_n, t'_n]$ with respect to $[t_{n-1}, t'_{n-1}]$ (t_n and t'_n being as indicated in paragraph 3) is $a'_n \alpha$. Its angle with the direction $[0, 1]$ taken as the x axis is thus $(a'_1 + a'_2 + \cdots + a'_n)\alpha$. The segment $[t_{n-1}, t_n]$ has length equal to $a_n q^n$ and if this length is non zero its direction makes an angle of α with the direction (t_{n-1}, t'_{n-1}). One deduces that, the origin being taken at the point [0],

$$(9) \qquad x + iy = \sum_{1}^{\infty} a_n q^n e^{i(a'_1 + a'_2 + \cdots + a'_n)\alpha},$$

this series being convergent when $\alpha < \pi/4$, that is $q < 1$.[29]

The base 4 expansion of t consists of numbers $b_n = 2a_{2n-1} + a_{2n}$. We note that, if in this expansion we replace $0, 1, 2, 3$ by $1, 0, 3, 2$, respectively, we get a number τ such that the direction $[t_n, t'_n]$ for the curve C_0 is parallel to the direction $[\tau_n, \tau'_n]$ (τ_n and τ'_n being related to τ in the same way that t_n and t'_n are to t) for the von Koch curve with the same value of α.

2. We will mostly use the geometric method to study the properties of C_0. The similitude that takes the entire curve to the arc $[0, 1/2]$ is the product of a rotation α around the origin and a homothety of ratio q. The point $[1/2^n]$, the nth iterated image of the point $[1]$, has as polar coordinates ρ^n and $n\alpha$. The different points $[1/2^n]$ thus lie on a logarithmic spiral surrounding the point $[0]$ which gives the behavior of the curve at this point. At the point $[1]$ the curve also has a spiral form surrounding this point and, due to the similitude of the curve and its two halves, at the point $[1/2]$ the approach arc and the retreat arc both follow a spiral pattern symmetric to each other with respect to the axis of symmetry of the curve and cross each other infinitely often at points of this axis. Because of the similitude between the curve and its parts, the double points comprise an everywhere dense set on the curve.[30]

3. We now turn our attention to points $[t]$ of non decimal rational height. By considering the curve as having order four, the values of $\varphi(h)$ are $2\alpha, 0, 0, -2\alpha$ and a necessary and sufficient condition for the number Ω of formula (7) to be zero, i.e., that the similitude associated to such a point be a homothety for any value of α, is that the period of

---

[29] For the von Koch curve the expression for $x + iy$ is the same but with alternating signs in the parenthesis.

We note that $x + iy$ is a function of q defined for $q < 1$ and has analytic continuation for certain values of t. If we replace a_n with $(1 - a'_n)/2$ then the series (9) becomes a sum of two entire series whose modulus is given but with randomly chosen argument. It is analogous to a type considered in the last paragraph of my memoir of 1935 cited above (see the footnote on that page), but is more simple. The arguments used there apply without difficulty and show that the general case is the one in which the series (1) does not have analytic continuation outside its circle of convergence. The contrary holds only on a set of measure zero.

[30] The interlacing of the different arcs leads one to think that there might be other double points located on the axes of symmetry of different branches of the curve. We have not proved this.

the base four representation of t have the same number of zeroes and threes. In particular, it is sufficient that this representation have none of these digits.

The simplest center of homothety is the point [1/3], the center of the homothety between the segments [0, 1] and [1/4, 1/3], and it is thus located at the intersection of the lines

$$y = x \tan 2\alpha \qquad y = (1 - x) \tan \alpha.$$

Its coordinates are thus

(10) $\quad x = \dfrac{1}{2} - \dfrac{1}{2\,(3\cos^2\alpha - \sin^2\alpha)}, \qquad y = \dfrac{\sin 2\alpha}{2\,(3\cos^2\alpha - \sin^2\alpha)},$

and the point [2/3] is symmetric to the preceding one with respect to the line $2x = 1$, the axis of symmetry of the curve.

The segment [1/3, 2/3], parallel to Ox, that joins these two points play a remarkable role. Let us first show that it contains an uncountable infinity of points of C and that, if $\alpha \geq \pi/4$, it belongs entirely to this curve.

Two triangles of the area S_2, denoted by $S_{2,1}$ and $S_{2,2}$ in Figure 2, are homothetic to S, the centers of homothety being the points [1/3] and [2/3], and the axis of homothety of the three triangles is thus the line containing the segment in question. From each of these two triangles of S_2 we derive two new triangles by repeating the operation that constructed them from S. We thus obtain four triangles belonging to S_4, homothetic to S but q^4 times as small, and the homothety axis of these triangles is still the same line. By iterating this operation indefinitely we get 2^n triangles in S_{2n} homothetic[31] to S: all these triangles have the same homothety axis so that, for n infinite, one has on this line an infinite number of limit points of the curve. Even without going to the limit one observes that when $n' > n$ and the triangle S_{2v} is constructed similarly to the triangle S_{2n}, its center of homothety is on the segment [1/3, 2/3] and also lies on the curve. In fact, it is the center of homothety of an arc of the curve and one of its parts. Iterating

[31] There are in all C_{2n}^n triangles in S_{2n} oriented like S. Each of these corresponds to a sequence $a_1, a_2, \ldots, a_{2n}$ with each of the digits one and zero occurring exactly n times. The ones that interest us here are characterized by the condition that, if we pair off the digits, there be a 0 and a 1 in each pair, i.e., that

$$a_1 + a_2 = a_3 + a_4 = \cdots = a_{2n-1} + a_{2n} = 1.$$

this homothety gives a nested set of arcs and in the limit they reduce to a single point on the curve, the center of homothety.

Furthermore, we observe that the points [1/3] and [2/3] are centers of homotethies of ratio q^2 taking the curve to its second and third quarters, respectively. The first of these homotethies takes the point [2/3] to a point of height 5/12 (since $t - 1/3$ becomes four times smaller from a point to its homologue). The second takes the same point [1/3] to a point of height 7/12. The points [5/12] and [7/12] are thus on the segment [1/3, 2/3] and split it according to the same rule which, for the von Koch curve constructed from the same value of q, allows one to put the points [1/4] and [3/4] on the segment [0, 1]. The similitude that the arc [1/3, 2/3] has with each of the arcs [1/3, 5/12] and [7/12, 2/3] allows one to continue: one leaves the intermediate segment [5/12, 7/12] alone and splitting each of the outside segment as before we put four new points with heights 17/48, 19/48, 29/48, 31/48 and so on. The construction is identical, modulo the fact that we started with a segment of shorter length (if $\alpha < \pi/4$), to the one which for the von Koch curve gives the points located on [0, 1]. The resulting set is similar to the set of points of the von Koch curve lying on [0, 1] and for $\alpha < \pi/4$ it is a discontinuous perfect set. For $\alpha = \pi/4$ it includes all the points of the segment [1/3, 1/2] with some points being double. For $\alpha > \pi/4$, as for the segment [0, 1] of the von Koch curve, all points of this segment are obtained an infinite number of times.

If t is expanded in base 4 then the 2^n triangles S_{2n} homothetic to S with centers of homothety lying on the segment [1/3, 2/3] correspond arithmetically to the 2^n intervals obtained by taking for the first n decimals only the digits 1 and 2. In the limit we see that the points [t] lying on the segment [1/3, 2/3] are those whose height can be written without using the digits 0 and 3. These heights correspond, by the transformation described above (exchanging 0 and 1 and exchanging 2 and 3), to the points of the von Koch curve lying on the segment [0, 1].

What we said about the segment [1/3, 2/3] also applies to all those that correspond to it under the similitude between the curve and its parts. In particular, we consider the similitude that gives the correspondence between the entire curve and its first half: the successive images of the point [2/3] have height 1/3, 1/6, 1/12, ..., and are the vertices of a polygonal spiral surrounding the point [0] (homothetic to the spiral containing the points of height 2^{-n}). By the symmetry of the curve a second analogous spiral having vertices of height 2/3, 5/6, 11/12, ... surrounds the point [1]. The union of these two lines and the initial segment $[1/2, 2/3]$ is a continuous line Π joining $[0]$ and $[1]$ whose vertices belong to the curve as well as an uncountable number of points on each side. If $\alpha \geq \pi/4$ this line is entirely contained in the curve.

4. We show now that the smallest convex region, denoted by S', containing the line Π also contains the curve C. Since its vertices belong to C, it will follow that it is indeed the smallest convex region containing S.

Let us first consider the line $\Pi_{1,1}$ and the region $S'_{1,1}$ defined from Π and S' by the similitude that gives the correspondence between the arc $[0, 1/2]$ and the entire curve and at each point $[t]$ the point $[t/2]$. A first part of $\Pi_{1,1}$, the image of the polygonal line $[2/3, 1/3, 1/6, \ldots]$, is the polygonal line $[1/3, 1/6, 1/12, \ldots]$ which is part of the preceding one and so of Π. The rest of $\Pi_{1,1}$ is the polygonal line $[1/3, 5/12, 11/32, \ldots]$ and can be considered as coming from the line $[2/3, 5/6, 11/12, \ldots]$ (which is a subset of Π and so of S') by the similitude that takes $[t]$ to $[(t+1)/4]$, that is, the entire curve at the arc $[1/4, 1/2]$. It is a homothety of ratio $q^2 < 1$ and center $[1/3]$ located on the contour of S'. The line derived by this operation from a line located in S' or on the contour of this region therefore lies in the interior of this region (the first side being the only one on the contour). The two parts of the line $\Pi_{1,1}$, and so the entire line, and the convex region $S'_{1,1}$ that contains it both belong to S'.

The same also holds for the region S'_1 consisting of the regions $S'_{1,1}$ and $S'_{1,2}$ surrounding the two triangles of S_1, since S' surrounds S. We iterate these operations as was done to get S_n by iterating the operation that takes S to S_1. At the nth operation we get an area S'_n made up of 2^n parts similar to S' where the correspondences are given, respectively, by the similitudes that give the correspondence between S and S_n. They contain these triangles, respectively, and so S'_n contains S_n. Moreover, we know that just as each of the parts of S'_1, and so this entire region, is contained in S', each part of the region S'_n, and so this entire region, is contained in S'_{n-1}. Therefore, each of the regions S'_n contains the following: for $\nu > n$, the region S_ν which is a part of S'_ν and, since it is a closed set, it contains C, the limit of S_ν (or of Γ_ν, a subset of S_ν). We have constructed a nested sequence of regions containing C. Since S'_n is made up of parts that are similar to S' and getting smaller as n increases and that each of these contains points of C, the intersection of all the regions S'_n contains only the points of

C. These regions close around C and define this curve at the limit just as the regions S_n do in the case of the von Koch curve.

5. In Figure 2 when $\alpha = \pi/6$ and in Figure 3 for $\alpha = \pi/4$ it is clear what role the regions S'_n play in the diagram of the curve.

In the case that we are going to study presently, that is $\alpha = \pi/4$, a part of the contour S'_n, for any n, is the line Π, i.e., the segment lying between the points [1/48] and [47/48]. The rest of the contour S' consists of the line segment joining these two points. For increasing values of n the regions S'_n close around the curve and the exterior of S'_n penetrates between the two endpoints of the line Π, first in S, then in derived triangles of S of higher and higher order (we call the triangles S_n the nth derivatives of S). It suffices to consider the area S'_2 to ensure that the curve C is entirely contained in a region that we will call the derived region of S and denoted by S''. The contour of S'' consists of the line Π between the points [1/192] and [191/192] where it intersects the segment [0,1] and the part of this segment lying between these two points. This region is the union of all the areas S_n (and so of S and all the derived triangles of S), its area is 8 times that of S and can be substituted for S' in the study of the curve C. The region S''_n replacing S'_n would then be the union of the triangles derived from S whose orders are less than or equal n.

If one considers, still for $\alpha = \pi/4$, the regions $S''_{1,1}$ and $S''_{1,2}$ (similar to S'' and containing the arcs [0, 1/2] and [1/2, 1] of C, respectively) we see (see Figure 3) that their intersection is a heptagonal region H whose contour consists of four sides of $\Pi_{1,1}$ and four of $\Pi_{1,2}$ (H being considered as an octagon with two sides the same). The two arcs [0, 1/2] and [1/2] have no common points outside the interior of H.

Due to the similitude between the curve and its parts, there exist an infinite number of heptagons similar to H. Their contours belong to the curve and this set consists of points on the curve for which either t or $2t$, when written in base 4, only has a finite number of 0's or 3's. The set given by these lines gives both points on the curve and on the boundary of regions containing different arcs of the curve. They quite quickly give an accurate picture of the curve as a point set without regard to order. On the other hand, the set given by the lines Γ_n or the areas S_p gives a better idea of the way a point traverses the curve.

In Figure 3 the shaded regions of decreasing size indicate only in regions not containing the curve, the segment [0, 1] and the lines $\Gamma_1, \Gamma_2, \Gamma_3, \Gamma_4$. The vertices of Γ_n are shown by large dots and numbered from 0 to 64 (the number given next to each of these represents $64t$ but

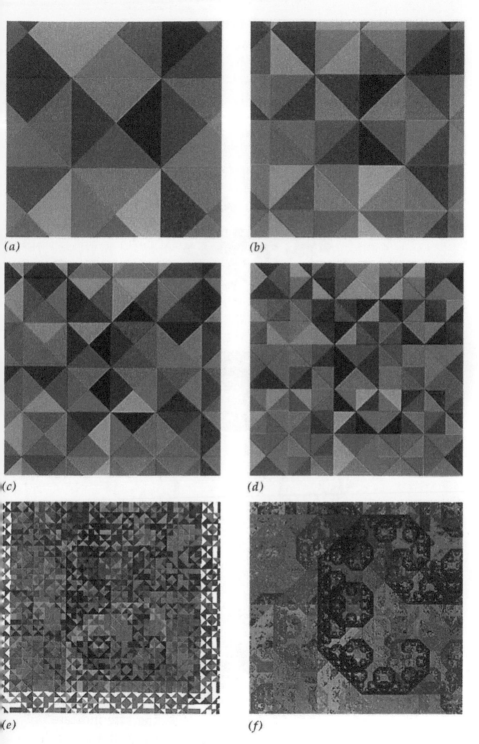

Plate 1. Tiling the plane by Lévy Dragons.

Plate 2. Parametrization shown by colors.

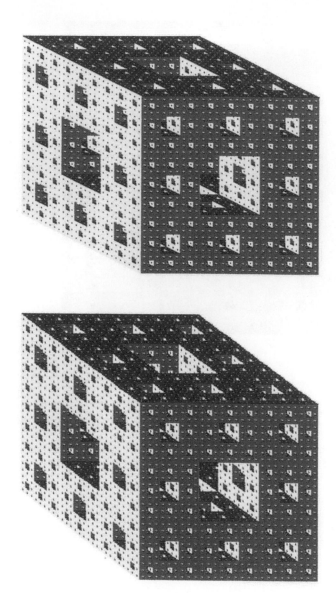

Plate 3. Menger's universal curve, cross-eye stereo pair.

Plate 4. The MacPinski tartan is warn on formal occasions in the remote regions of the Highlands of Scotland bordering Poland.

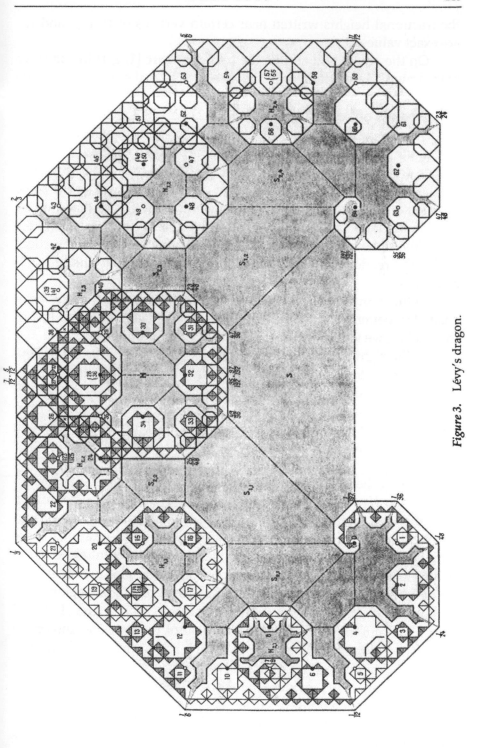

Figure 3. Lévy's dragon.

the fractional heights written near certain vertices of Π, $\Pi_{1,1}$, and $\Pi_{1,2}$ are exact values of t).

On the arc $[0, 1/2]$ and on the part of the arc $[1/2, 1]$ inside Π we have drawn all the triangles of the area $S_{1,0}$. To make their order more transparent, we have shaded in each of the five disjoint parts bounded by the points of height $1/96, 1/48, 1/24, 1/12, 1/6, 1/3$ the first half of the triangles located within them (thus, the ones for which the part before the points marked $1, 2, 4, 8, 16$ is near the third half of the approach arc), and the second half of the triangle in each of these regions has been left blank. By observing that one always winds in the same direction around the empty regions and that the figure made up by 16 triangles lying between two consecutive vertices of Γ_6 is always the same (the same holding for the 32 triangles lying between two vertices of Γ_5), the reader will easily follow the course taken by the lines Γ_{10} and Γ_{11}.

In the heptagon H all triangles making up the area S_{10} have been shaded to better show the formation of quite dense groupings. In 7 of this section we will return to the study of these groupings.

In the right half of the figure we have highlighted the set of heptagons similar to H. In the left half we have drawn these lines only to the extent that they could show at the scale of the drawing the boundaries of regions occupied by the curve (separating them, either from their exterior or from holes within them). In the figure these have been drawn by boldface lines in order to distinguish them from the contours of the triangles S_n.

Among the heptagons similar to H we consider those that are homothetic to H with respect to the points $[1/3]$ and $[2/3]$. Their union with the line segment $[1/3, 2/3]$ lies in a polygonal region with an infinite number of sides and entirely contains in its interior the arc $[1/2, 2/3]$ of C. The arcs of C subtended by the different sides of the polygonal line Π all lie in the interior of regions similar to the one that we have just defined. The figure clearly shows that those regions containing two consecutive arcs such as the arcs $[1/6, 1/3]$ and $[1/3, 2/3]$ have no common point other than the end of the first one and origin of the second one. All these points (which are vertices of Π) are thus accessible from both sides of the curve.

We also note in the figure the shape of some of the surrounded regions that do not contain the curve. The grey shading indicates exactly those regions not containing the curve. Naturally, the largest of the regions surrounded entirely by the curve is H. Four vertices of H are accessible from the interior of this region but the path that, if one only allows half of the curve, would allow one to reach the vertex [11/24] (this height is not marked in the figure since this point coincides with the vertex [40/64], a vertex of Γ_6), is blocked by the contour of $H_{2,3}$ (the transform of H by the similitude that gives the correspondence between S and the third triangle of S_2 with respect to the point [2/3]).

As well, we have shaded in grey the empty regions outside the six heptagons transformed from H by the similitudes that give the correspondence between the entire curve, its two halves, and its four quarters. We note that four of these empty regions are similar to the one that is in H. On the other hand, two others inside $H_{2,2}$ and $H_{2,3}$ are crossed by the contour of H.

6. We will return to the case $\alpha = \pi/4$ in order to determine the area enclosed by the curve. First, let us make some remarks about the case when α is unequal to $\pi/4$.

The points of height $1/2^n$ lie on a logarithmic spiral having the equation

$$\rho = e^{-m\omega}, \qquad \left(m = \frac{-1}{\alpha} \log q = \frac{1}{\alpha} \log(2 \cos \alpha) \right),$$

and all the spirals that occur in the construction of the curve and end at a vertex of one the Γ_n are similar. However, m increases indefinitely with $1/\alpha$ so if α is small, each spiral rapidly tends to its limit point. Already for $\alpha = \pi/6$ it was not possible for us to draw in the scale of the diagram an entire circuit of of the line Π around each of its endpoints, nor were we able to draw the polygon H completely (having twelve sides in this case). On the contrary, if α tends to $\pi/3$, m tends to zero and when $\pi/3 - \alpha$ is quite small, the figure formed by a certain number of consecutive sides of the line Π, even though it is an open polygonal line, differs very little from a regular hexagon traversed several times.

Moreover, we should note that the point [1/3] is defined by the intersection of the lines [0, 1/3] and [1, 1/3] which form an angle of $\pi - 3\alpha$. If $\pi/3 - \alpha$ is small then this point is distant and the line Π covers a large area and, in the limit, the region S' that it surrounds will cover the whole plane. If one considers that, as soon as $\alpha > \pi/4$, the area of S_n increases indefinitely with n in such a way that some parts of the plane are covered an infinite number of times by the curve (whose different arcs must necessarily cross each other), one realizes that it is quite difficult to give a visual rendering of the curve. We merely indicate that, if for $\alpha = \pi/4$ the different arcs separated by the vertices of Π do not have any common points other than these vertices, this circumstance quite quickly ceases to occur (as n increases). We first see that two contiguous arcs can cross and that non neighboring arcs can also cross. As well, since end points of the line Π describe near [0] and [1] larger and larger spirals, it follows that the curve C has an indefinitely increasing number of double points located on the axis of symmetry.

We now consider the case when $\alpha < \pi/4$. One must not forget that the line Π, even though it is drawn in boldface in Figure 3, does not belong to C_1. The point of Π that belong to C comprise a perfect discontinuous set containing all the vertices.[32]

As α decreases from the starting value π the different arcs of the curve that cross over each other for $\alpha = \pi/4$ start to progressively come apart. As soon as $\alpha < \pi/4$ the point [5/12], a vertex of the line $\Pi_{1,1}$, is outside the region occupied by the arc [1/2, 1]. The regions S'_n then close around the curve from both sides (and not just from one side) as n increases. From above, passing from S' to S'_1 already removes a triangle Δ whose side lying on Π is the segment [5/12, 7/12]. By successively removing analogous triangles we obtain a curve C that is a part of the boundary of the region occupied by C and which is greatly analogous to the von Koch curve.

This analogy is especially remarkable in the case of Figure 2 ($\alpha = \pi/6$). The equal sides of the triangle Δ coincide with the sides [5/12, 11/24] and [13/24, 7/12] of the line Π_1. Passing from Π to Π_1 has therefore substituted the segment [1/3, 2/3] with a line Γ'_1 having four sides which are precisely the sides of Π. The following operations will have the effect of replacing each of the sides by a line directly similar to Γ_1 and similarly for each of its sides and so on indefinitely. The curve $\bar{C}$ is therefore what we have called a line C_0 of order 4 and would be the same curve studied in paragraph 9 (for $\alpha = \pi/12$) if the line L'_1 had its four sides equal. Everything that we have said about these lines applies and proves that the outside boundary of the region surrounded by our curve gives detours that the figure cannot hint at.

As α decreases from $\pi/4$ to zero, the vertex A of Δ located on the axis of symmetry traverses the line Π_1 from the point [5/12] to its endpoint. All the sides are traversed one after another in a continuous way. Moreover, the set of values of t corresponding to the points of C lying on one of these sides is independent of α. A point [t] defined by such a value moves in a continuous way on the side of Π_1 on which it

---

[32]In the case of Figure 3 ($\alpha = \pi/6$), all the angles are multiples of $\pi/6$. The triangle [5/12, 1/2, 7/12] is equilateral and the triangles [1/8, 5/12, 1/2], [1/2, 7/12, 2/3] are isosceles. The set in question is constructed by taking away on each side of Π exactly the middle third and repeating on the remaining thirds, and so on indefinitely.

is located and, as α varies, coincides at least once with A. This point A will sometimes be a vertex of Π_1 or one of the other lines Π_n, sometimes a point of C with height not of the form $h/(3 \cdot 2^n)$, and sometimes a point outside C. In the first case, the arc of $\bar{C}$ between the points [1/3] and [2/3] will be a curve C_0 of finite order (> 4 if $\alpha \neq \pi/6$). In the second case, it will be a curve C_0 of infinite order and in the third case its structure will be more complicated. But, in all cases the fundamental fact remains: it is constructed by iterating the same construction at smaller and smaller scales which gives, with regards to the behavior of the curve, the consequences developed for plane curves C. In particular, there are on each arc of this curve, points which are centers of similitude for nearby parts of the curve and, since the similitude is not a homothety, this implies the spiral shape for the approach arc as well as the retreat arc.

To give an idea of the behavior of the curve C itself one must imagine the double points located on $\bar{C}$. They comprise an everywhere dense set. Below each of these double points (we mean by this the side of the region surrounded by $\bar{C}$ and the segment $[0, 1]$) we will attach various loops of the curve, all similar to each other with the largest one inside the polygon H. It can happen that two of these loops cross each other. In a general way the different loops become more apart from each other as α becomes smaller.

Thus, for $\alpha = \pi/4$ the heptagons H and $H_{2,2}$ are homothetic to each other with respect to the point [1/3] and have the same area. The deformation of the curve being continuous it follows that, for $\pi/4$ small enough, the loops surrounding the points [1/2] and [3/8] respectively, cross each other. Figure 2 shows that, for $\alpha = \pi/6$ they are already quite far from each other. As soon as these two loops are separated the point [5/12], and so an infinite number of vertices of the lines Π_n other than Π, become points of C accessible from both sides. For $\alpha = \pi/6$ it can be shown that two of the loops in question do not cross so that points of C accessible by a path coming from infinity without crossing the curve or the segment $[0, 1]$ are accessible from both sides. It does not seem that this property (shown quite easily by taking into account that A is a vertex of Π_1) persists as α decreases. In fact, no matter how small α is, the interior of each of the loops in question contains similar loops that leave it as α continues to decrease. It must therefore happen that at a certain point the contours of two such loops cross and for all the points of $\bar{C}$ to remain accessible from both sides one has to have that the point where the interior loop is attached to the contour of the outside loop is the last to disengage from this loop. This does not seem to be the case.

7. If $\alpha < \pi/4$ we know that the curve C has zero planar measure. We will show that *for $\alpha = \pi/4$ it fills an area equal to that of the triangle S.*

In fact, let us consider a tesselation Q_0 of the plane by squares. By drawing the diagonals of the squares of Q_0 one gets another tesselation Q_1 that we will call the derivative (or first derivative) of Q_0. By iterating this operation we obtain the successive derivatives $Q_2, Q_3, \ldots,$ $Q_n, \ldots$. Conversely, Q_{n-1} will be, with respect to Q_n, the primitive tesselation. It is clear that, if all the derivatives of a given tesselation are well defined, the converse is false: Q_1 is the derivative of Q_0 as well as the derivative of the tesselation given by taking as vertices the centers of the squares of Q_0. Similarly, Q_n is the derivative of 2^n tesselations with distinct vertices such that the set of their vertices coincides with the vertices of Q_n. To distinguish Q_0 among all these nth primitives of Q_n, it suffices to know one vertex, for example, the origin which is also a vertex of $Q_1, Q_2, \ldots, Q_{n-1}$, and this allows us to go, without ambiguity, from Q_n to Q_0.

To each tesselation Q_n we associate a triangulation R_n consisting of the right isosceles triangles $S^{(v)}$ having as hypotenuse the sides of the squares of Q_n. The Q_n determine R_n and conversely. We will say that R_n is the first derivative of R_{n-1} and the nth derivative of R_0. We go from R_{n-1} to R_n by dividing each triangle of R_n into two equal triangles, this operation being well defined. Conversely, since R_n is related to Q_n and R_0 to Q_0, there are 2^n ways to go from R_n back to R_0. This ambiguity can be removed as soon as we know that the origin is a vertex of Q_0 (and so a vertex of an acute angle for eight of the triangles of R_0).

However, one can go from R_0 to R_1 just as well by replacing each triangle of R_0 by its two halves as by replacing it by the symmetrical parts of these halves with respect to the hypotenuse, i.e., by its first derivatives, in the sense of **5** of this section. By iterating this result n times we see that the set of regions $S_n^{(v)}$ derived from the triangles $S^{(v)}$ in the same way as S_n was from S (i.e., the set of nth derivatives of all the triangles of R_0) coincides with the set of triangles of R_n. They form a tesselation of the plane and one can, without ambiguity (if it is known that the origin is a vertex of Q_0) go from a triangle $S_n^{(v)}$ back up to the triangle $S^{(v)}$ in R_0 of which it is the nth derivative. It follows from this that each of the regions $S_n^{(v)}$ (in particular S_n) consists of

2^n effectively distinct triangles and its area is equal to the area of the initial triangle $S^{(v)}$ (so $1/4$ for our triangle S).[33]

The area s' of paragraph 6, the limit of the areas of S_n for infinite n, is thus equal to the area of the initial triangle. By the final result of paragraph 6, the area measure m of the curve C is at least equal to this area.

Let us now consider in the plane a point M not lying on any side of the tesselation Q_n (this restriction is unimportant in what follows, since it only excludes a set of measure zero). For each value of n it is inside a well defined $S_n^{(v)}$, that is, $v = v_n$ is a well defined function of n. Moreover, since M is inside a triangle $S^{(v_0)}$, it can only belong to derived regions of $S^{(v_0)}$ itself and the three triangles of R_0 having a common side with it, or the eleven other triangles having a common vertex with it. The numbers nu_n thus have at most fifteen possible values,[34] and at least one of these values occurs an infinite number of times. Let v be such a value: M belongs to the curve $C^{(v)}$ constructed from $S^{(v)}$ just as the curve C was from S, and so is a limit of $S_n^{(v)}$ for n infinite. Thus, M belongs to $C^{(v)}$, that is, the set of curves $C^{(v)}$ covers the whole plane.[35]

We consider the $4p^2$ curves $C^{(v)}$ constructed from the $4p^2$ triangles which form a square of side p. They lie inside a square of side $p + 1$ and fill a square of side $p - 1$ (which all other curves $C^{(v)}$ cannot enter). Since the sets made up by these different curves are all identical and measurable, their common measure m must be at least $(p - 1)^2/4p^2$ for any p, and so $1/4$ (a result obtained above). Similarly, the set of points belonging to C but not to any other curve $C^{(v)}$ is measurable and its measure μ is at most $(p + 1)^2/4p^2$ for any p and so at

---

[33]These results suggest the idea of a color drawing in which each triangle would have a color depending on v. It would indicate the way in which different curves meet to cover a region of the plane and it would be enough to conceive the same figure at a smaller scale to see how different arcs of the same curve cover a region. One would need at least sixteen colors for two curves to be always distinguishable by their color, but a drawing using only eight colors would still give a quite precise idea of the general shape of the regions $S_n^{(v)}$ and the curves $C^{(v)}$. We do not pursue this here.

[34]We note that these fifteen values are effectively possible in any part of the triangle. The derived region of S having an area equal to eight times that of S, the plane, and in particular each triangle of a square tesselation, is covered eight times, on average, by the derived regions of the tesselation.

[35]The conclusion also holds for the points excluded above: the number of $C^{(v)}$ penetrating a finite region of the plane remains fixed so the set of points they run over is a closed set. It suffice to show that they run almost everywhere in order to show that they run everywhere. The proof in the text is valid for the points that we have excluded (modulo some details). The essential point is that for each n the point M must be in at least one $S_n^{(v)}$ and that the number of values of v to consider be finite. This holds in all cases.

most 1/4. Therefore, if the set of points belonging to more than one curve $C^{(v)}$ has measure zero (we will see that this is the case), we have $m = \mu = 1/4$.

The set E representing the intersection of two curves $C^{(v)}$ and $C^{(v')}$ is related to the set E' consisting of points M for which each of the numbers v and v' occurs infinitely often in the sequence of v_n's (that is, there are an infinite number of values of n for which $S_n^{(v)}$ contains M and similarly for $S_n^{(v')}$). Any point M in this set clearly belongs to both $C^{(v)}$ and $C^{(v')}$ but the converse is false. Thus, each vertex of the tesselation Q_n belongs to sixteen curves $C^{(v)}$ but for eight of these curves it is an endpoint (i.e., of height 0 or 1) and belongs to $S_n^{(v)}$ for any value of n. For the eight others it is a point of height 1/3 or 2/3 and belongs to none of the $S_n^{(v)}$.[36] So, if $C^{(v)}$ is any of these sixteen curves and $C^{(v')}$ one of the latter eight, the point in question belongs to the set E but not to the set E'. We will see that E has zero measure (that is to say $4m = 1$) and so the same holds for E'. On the other hand, one could not simply deduce the first of these results from the second.

We were able to determine these measures only with the help of a detailed analysis of Figure 3.[37] However, the union of the fifteen curves $C^{(v)}$ covers a triangle of Q_0 and one can thus find an area filled by a curve $C^{(v)}$. All these curves are equal, so the fact that one can find an arbitrarily small arc of C similar to C and containing any given point of the curve shows that one can define C as the union of a certain number of similar areas and their limit points. A pointilist painter could imagine a representation of the curve by spots representing the largest of these areas.[38,39]

---

[36]All the points of the polygonal line Π other than its endpoints (and with the same restriction, all those of the line Π_n that are on the boundary of a region not occupied by C) belong to C as well but to none of the regions $S_n^{(v)}$. In the text we have chosen to give the example of points of height 1/3 and 2/3, vertices of a square of Q_0 to show that a point can belong to sixteen curves $C^{(v)}$. The centers of the squares Q_0 each belongs to twelve of these curves.

[37]Because of this, the accuracy of the figure is quite important, especially in the neighborhood of the area denoted below by $\mathcal{A}_1$, located at the cross above the point marked 26. We will indicate to the reader how to verify the precision of the diagram. All the arcs of C bounded by two consecutive vertices of Γ_0 (numbered in Figure 3) are equal. It suffices to draw on a transparency the figure formed by the sixteen triangles of S_{10} comprising such an arc and to verify that, by placing the endpoints of this figure over the points 23 and 24, 25 and 26, 26 and 27, 35 and 36, one has covered the six triangles whose derivatives fill $\mathcal{A}_1$. On the other hand, the same transparency placed over the points 26 and 25, 26 and 35, 27 and 26, does not cover the blank triangles and covers successively all those that are near $\mathcal{A}_1$. The diagram in Figure 4 can also be easily verified by this method.

[38]Results below will show the small size of the spots needed to represent, for example, a quarter of the area occupied by the curve. We could only mark points whose dimensions were each unrelated to the area that they represent.

[39]These "areas" are discussed in the commentary, below. –Ed.

It is now indicated to look for these areas in Figure 3.[40] However, we are assured of finding some only if there exists an area A that is not only completely covered by the curve C, but also does not contain in its interior any point of the other curves $C^{(v)}$ (constructed via the triangulation Q_0 which contains the triangles S). So the lines similar to Π, constructed from the triangles of the triangulation Q_n near the area A and not belonging to S_n (that is, located with respect to triangles just as Π is with respect to S), and bounding the region of the plane where the curves $C^{(v)}$ other than C can enter, will exhibit, if n is large enough, an area that can be covered only by C and which is effectively covered by C. In particular, the necessary and sufficient condition for a triangle of Q_n to be entirely contained in such a region is that this triangle and the fourteen others whose derivatives intersect nontrivially belong to S_n. This comes down to saying that the vertices of the triangle in question are in the interior (in the strict sense) of S_n. If an area like A exists then for some n onwards it contains a triangle S_n in its interior as is obvious by consideration of the set S_n, by the criterion just alluded to. It can thus be found in an effective way assuming, of course, that n is not so large as to exhaust the patience of the researcher.

This search can be facilitated by some simple remarks. The largest of these areas (whether one measures its area, its largest dimension, or the largest derived triangle of S that lies inside it) cannot entirely belong to the arc $[0, 1/2]$ nor to the arc $[1/2, 1]$, for there would be, due to the similitude between these arcs and the entire curve, an area with the same property and larger by a linear factor of $\sqrt{2}$. For the same reason, one would not look in the small heptagon similar to H which contains the arc $[95/192, 97/192]$. We need only examine the upper part of the heptagon H and, due to the symmetry of the figure, it suffices to examine one half of it.

We have not found any triangle of the area S_1 having the indicated property. On the other hand, S_{13} contains four groups of triangles, pairwise symmetric and having this property.

The two most important are naturally symmetric to each other and located below the points marked 26 and 38 in Figure 3, in the triangles of Q_2 marked by crosses. Figure 3 shows, at 8 times the scale, the triangle of Q_0 located below the point $[26/64]$. This figure is in negative in the sense that the shaded triangles of Q_{15} are those that

---

[40]Figure 3 was reproduced exactly by R. W. Gosper [8]. Gosper named it "the C curve" and gives a figure identical to the one here. It is one of the first fractal images generated on a computer. –Tr.

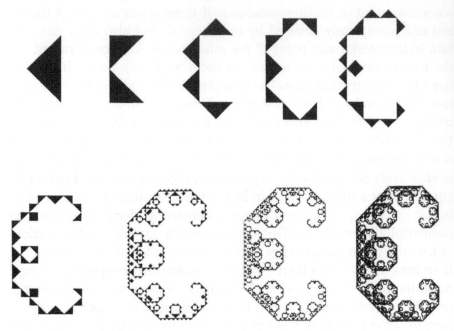

Figure 4.   A construction of Lévy's dragon.

do not belong to S_{15} and the bold lines represent the parts of the lines
Π surrounding the derived regions of these triangles. None of these
regions penetrates the shaded area which is thus completely covered
by C and does not contain in its interior any point of another curve
$C^{(v)}$. We note that it contains ten triangles and two halves of triangles
of S_{15}. Its area is thus $11 \times 2^{-15}s = \frac{11}{8}\frac{s}{4096}$, i.e., about $s/3000$ ($s =$
$1/4$ being the area of S). We denote this area by A_1 and its symmetric
counterpart by A_1'.

The point marked 26 is a quadruple point of C, its heights being
$26/64, 35/96, 53/96$, and $55/96$. The area A_1 is crossed by the approach arcs
of the points $[26/64]$ and $[53/96]$ and by the retreat arcs of the points
$[26/64]$ and $[35/96]$. But we note that it is crossed by arcs of C derived
from six triangles of S_{10} and, if one considers the arcs subtended by
the sides of Γ_{11}, one finds eight different arcs crossing this region and
having their endpoints outside it. The union of these arcs is needed to
cover it completely.

On the subject of two other areas analogous to those that we have
found, denoted by $A-2$ and A_2', we will merely indicate the following:
the area A_2 is inside the 563rd triangle of S_{10} (the third after the point

marked 35 in Figure 3). It contains a square made up of two triangles of S_{15} and fractions of three other triangles. Its measure is three eighths that of A_1. Naturally, A'_2 is symmetric to A_2.

Knowledge of one of the areas A_1 or A_2 suffices either as a starting point for a construction of the curve by iterating the construction of the areas similar to A_1 or A_2 that they contain[41] or by the line of reasoning that we pursue presently. The key to this line of reasoning is knowing that there exists an area (A_1 to fix our ideas) not penetrated by any of the curves $C^{(v)}$ other than C. We will show that any area filled by C has this property.

If, in fact, the area B has in its interior a point M of a curve $C^{(v)}$ other than C, one can find in any neighborhood of M, and thus in B, an area similar to A_1 not containing points of another $C^{(v)}$ other than $C^{(v)}$ (due to the similitude between C and its parts $C^{(v)}$). Thus the area B is not filled by C. Q.E.D.

In order to state our conclusions, there only remains to show that the set $\mathcal{E}$ of points belonging to more than one $C^{(v)}$ has measure zero. In fact, let us consider the triangle S (a similar argument works for the other triangles of R_n), removing the two areas equal to A_1 that belong to the curve symmetric to C with respect to the point $1/2$. Instead of the initial area s there remains a region of area $(1 - 2\varepsilon)s$, where ε has the value 11×2^{-15} as indicated above. Moreover, this region is the union of $2^{16} - 44$ triangles of Q_{16} that can be grouped as 2 triangles from Q_2, 2 from $Q_3, \ldots$, 2 from Q_9, 4 from Q_{11}, 4 from Q_{12}, 2 from Q_{13}, and 4 from Q_{16}. In each of the triangles in question (it is irrelevant whether one considers those of Q_{16} or the 30 groupings above) we remove the two areas located with respect to it just as the first two areas removed are with respect to S and iterate this indefinitely. After the nth operation there remains an area $(1 - 2\varepsilon)^n$ that tends to zero for infinite n. Moreover, the areas removed do not contain in their interior any point of $\mathcal{E}$ and so for any n this set lies inside an area $(1 - 2\varepsilon)^n$ obtained after the nth operation and the contours of the removed areas. It thus has measure zero. Q.E.D.

Thus: *The plane contains an infinite number of areas similar to A_1 each of which is entirely covered by a single curve $C^{(v)}$ and does not contain in its interior any point from the other curves. If one removes these areas and their contours there remains a set $\mathcal{E}$ of measure zero.[42] Those areas that belong to a given curve $C^{(v)}$ cover a total area exactly equal to s (with the convention*

---

[41]This process does not simply give one of the areas from the other—one needs an infinite number of areas similar to A_1 to form A_2 and conversely.

[42]This set indeed exists and in particular contains the sides of the original tesselation.

that areas covered more than once are only counted once). Moreover, these areas can be grouped in groups (such as A_2), and can cover over each other. This is necessarily the case since the sum of the areas of all the region derived from A_1 by the similitude between the curve and its parts is infinite.

8. We end the study of the case $\alpha = \pi/4$ with a final theorem: *If a set of values of t lying between 0 and 1 is measurable then the set of points of the curve corresponding to it is area measurable and, conversely, the area measure of this set of points is one quarter of the measure of the former set: more generally, for any set of values of t lying between 0 and 1 there corresponds a set of points whose outer and inner area measures are one fourth of the corresponding measures of the former set.* [43]

This follows from the fact that C is divided by the vertices of Γ_n into 2^n equal parts each of which covers an area equal to $s/2^n = 1/2^{n+2}$ and the set of points belonging to two parts has measure zero, so applying the preceding results yields the statement. Replacing Q_0 by Q_n so that the curves $C^{(v)}$ are replaced by fractions of curves going from one vertex of Q_n to the next.

Our measure theory results follow without difficulty: any statement (measure, inner measure, outer measure) concerning a set of points can be translated into a corresponding statement for the curve, as indicated above. It seems to us that this is a correspondence principle too well understood to discuss here.

Conversely, if we consider an open square covered by the curve C, any point inside this square corresponds to at least one value of t in an interval (t', t'') such that the arc $[t', t'')$ belongs entirely to the square. The set of points of the curve in the interior of the square is then the union of a finite or countably infinite number of arcs like $[t', t'')$ and the square in question corresponds to a measurable set of values t. Measure theory shows us that, inversely to how we proceeded at first, one can go from the area measure of any set of points of the curve to the measure of the set of value of t to which they correspond. Everything that holds in one of these measure theories holds in the other and the theorem follows.

The same result applies also, as is well known, to the von Koch curve in the case $\alpha = \pi/4$ when it fills the area S.

---

[43]The parameterizing function $t \mapsto [t]$ is *measure preserving* from one-dimensional Lebesgue measure to 1/4 of two-dimensional Lebesgue measure. −*Ed.*

Chapter III. Results for three dimensional space.

11. The set E for the curve C.   Denote by $[t_n, t'_n]$ the hth side of Γ_n, i.e., if $a_1, a_2, \ldots, a_n$ are the first n decimals of t_n then

$$h - 1 = a_1 p^{n-1} + a_2 p^{n-2} + \cdots + a_n.$$

We will consider it oriented with the positive direction being in the sense of increasing t.

Denote by $D_n^{(h)}$ (or more simply D_n, or D when the indices are not explicitly needed) the half line having as origin a fixed point O, direction parallel to the side $[t_n, t'_n]$ of Γ_n, and orientation as above. Also denote by $A_n^{(h)}$ (or A_n or A) the point where these half lines intersect the sphere with center O and radius 1. So D_0 and A_0 are the half line and point corresponding to the segment $[0, 1]$.

The aim of this paragraph is to study the set E of all points A. One can define this set without mentioning the curve C as follows: from the point A_0 one derives, by a first operation, p points $A_1^{(1)}, \ldots, A_1^{(p)}$ distributed in an arbitrary way on the sphere, but in such a way that they correspond to a line Γ_1, i.e., that the vector

(11)
$$\lambda_1 O \Lambda_1^{(1)} + \lambda_2 O \Lambda_1^{(2)} + \cdots + \lambda_n O \Lambda_1^{(p)}, \qquad (\lambda_1 \geq 0, \ \lambda_2 \geq 0, \ldots, \lambda_p \geq 0),$$

must be able to represent $O\Lambda_0$. To each of the points $A_1^{(i)}$ one associates either a rotation of the sphere onto itself or the product of such a rotation by a symmetry about the center O taking A_0 to this point. For each of the p points $A_1^{(i)}$ this operation gives a correspondence to p new points that are its first derivatives and that, as a whole, represent p^2 second derivatives of A_0. The first operation (without introducing a new random element) allows one to construct from these points $A_2^{(h)}$ new points that are the second derivatives of $A_1^{(i)}$ and so the points $A_1^{(h)}$, and so on.[44]

In particular we propose to study the different possible shapes of the set E when it is finite. First let us briefly indicate some results relating to the general case. These will be immediate applications of known principles of group theory and that of probability in chains.

If the set E is finite or everywhere dense on the sphere, or if the lines D are everywhere dense in a cone of revolution (reducing to a

---

[44] One easily verifies that the points $A_n^{(h)}$ obtained in this way are those defined at the beginning of this chapter, even if the similitudes involved do not preserves the direction of t.

plane if C is a plane curve) and the set E is everywhere dense on two circumferences symmetric to each other with respect to the center of the sphere (the hypothesis that all the points A lie on the same circumference other than a great circle is excluded by the fact that (11) must represent $O\Lambda_0$).

Let e_0 be the set of points of $A_n^{(h)}$ of E from which return to A_0 is possible, i.e., such that the set of points $E_n^{(h)}$ derived from $A_n^{(h)}$ contains A_0. This set might contain just a single point, or a finite number, or an infinite number of points. It can also coincide with E or just be a subset of E. In the latter case, a sequence of points, each derived from the previous, can escape from e_0 and it is obvious that once outside e_0, one cannot reenter it. At every point $A_n^{(h)}$ there corresponds a set $e_n^{(h)}$ consisting of those points derived (in some order) from $A_n^{(h)}$ for which return to $A_n^{(h)}$ is possible. All these sets are equal and, as escaping each one is possible (as was possible from e_0), but not to reenter once outside, and there are an infinite number of them (a number of cases must be distinguished depending on the number of distinct $e_n^{(h)}$ that one can reach in a single operation starting from e_n, this number possibly being finite).

All the sets $E_n^{(h)}$ are subsets of E and are equal to E. It is clear that if the set E is finite then the $E_n^{(h)}$ coincide with E, in other words, that from any point of E return to A_0 is possible and e_0 coincides with E.

From the point of view of probability theory, if one chooses a point $A_n^{(h)}$ (n being very large and h random) by choosing at random integers $a_1, a_2, \ldots, a_n$ (the p^n possibilities being equiprobable) then one can assert the following: if e_0 does not coincide with E, the probability of finding a point of e_0 (or from any of the sets $e_\nu^{(h)}$, remaining constant as n varies) tends to zero for infinite n. If E is infinite the probability tends to distribute itself in a uniform way on the derived set of E (i.e., if this set is the sphere, to a surface of area $4\pi\sigma$ there corresponds a probability tending to σ as n goes to infinity and if the derived set of E contains two circumferences, the probability associated to each circumference tends to $1/2$ in a uniformly distributed way). If the set E is finite it can happen that the set of points $A_n^{(h)}$ (n given, h indeterminate) varies with h. In this case it varies periodically, for sufficiently large n, but in all cases (independent of n) the probability tends to a uniform distribution among all possible values.

We now concentrate on the special case when E is finite. This set is then invariant under certain transformations (rotations or products of rotations by a symmetry) that constitute a transitive group (that is, any vertex can be mapped to any other vertex).

In the planar case, E is necessarily made up either of vertices of a regular polygon or of two regular polygons obtained from each other by a rotation through an angle around O. The latter case is realized, for example, for the curve C_1 of order 2 that fills the area of a right non isosceles triangle.

In the three dimensional case these are the different possible shapes:

1. The set E can be the set of vertices of a regular polyhedron that can be assumed to be convex. The five regular convex polyhedra thus give five solutions to our problem.

2. The points of E can lie on two circumferences symmetrical to each other with respect to O (the distance of their plane to O being arbitrary). There are thus four possible types of shapes:

a. A prism having as section a regular polygon with an arbitrary number of sides.

b. A prism having as section a polygon whose vertices are those of two regular polygons with the same number of sides and inscribed in the same circle (in this case as in the previous one the height of the prism can degenerate to zero and one has the solution occurring in the plane case). Note that this case also includes the particular case of the rectangular parallelopiped with unequal edges.

c. A polyhedron with $2n$ vertices belonging to two regular polygons of n sides each and located on the two circumferences considered above (having the same axis Oz and the same radius) but out of phase with each other (by a rotation around the center) in such a way that the polyhedron is not a prism. The lateral surface thus defined consists not of n rectangles but of $2n$ triangles (possibly isosceles or even equilateral but having unequal sides in general). About each vertex of the polyhedron there are three triangles and a polygon of n sides (this case, if $n = 3$, includes the regular octahedron and also the polyhedron having as vertices six vertices of a cube, pairwise symmetric about the center; each these solutions can be obtained from the other by changing z into λz for λ a suitable constant).

d. A polyhedron with $4n$ sides also lying on two circumferences with the same axis and radius. On each of these are located the vertices of a polygon with $2n$ sides defined as in b, but with the two polygons out of phase by an angle π/n in such a way that the smallest sides of each of the polygons is parallel to the largest sides of the other. The lateral faces of this polyhedron are thus $2n$ equal isosceles trapezoids. About each vertex of this polyhedron there is one of the basis polygons and two trapezoids (the three angle are thus $\frac{n-1}{n}\pi$ and two supplementary angles).

3. We now enumerate the convex polyhedra that are solutions to our problem and whose faces are all regular polygons not all having the same number of sides. This is possible due to the fact that all the points of E playing the same role does not mean that the faces of the convex polyhedron having these points as vertices must play the same role. A systematic enumeration of these polyhedra is easy since one can only group angles with sum less than 2π around a vertex. It must be observed that some plausible solutions are excluded for a simple reason: one can, for example, group around a vertex a triangle, a square, and a hexagon, but, in order to continue the construction, it is necessary that the polygons having a common side with the triangle alternate between a square and a hexagon, which is impossible. More generally, a polygon with an odd number of sides can only have edges of two different kinds and these have to alternate.

Not repeating the solutions mentioned above (regular polyhedra, solutions to 2.a, for which lateral faces can be squares, and solutions to 2.c, for which the lateral faces can be equilateral triangles) one is led to the following enumeration:

P_1–About each vertex two triangles and two squares (alternating around the vertex), 12 vertices in all, that are centers of the edges of a tube (or of a regular octahedron), 6 squares in the plane of the faces of the cube, and 8 triangles in the planes perpendicular to the diagonals of the cube (that are also the planes of the faces of the regular octahedron). The polyhedron P_1 can be thought of as being obtained from a cube by removing 8 pyramids with triangular bases, or as obtained from the octahedron by removing 6 pyramids with square bases. One can also produce it by observing that the diameters of the circumscribed sphere having the vertices as endpoints are parallel to the edges of a regular octahedron. Naturally, the symmetry group is that of the cube (or of the octahedron).

P_2–About each vertex two triangles and two pentagons (alternating), 30 vertices in all, 60 triangles, and 30 edges. This polyhedron is related to the regular dodecahedron and octahedron just as the previous one was to the cube and octahedron.

P_3–About each vertex a triangle and two hexagons, 12 vertices in all, these being the points that divide into three equal parts each of the

6 edges of a regular tetrahedron, 4 triangles, 4 hexagons, and 18 edges (these are the edges that belong both to the cube and to two hexagons, the others to a triangle and a hexagon).

P_4–About each vertex a triangle and two octagons, 24 vertices in all, that are the points dividing each of the 12 edges of a cube into three segments proportional to the numbers $1, \sqrt{2}$, 6 octogons, 8 triangles, and 36 ($= 12 + 24$) edges.

P_5–About each vertex a triangle and two decagons, 60 vertices in all, that are the points dividing each of the 30 edges of a regular dodec-ahedron into three segments proportional to the numbers $1, 2\cos \pi/5$, 12 decagons, 20 triangles, and 90 ($= 30 + 60$) edges.

P_6–About each vertex a square and two hexagons, 24 vertices in all, that are the points dividing into three equal parts each of the 12 edges of an octahedron, 6 squares, 8 hexagons, and 36 ($= 12 + 24$) edges.

P_7–About each vertex a pentagon and two hexagons, 60 vertices in all, that are the points dividing into three equal parts the 30 edges of a regular icosahedron, 20 pentagons, 12 hexagons, and 90 ($30 + 60$) edges.

Thus, one does not find any polyhedra other than those that are obtained in a simple way from the regular polyhedra. This could have been expected by observing, for example, that the presence of a pen-tagon implies the symmetry of the regular dodecahedron. More gen-erally, all the possible combinations that would seem, at first glance, not to have the symmetry of a regular polyhedron, are either excluded by the remarks made at the start of this enumeration, or lead to one of the polyhedra listed in 2. above.

4. Naturally, new solutions to our problem can be found if we give up the condition that all the faces of the polyhedron be regular polygons. For the solutions indicated in 2. this happens only in very special cases. Similarly, it is easy to find the polyhedra related to those of 3. that are invariant by the transformations of a transitive group but have faces which are not regular polygons.

Each of these polyhedra is a volume of intersection of two regu-lar polyhedra P and P' having the same symmetry type. However, to obtain one of the polyhedra just considered, not only must these two polyhedra be concentric, but their dimensions (for example, the dis-tances from their faces to their center) must have, in each of the seven cases, a well defined relationship. If we modify this relationship, the symmetry group persists in such a way that we still get a polyhedron invariant under the transformations of a transitive group. Each of the polyhedra P_i, $(i = 3, \ldots, 7)$ transforms itself in this way into a new polyhedron P_i' having the same angles but with edges no longer equal, the two types of edges mentioned above for the polygons P_3 to P_7 having different lengths. Those for which polygons contain edges of both types (hexagons for P_3', P_6', and P_7', octagons for P_4', and decagons for P_5') are thus of the type already mentioned in 2.b and d.

The same operation applied to the polyhedra P_1 and P_2 does not give anything new. In the case of P_1 it is in fact the limiting case of P_4' and P_6'. The case P_2 is the limiting case of P_5' and P_7'.

We can now show that the preceding enumeration is complete. In order not to overly extend this paper, we will not give a complete proof of this. We will instead indicate the principles on which this proof rests. If our polyhedron contained a non equilateral triangle with angles α, β, γ ($\alpha \neq \beta$, with γ possibly equal to α or β) then, since all the vertices of the polyhedron play the same role, it would be necessary to group three triangles around a single point. One thus recovers the solution of 2.c and there are no other that contain non equilateral triangles as faces of the polyhedron. Similarly, we see that it is impossible to find pentagons or heptagons that are not regular. For polygons with an even number of sides the discussion is more involved. A quadrilateral could plausibly be a diamond without it being necessary to place more than two at each vertex. But, if one tries to define a polyhedron obtained from a right parallelopiped just as P_1 is obtained from the cube, then one finds that the vertices do not all play the same role, just as the cube cannot be replaced by a rhomboid. Finally, we are able to exclude quadrilaterals that are not rectangles or isosceles trapezoids, and an analogous study of the other cases ends up showing that our problem has no solutions other than the ones already mentioned.

12. Space measure and plane measure. The results obtained in Chapter II on the area of a plane curve C extend without difficulty to the volume occupied by a space curve C. Instead of

$$(12) \qquad k = \ell_1^2 + \ell_2^2 + \cdots + \ell_p^2,$$

one simply introduces the expression

$$k' = \ell_1^3 + \ell_2^3 + \cdots + \ell_p^3,$$

and the results are as follows: if $k' < 1$, then the measure of the set of points of C is zero. If $k' = 1$ then it can fill a volume, and if $k' > 1$ then it can fill this volume an infinite number of times. The Hilbert curve that fills a cube is a curve of order 27 for which $k' = 1$.

The expression k is related to a two dimensional measure, even if the curve is not a plane curve, by a projection. Denote by s' the maximum of the projection of the smallest convex volume containing C to a plane as the orientation of the plane is allowed to vary. At the nth operation in the construction of the curve, it is the union of p^n arcs whose projection to any plane lie respectively inside convex areas with sum $k^n s'$. Therefore, if $k < 1$, the projection of the curve onto any plane has zero two dimensional measure. If $k = 1$, this measure can be positive (in any case, this is true in the case of plane curves).

13. Miscellaneous examples.[45] 1. A first example consists of symmetric curves C of order 2. One starts with the same isosceles triangle as in the von Koch curve or the curve C_0 of paragraph 10 but in the second operation, instead of constructing a similar triangle on each side of Γ_1 and in the same plane as the first triangle, one adds the triangle at an angle β with the initial plane. For $\beta = 0$ one gets the curve C_0 and for $\beta = \pi$ one obtains the von Koch curve. Other values do not yield plane curves.

The surface obtained by varying β (with Γ_1 fixed) is quite curious. The points $[1/4]$ and $[3/4]$ describe circumferences and the points $[h/2^n]$ describe epicycloids whose type becomes complicated as n increases. The segments $[0, 1/2^n]$ (β fixed and n variable) are distributed in a regular way on a cone of revolution whose half angle at the vertex decreases from $\pi/2$ to $\alpha/2$ as β increases from 0 to n.

For $\alpha = \pi/4$ and β decreasing from π to 0 there is a continuous deformation of the Peano curve filling a right isosceles triangle to the curve represented in Figure 3. As soon as β is less than π the points $[1/4]$ and $[3/4]$, which were identified, become separated and the curve does not fill an area (this does not prevent the possibility that a suitable definition of a two dimensional measure of a set of points in space might assign it a positive measure).

As for the surface obtained by varying β (with $\alpha = \pi/4$), it would be interesting to know if it fills a volume. We have not resolved this question.

---

[45]Several of these examples would be good candidates for computer graphics to aid in visualization. —Ed.

2. Other curves C having $k = 1$ and which have quite curious properties are the curves of order 3 for which the three sides of Γ_1 are equal (and so equal to $1/\sqrt{3}$) and parallel respectively to the three edges of a rectangular tetrahedron. These sides and the segment $[0,1]$ are thus three edges and a diagonal of a cube. One can force the condition that each of the three parts of Γ_2 separated by a vertex of Γ_1 has an edge parallel to $[0,1]$. For each of these there are six possible solutions (depending on the order of the sides given that we only know that they are edges of a well defined cube), so there are 216 solutions in all. Among these solutions we mention those that are symmetric with respect to the axis of symmetry of Γ_1 (the line perpendicular to the segment $[0,1]$ and to the second side of Γ_1). There are twelve, the numbers of the sides of Γ_2 that are parallel to $[0,1]$ possibly being $1, 5, 9$, or $2, 5, 8$, or $3, 5, 7$. Each of these possibilities leads to four solutions, the set of the first three sides of Γ_2 possibly being directly or inversely similar to Γ_1, and similarly for the last three.[46]

All the curves obtained in this way have the following property in common: each line Γ_{2n} contains 3^n sides parallel to the segment $[0,1]$, each of length $1/3^n$ and their projections on the segment $[0,1]$ cover it entirely.

In particular, let us consider the curves C for which the first and last sides of Γ_2 are parallel to $[0,1]$ and thus lie on this segment. There are 24 of these, four of which have the symmetry alluded to just before. For these curves the points located on the segment $[0,1]$ comprise a perfect discontinuous set obtained by dividing this segment into three equal parts, removing the middle part, and iterating this procedure on the remaining parts. This is exactly the same set as in the case of the von Koch curve for $\alpha = \pi/6$. Similar sets can be found on all the sides of the line Γ_n and, for each of these lines, the ordinates of these points are the numbers for which $3^n t$ expressed in base 9 only contains the digits 0 and 8.

The curves obtained in this way, notably the four curves that have an axis of symmetry, seem to be quite analogous to the von Koch curve. By supposing instead that the sides of rank $3, 5, 7$ of Γ_2 are the

---

[46]For these symmetric curves the point $[1/3]$, the center of a similitude that gives the correspondence between $[0,1]$ and the second side of Γ_1 and of a homothety that gives a correspondence between $[0,1]$ and the fifth side of Γ_2, lies on the axis of symmetry and has two possible locations, these being conjugate with respect to the segment of this axis lying between the center of $[0,1]$ and the center of the second side of Γ_1. There corresponds to these, for the fifth side of Γ_2, two possible positions that can be defined by the following remark: by projecting to a plane perpendicular to $[0,1]$ the point set of these lines and the endpoints of the second sides of Γ_1 are the vertices of a square.

ones parallel to $[0, 1]$, one would get curves analogous to the curve studied in paragraph 10.[47]

Naturally, it would be easy to develop the properties of the 216 curves that we have just defined and in particular, the 12 symmetric ones. We will not pursue this further.

14. Surfaces analogous to the curves C. There exists surfaces that, like the curves C_1, consist of parts similar to the whole.

Let us first give a simple example.[48] Start with an equilateral triangle Σ_0 and decompose it into four equal triangles and replace the central triangle by the three other faces of a regular tetrahedron having it as a base. The triangle Σ_0 is thus replaced by a surface Σ_1 composed of the union of six triangles similar to Σ_0 and bounded by the perimeter of Σ_0. By iterating on each triangle of Σ_1 and so on, one obtains in the limit a surface composed of six parts similar to the whole. We will say that it is a surface of order six.[49]

More generally, one obtains a surface S of order p, i.e., composed of p parts similar to the entire surface by a construction analogous to the one for the curves C with condition that the initial surface Σ_1 be composed of p parts similar to each other and whose contours are similar to the entire surface Σ_1 itself. Moreover, it can happen (if the contour of Σ_1 has symmetries that Σ_1 does not have, which surely happens if this contour is planar) that there are a number of surfaces S defined starting from the same surface Σ_1. The definition of one of these surfaces will also have to include the similitude giving the construction starting from one of the parts of Σ_1.

Here are some remarks which hold for all these surfaces:

1. If the ratios of similitude $\ell_1, \ell_2, \ldots, \ell_p$ between the contours of the p parts of Σ_1 and Σ_1 are all less than one, then the above construction indeed converges to a surface S without tangent planes (excluding the case in which it is itself a plane).

---

[47] If one projects to a plane perpendicular to $[0, 1]$, then Γ_1 projects following an equilateral triangle and on this projection the four shapes of Γ_2 corresponding to these symmetric curves are clearly distinguishable from each other. For one of these the six sides of Γ_2 not parallel to $[0, 1]$ project into the interior of the triangle of projection of Γ_1. It is the one that gives the best analogy to the von Koch curve. On the other hand, it is by supposing that these projections lie outside the triangle that one obtains the best analogy with the curve of paragraph 10.

[48] Some remarks on this example are given as "a fractal fairy-tale" in the commentary. —Ed.

[49] The surface Σ_1 being given, there are clearly 64 possible surfaces S depending on which side of the triangles of Σ_1 the construction is iterated. They make up four surfaces having the symmetry of an equilateral triangle, 36 surfaces having only a single plane of symmetry (consisting of 12 types each repeated three times), and 24 surfaces without symmetry (consisting of 8 types, pairwise symmetric, and each repeated three times).

2. The condition $k' \geq 1$ (k still as defined in formula (12)) is necessary for S to be a set of points of positive measure.

3. The use of a single parameter t defined by the decimals of its base p expansion applies as well to the representation of the successive faces of Σ_n, and so to the points of S not lying on the contours of these faces. Of course, this representation does not conserve continuity. But in the study of points of rational non decimal height they serve the same function as in the case of curves. Each of these points is invariant under a similitude (which can be a homothety in special cases) that does not change the surface in the neighborhood of this point. One can therefore find a sequence of elements that are successive images of one of these points under this similitude and that converges to the point in question.

We now indicate the generalizations of the example given at the start. This example essentially consisted of the division of a plane region into parts similar to the whole region and the use of a regular polyhedron allowing one to "leave the plane."

It is clear that the regular tetrahedron can be replaced by any polyhedron having all its faces equal. If one also imposes the condition that these equal faces play the same role, that is, that the polyhedron is invariant under rotations or under rotations followed by symmetries taking any face to any other face, then one finds oneself asking the correlation problem dealt with at the beginning of this chapter. The solution is therefore immediate and we feel it useless to once again give an enumeration of the solutions. We merely note that, on the whole, one finds, on the one hand, polyhedra whose vertices are obtained by joining those of two regular polyhedra having the same centers of symmetry and, on the other hand, polyhedra related to those of paragraph 11. 2, having two opposite vertices S and S' and $2n$ vertices lying on two regular polyhedra either in the mediating plane of SS' or in two distinct planes perpendicular to SS' and symmetric to each other with respect to the center of SS'. In particular, one finds cases in which all the faces are equal triangles, the shape of the triangle being arbitrary.

Of course, these polyhedra are just a part of those that can be obtained by removing the condition that all the faces must play the same role. For example, if one asks for a polyhedron whose faces are equal equilateral triangles, then one can stellate a regular octahedron by one or more regulàr tetrahedra and, more generally, effect any construction combining previously obtained polyhedra having as faces equilateral triangles playing the same role.

To go from a figure to a smaller but similar figure one merely combines the preceding results with a suitable dissection of the plane. Thus, a triangle can be divided into p^2 similar triangles and one can even group p'^2 of these (where $p' < p$) into a single triangle. One of the partial triangles constructed in this way is replaced by the rest of the surface of one of the above polyhedra having this triangle as a face (determining this polyhedron is hardest if the triangle is equilateral but this is nevertheless possible). One can do this in a similar way for a number of partial triangles and continue by combining the different possibilities, but only a finite number of times. One ends up with a surface Σ_1 from which the analogous construction can be repeated, without any arbitrary component, so that the operation effected on each face of Σ_1 in order to define Σ_2 is the same as the one that defined Σ_1.

The division of a square into smaller squares and of a diamond into smaller diamonds gives rise to analogous constructions. It is just as easy to give examples of polyhedral surfaces Σ_1 having all their faces lying on pairwise parallel planes or rectangular planes and having rectangular contours. These rectangles are not arbitrary but have edge lengths with rational ratio or that are solutions to certain algebraic equations. One can even find surfaces Σ_1 bounded by skew quadrilaterals. We will not lengthen this list further. It would be interesting to find a method allowing a systematic treatment of all these surfaces.

The self-similar curves discussed here are sometimes known as **dragon curves**. Of course the Cantor set (see Selection 2) is a self-similar set, but it is not a connected curve, so it does not fall under Lévy's scheme. Selection 13 (Moran) discusses the Hausdorff dimension of many sets, including self-similar sets. Self-similarity was later treated in general terms, for example by Hutchinson [9]. Selection 19 (Mandelbrot) concerns another important generalization, **statistical self-similarity**.

Figure 3 was published as a large size 22 × 34 cm fold-out page in the original version of this paper. If you get a chance, take a look at it. The 1939 *Journal de l'École Polytechnique* may still be in good condition in some libraries. Imagine how this diagram would have been drawn in the days before computer graphics.

I will use the name **Lévy's dragon** for the limiting curve C_0 of Figure 3. There are many ways to construct it. For example: Begin with an isosceles right triangle T_0 (Figure 4, p. 222); replace it with two triangles, shrunk by factor $\sqrt{2}$, so that the hypotenuse of each new triangle is one of the legs of the old triangle. Continue in this way: at each stage, replace each triangle in T_k by two smaller triangles in T_{k+1}. These approximations T_k converge to the dragon C_0. Figure 4 shows T_0 (the original triangle), approximations T_1, T_2, T_3, T_4, T_5, T_8, T_{12}, and T_{16}. The last one is indistinguishable from the limit C_0 at this resolution.

The argument (Paragraph 10 Item 7, near footnote 39) showing that the Lévy dragon C_0 has nonempty interior is a tricky one. Lévy notes that it might "exhaust the patience of the researcher" (*lasse la patience du chercheur*).[50] With our modern knowledge, it may be done in an easier ("softer") way (see below). But, as is typical with such soft arguments, the information obtained is less precise.

C. Bandt [5] shows, as observed by Lévy, that the dragon C_0 **tiles the plane**: there is a sequence of sets C_j, $j \in \mathbb{N}$, congruent (isometric) to C_0 that are **non-overlapping** (the sets C_j intersect only in their boundaries) and cover the plane ($\mathbb{R}^2 = \bigcup_j C_j$). Plate 1 illustrates this tiling. (See footnote 33.) First tile the plane using isosceles right triangles as in (a). Then replace each tile by its two successors; the plane is still covered, each triangle of the new tiling is used exactly once. This continues to hold in all successive stages, by induction. When we go to the limit, the Lévy dragons still cover the plane.

The Baire Category Theorem [14], (3.55) tells us that if the plane is covered by a countable union of closed sets, then at least one of the sets has an interior point; thus C_0 has an interior point. By the self-similarity, each neighborhood in C_0 contains an image of C_0 under an appropriate similarity transformation, so each neighborhood in C_0 contains interior points of C_0. That is, C_0 is the closure of its interior. This is a property that the usual "tiles" for the plane have. But, of course, the interior of C_0 is not a connected set, even though C_0 itself is connected.

Curves constructed as suggested here ("dragon curves") are associated with a natural parametrization by the unit interval: in Lévy's notation, the map is from the number t in the interval to the geometric point $[t]$. One graphic way to illustrate this parametrization is using an animation: the parameter value stands for time. Unfortunately, this low-tech book cannot show this to you. Another graphic way to show the parametrization involves color. This is illustrated in Plate 2, using four fractals that correspond to possible variants in Lévy's definitions. The unit interval at the top shows the color associated with each value between 0 and 1. Each pixel in the pictures below is colored with the color corresponding to the highest value of the parameter in the corresponding square. [These pictures look quite striking on a 24-bit color monitor, where each pixel may be assigned a different color.]

Questions. Lévy's dragon C_0 is not a fractal in the sense of Mandelbrot: its Hausdorff and topological dimensions are both 2. Is the boundary of C_0 a fractal (is C_0 a "fractile")? What is the Hausdorff dimension of the boundary? Is the Hausdorff dimension of the boundary of a set that tiles the plane automatically less than a certain value, or can it be as large as 2? [See special note at end of chapter.]

---

[50] It might be a challenging computer exercise to automate this kind of tedious search. In [7] I stated that the overlap sets of "Barnsley's wreath" contain copies of the entire wreath. For the unpublished verification of this I carried out such a painstaking search; it involved drawing page after page of little triangles, and keeping track of the combinatorics of their overlaps. I know from experience how it might "exhaust the patience of the researcher". Some other fractal sets which I attempted to examine in the same way did indeed exhaust my patience, and thus were never completed. –Ed.

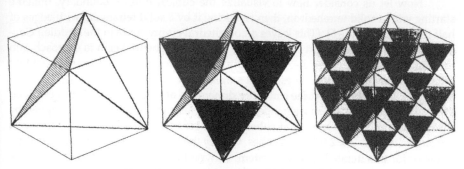

Figure 5.   A construction of Lévy's cube.

A fractal fairy-tale. Some modern developments concern one case of the self-similar surfaces discussed in Paragraph 14. Begin with an equilateral triangle in a horizontal plane. Subdivide it into four smaller triangles by bisecting each of its sides. Think of a regular tetrahedron extending above the plane with the central small triangle as its base. Replace that central triangle by the other three faces of the tetrahedron. This basic step, replacing a triangle by six triangles with sides of half the length, should be repeated indefinitely. Always place the tetrahedron for the new faces on the same side of the surface.

What is the limit surface like? Most people expect it to look much like von Koch's curve, but three-dimensional. Surprisingly, from the top, you can see only three plane faces, isosceles right triangles, with hypotenuses along the sides of the original triangle. From the bottom you see a very complicated system of many rooms separated by walls penetrated by many doors into smaller rooms, and so on. The top and the bottom were described by Mandelbrot [12], pp. 139–140, but the mention of the plane features of the top were stated so briefly that hardly anyone noticed. Scot Morris [13] calls the expectation that the top would be an extremely elaborite crinkly surface "a fractal fairy-tale". He relates how the flat nature of the faces was redis-covered in the mid 1980's. The cover of [16] shows this construction; Mathematica programs to draw the pictures begin on page 238.[51] Among the people involved in the story, only Mandelbrot seems to have been aware that Lévy had described the construction many years before.

A more symmetric three-dimensional figure is obtained if we begin with a regu-lar tetrahedron, and then apply Lévy's construction to each of its faces, with the new faces always bulging outwards. The resulting limit set is on the outside a cube, on the inside riddled with holes and connecting doorways. I will call this limit set "Lévy's cube". We can think of the initial tetrahedron as inscribed in a cube: the edges of the tetrahedron are diagonals of faces of the cube (Figure 5). After k steps, we have a sur-face consisting of $4 \cdot 6^k$ triangles, with edges of length $(1/2)^k$, if the original tetrahedron had edge length 1. The similarity dimension of the surface is $\log 6 / \log 2 \approx 2.585$. The open set condition is not hard to verify (think of the surrounding cube), so this is also the Hausdorff dimension.

---

[51] Thanks to the translator for pointing out this reference.

Now let us consider how to visualize the cubical outside boundary. Imagine starting with a solid tetrahedron, then replacing it by 8 solid tetrahedra with edges of half the length (Figure 5). (This leaves an octahedral empty space in the middle.) Continue in this way: in the next step, some of the new tetrahedra begin to encroach into the empty space in the middle. The similitudes that describe this process constitute an iterated function system. These eight similitudes have ratio 1/2. Solving the equation $8 (1/2)^s = 1$ shows that the similarity dimension of Lévy's cube is 3. But the open set condition is satisfied, as well: indeed, the original surrounding cube is mapped by these eight similitudes onto eight cubes with edges of half the length; they partition the large cube into $2 \times 2 \times 2$ small cubes. Clearly the large cube (with its interior) is an attractor for this iterated function system. A hyperbolic iterated function system has a unique nonempty compact attractor [9], so this cube must also be the limit set of the process consisting of the tetrahedra. —Ed.

Bibliography

[1] H. von Koch, *Sur une courbe continue sans tangentes obtenus par une construction géométrique élémentaire*, Arkiv för Mat., Astron. och Fys. **1** (1904), 681–702.

[2] Helge von Koch, *Une méthode géometrique élémentaire pour l'étude de la théorie des courbes planes*, Acta Mathematica **30** (1906), 145–174.

[3] E. Cesàro, *Remarques sur la courbe de von Koch*, Atti della R. Accad. Sc. Fis. Mat. Napoli **12** (1905), 1–12.

[4] P. Lévy, *Sur la sommabilité des séries aléatoires divergentes*, Bull. Soc. Math. France **63** (1935), 1–35.

[5] C. Bandt, *Integer matrices and fractal tilings of* $\mathbb{R}^n$, Proc. Amer. Math. Soc. **112** (1991), 549–562.

[6] P. Billingsley, *Probability and Measure*, Wiley, New York, 1979.

[7] G. A. Edgar, *A fractal puzzle*, The Mathematical Intelligencer **13** (1991), no. 3, 44–50.

[8] R. W. Gosper, *The "C" curve*, HAKMEM (M. Beeler, R. W. Gosper, and R. Schroeppel, eds.), AIM 239, M.I.T. Artificial Intelligence Lab, Cambridge, February 1972, pp. 65–66, (item 135).

[9] J. E. Hutchinson, *Fractals and self similarity*, Indiana Univ. Math. J. **30** (1981), 713–747.

[10] J.-P. Kahane, *Some Random Series of Functions*, Heath, Lexington, 1968.

[11] Paul Lévy, *Quelques Aspects de la Pensée d'un Mathématicien*, Paris, 1970.

[12] B. Mandelbrot, *The Fractal Geometry of Nature*, Freeman, San Francisco, 1982.

[13] Scot Morris *A fractal fairy tale*, OMNI 11 (November 1988), no. 2, 139–140.

[14] K. R. Stromberg, *An Introduction to Classical Real Analysis*, Wadsworth, 1981.

[15] S. J. Taylor, *Paul Lévy*, Bull. London Math. Soc. 7 (1975), 300–320.

[16] Stan Wagon, *Mathematica in Action*, Freeman, New York, 1991.

Special note: The dimension of this boundary (and many other boundaries of self-similar tiles) was computed by Duvall & Keesling (*The dimension of the boundary of the Lévy Dragon*, International Journal of Mathematics and Mathematical Sciences **20** (1997) 627-632.) The dimension is approximately 1.93400718298829; it was determined from the largest eigenvalue of a certain primitive 734 X 734 matrix. The Strichartz & Wang (*Geometry of self-affine tiles I*, preprint) obtained the same result using an 11 X 11 matrix. Keesling (*The boundary of a self-similar tile in* $\mathbb{R}^n$, preprint) says that the boundary of a tile must be strictly less than 2, but may be as close to 2 as we like.

Received 20 March 1945

This selection contains a discussion of the relation between Hausdorff dimension and existence of finite measures with metric properties. If there is a finite nonzero measure φ concentrated on the compact set $E \subseteq \mathbb{R}^q$, and there are positive constants k and p such that, for every q-dimensional cube R, the inequality $\varphi(R) \leq k(\text{diameter } R)^p$ holds, then the Hausdorff dimension of E is $\geq p$.

Among the applications of this result is the computation of the Hausdorff dimension of a "self-similar" fractal satisfying an "open set condition" (Theorem III). Another application of this result is made to the Hausdorff dimension of a Cartesian product $A \times B$ of sets.

Patrick Alfred Pierce Moran (1917–1988) was born in Sydney, Australia. He obtained a Ph.D. at Cambridge University under Besicovitch. He returned to Australia in 1952, to occupy a newly-created Chair of Statistics at the Australian National University. Moran's work was mostly concerned with pure and applied probability. [10]

THIRTEEN

Additive Functions of Intervals and Hausdorff Measure

P. A. P. Moran

Notation and terminology notes: the p-dimensional Hausdorff measure of a set E is written $p\text{-}mE$. Union of sets is called a sum: $A_1 + \cdots + A_k + \cdots$ means $A_1 \cup \cdots \cup A_k \cup \cdots$; intersection of sets is called a product: $A.V$ means $A \cap V$. But note that the cross $\times$ is the Cartesian product: if $A, B \subseteq \mathbb{R}$, then

$$A \times B = \left\{ (x, y) \in \mathbb{R}^2 : x \in A, y \in B \right\}.$$

An overline (vinculum) is used to indicate grouping:

$$A - \overline{A.V} \qquad \text{means} \qquad A - \left(A \cap V \right).$$

A **half-open interval** in the line $\mathbb{R}$ is a set of the form

$$[a, b) = \left\{ x \in \mathbb{R} : a \leq x < b \right\},$$

where $a, b \in \mathbb{R}$ and $a < b$. Similarly a **half-open interval** in $\mathbb{R}^q$ is a set of the form

$$[a_1, b_1) \times [a_2, b_2) \times \cdots \times [a_q, b_q),$$

where $a = (a_1, \cdots, a_q)$ and $b = (b_1, \cdots, b_q)$ are points in $\mathbb{R}^q$, and $a_i < b_i$ for all i. –Ed.

Consider bounded sets of points in a Euclidean space R^q of q dimensions. Let $h(t)$ be a continuous increasing function, positive for $t > 0$, and such that $h(0) = 0$. Then the Hausdorff measure $h - mE$ of a set E in R^q, relative to the function $h(t)$, is defined as follows. Let ε be a small positive number and suppose E is covered by a finite or enumerably infinite sequence of convex sets $\{U_i\}$ (open or closed) of diameters d_i less than or equal to ε. Write $h - m_\varepsilon E =$ greatest lower bound $\sum_{\{U_i\}} h(d_i)$ for any such sequence $\{U_i\}$. Then $h - m_\varepsilon E$ is non-decreasing as ε tends to zero. We define

$$h - mE = \lim_{\varepsilon \to 0} h - m_\varepsilon E .$$

$h - mE$ may be zero, positive and bounded, or infinite, and is clearly an outer Carathéodory measure. It is usual to impose the further condition on $h(t)$ (in the case of linear sets) that it be concave from below. If this is so $h(t)/t$ is non-increasing. Therefore in the case where $q > 1$ we shall suppose that $h(t)$ satisfies the following:

Condition A: *If q is the dimension of the space, $h(t)/t^q$ is non-increasing at t increases.* This condition is satisfied if $h(t) = t^p$, where $p = \leq q$, and in this case we get the ordinary theory of sets of fractional dimensions. In what follows some of the theorems will depend on the fact that $h(t)$ satisfies condition (A) and some will not.

The main purpose of this paper is to establish necessary and sufficient conditions that a set E in R^q be such that $h - mE$ is greater than zero, in terms of the existence of a certain additive function of intervals. This theorem is an extension to q dimensions of a theorem due to Gillis (1), which involves the existence of a certain function of a real variable, the ideas of which really go back to Hausdorff. By means of this theorem we shall be able to discuss a number of problems in the theory of measure. By its aid we determine the dimensional number of a class of sets generalizing Cantor's ternary set, and in the third section of the paper we prove some theorems on the measures of plane sets which are the products of linear sets of fractional dimensions. Similar theorems can be established for cylinder and product sets in three dimensions.

Theorem I. *Given a closed bounded set E in R^q of finite h-measure, a necessary and sufficient condition for $h - mE > 0$ is that there exists an additive function $\varphi(R)$ of half-open figures* R such that (1) for any figure R,*

---

*A half-open figure is a set expressible as a finite sum of half-open (e.g., open on the right) q-dimensional intervals.

$\varphi(R) \geq 0$; (2) if R contains E, $\varphi(R) \geq b > 0$, where b is some fixed constant; (3) there is a finite non–zero constant such that, if δ is the diameter of R, then $\varphi(R) \leq kh(\delta)$.

Then, in fact, $h - mE \geq b/k$.

Proof of sufficiency. We shall show that if $\{U_i\}$ is any sequence of open sets of diameters δ_i, covering E, we must have

$$\sum_{\{U_i\}} h(\delta_i) \geq b/k.$$

By the Heine–Borel theorem we can take the number of such sets U_i to be finite and moreover we can suppose them to be convex. We can then consider each to be enclosed in a half–open figure U_i' of diameter δ_i' so near δ_i that

$$h(\delta_i') < (1 + \varepsilon)h(\delta_i),$$

where ε is a given small positive number. Therefore for each figure U_i' we have

$$h(\delta_i') \geq \frac{1}{k}\varphi(U_i'),$$

and so

$$\sum_{\{u_i\}} h(\delta_i) \geq \frac{1}{1+\varepsilon}\sum_{\{U_i\}} h(\delta') \geq \frac{1}{k(1+\varepsilon)}\sum_{\{U_i\}} \varphi(U_i') \geq \frac{1}{k(1+\varepsilon)}\varphi\left(\sum_{\{U_i\}}\right),$$

because $\sum \varphi(U_i') \geq \varphi\left(\sum U_i'\right)$ (the U_i' may be overlapping). Moreover, $\sum U_i'$ contains E_1 and so $\varphi\left(\sum U_i'\right) \geq b$. Therefore

$$\sum_{\{U_i\}} h(\delta_i) \geq \frac{b}{k(1+\varepsilon)},$$

and so

$$h - mE \geq b/k.$$

It will be seen that in the proof of the sufficiency of the condition of the theorem, no use has been made of condition (A).

Proof of necessity. The necessity of the condition is not used in later applications but is proved here for the sake of completeness. We first prove a lemma which is a simple generalization of a result due to A.S. Besicovitch (2). Given a set E in R^q, we define the upper h–density of E, $h - \bar{D}(a)$, at a point a as the upper limit, as r tends to zero, of the expression

$$\frac{h - m[E \cdot C(a, r)]}{h(2r)},$$

where $C(a, r)$ is an open sphere of centre a and radius r.

Lemma I. *If E is h–measurable and $h - mE$ is finite, then*

$$h - \bar{D}(a) \leq 1$$

at all points of E except possibly for a set of h–measure zero.

This result has been proved when E lies on a straight line and $h(t) = t^s$, $(0 < s \leq 1)$, and when E lies in a space R^q and $h(t) = t^p$, where p is integral (A.S. Besicovitch (2) and (5), R.L. Jeffrey (4)). By using condition (A) we can make the proof in the general case follow very closely that of Besicovitch. Given a set A, we denote by $U(A, \rho)$ any sequence of open sets $\{U_i\}$, of diameters $d_i < \rho$, covering A, and we denote by $V(d_0)$ any sequence of open sets of diameters d_i less than d_0. Then we need the following lemma.

Lemma II. *Given a measurable set A for which $h - mA < \infty$ and any positive number ε, there exists a number $d_0 = d_0(A, \varepsilon) > 0$, such that for any $V(d_0)$,*

$$h - m[A \cdot V(d_0)] \leq \sum_{V(d_0)} h(d_i) + \varepsilon . \tag{1}$$

Proof of Lemma II. By the definition of h–measure, there exists a number $d_0 = d_0(A, \varepsilon)$, so that for any $U(A, d_0)$

$$\sum_{U(A, d_0)} h(d_i) > (h - mA) - \varepsilon/2 , \tag{2}$$

where ε is any prescribed positive number. Since any $V(d_0)$ is an open set, $(A \cdot V)$ is measurable, and we can write

$$h - mA = h - m(A \cdot V) + h - m(A - \overline{A \cdot V}) . \tag{3}$$

So we conclude

$$\sum_{U(A, d_0)} h(d_i) \geq h - m(A \cdot V) + h - m(A - \overline{A \cdot V}) - \varepsilon/2 . \tag{4}$$

Now, given any $V(d_0)$, we can find a $U_1(A - \overline{A \cdot V}, d_0)$ so that

$$\sum_{U_1} h(d_i) \leq h - m(A - \overline{A \cdot V}) + \varepsilon/2 . \tag{5}$$

Putting this in (4), we can write

$$U(A, d_0) = V(d_0) + U_1(A - \overline{A.V}, d_0), \qquad (6)$$

and, since

$$\sum_{U(A,d_0)} h(d_i) \leq \sum_{V(d_0)} h(d_i) + \sum_{U_1} h(d_i),$$

we have

$$\sum_{V(d_0)} h(d_i) + h - m(A - \overline{A.V}) + \varepsilon/2 \geq \sum_{V(d_0)} h(d_i) + \sum_{U_1} h(d_i)$$

$$\geq \sum_{U(A,d_0)} h(d_i),$$

and therefore

$$\sum_{V(d_0)} h(d_i) \geq h - m(A.V) - \varepsilon,$$

which is the required result. We now prove Lemma I. It is easy to show that, given any positive number u, the set of points at which the upper density is greater than u is measurable. Let A' be the set where $h - \bar{D}(a) > 1$ and suppose $h - mA' > 0$. Then there exists a positive number b such that $h - mA'' > 0$, where A'' is the set at which

$$h - \bar{D}(a) > 1 + b.$$

Write

$$\varepsilon = \min\left(\frac{1}{2}h - mA'', \frac{1}{18}b(h - mA'')\right).$$

In the case of a space of $q > 2$ dimensions we replace the factor $\frac{1}{18}$ by $1/(2.3^q)$. Let ρ_1 be a positive number such that the inequality of Lemma II above is satisfied for ε, the set A, and any $d_0 = \rho_1$. Write

$$A'' = A_1 + \ldots + A_k + \ldots,$$

where A_k is the set of points a of A'' about which it is possible to draw an open circle $C(a, r)$ of centre a, radius r, where

$$\frac{\rho_1}{k+3} \leq 2r < \frac{\rho_1}{k+2},$$

such that

$$\frac{h - m(A.C(a,r))}{h(2r)} > 1 + b.$$

A_k is measurable and $C(a, r)$ is called a density circle of class k. About any point of A_1 we draw a circle of radius and class 1, and a concentric circle of radius $3r$. Then about any point of A_1 outside these two circles we describe two concentric circles in a similar way. We continue this process, at each stage taking a point outside those circles already drawn such that the density circle of the lowest possible class can be drawn, together with a concentric circle of three times the radius. We thus obtain a finite or enumerably infinite* set C of non–overlapping circles and a set C_1 of concentric circles such that C_1 covers all the points of A''. Now $h(r)/r^2$ (or in the general case $h(r)/r^q$) is a decreasing function of r. Therefore, for any positive r,

$$h(2r) \geq \frac{1}{9}h(6r),$$

and consequently

$$\sum_C h(2r) \geq \frac{1}{9} \sum_{C_1} h(6r).$$

In the general case $\frac{1}{9}$ is replaced by $1/3^q$. The radius of any circle of C_1 is $3r < \rho_1$, and so

$$\sum_{C_1} h(6r) \geq (h - mA'') - \varepsilon$$

$$\geq \frac{1}{2}(h - mA'').$$

Therefore

$$\sum_C h(2r) \geq \frac{1}{18}(h - mA'').$$

Now the circles of C do not overlap, and so

$$h - m(A'' . C) > (1 + b) \sum_C h(2r) \geq \sum_C h(2r) + \frac{b}{18}(h - mA'')$$

$$> \sum_C h(2r) + \varepsilon. \tag{7}$$

But, taking C as the set $V(\rho_1)$, we have

$$h - m(A'' . C) \leq \sum_C h(2r) + \varepsilon. \tag{8}$$

---

*$h - mA$ being finite, only a finite number of non–overlapping density circles, of any given order, can be drawn.

(7) and (8) being contradictory, we conclude that $h - mA'' = 0$. This proves Lemma I.

We now prove the necessity of the condition in Theorem I. Let E_1 be the part of E where $h - \bar{D}(a) > 1$. Then E_1 can be enclosed in an open set F_1 so that

$$h - m(E \cdot F_1) \leq \frac{1}{2}(h - mE).$$

We write $E_2 = E - (E \cdot F_1)$. Then E_2 is closed and $h - mE_2 > 0$. Also $h - \bar{D}(a) \leq 1$ for all points of E_2, the density being taken relative to E_2. For any point p outside E_2

$$h - \bar{D}(p) = 0,$$

because E_2 is closed. Then, given any point a of E_2, there exists a number $r_0(a)$ such that, if $r \leq r_0(a)$,

$$\frac{h - m[E_2 \cdot C(a, r)]}{h(2r)} \leq 2.$$

For any given positive number r_0, let $G(r_0)$ be the part of E_2 for which this inequality is true for $r \leq r_0$. Then $G(r_0)$ is a measurable set and as $r_0 \to 0$, $G(r_0) \to E_2$. We can therefore choose r_0 so small that $h - mG(r_0) > 0$. Now given any half–open figure R we define

$$\Phi(R) = h - m(G(r_0) \cdot R).$$

This is clearly an additive function of figures and the reason for the condition that the latter be half open is made clear. For then two abutting figures or intervals have no points in common, even on their boundaries, whereas without this condition $\Phi(R)$ would not necessarily be additive. For example, if the dimension of E was one, we could have a set of positive measure lying entirely in the boundary between the figures. $\Phi(R)$ would then not be additive. Now suppose R contains points of $G(r_0)$ and its diameter is less than $\frac{1}{4}r_0$. R can then be enclosed in a circle of radius equal to the diameter of R whose centre is some point of $G(r_0)$. Then we have

$$\Phi(R) = h - m[G(r_0) \cdot R] \leq 2h(2 \times \text{diameter } R) \leq 8h(\text{diameter } R).$$

On the other hand, if R does not contain points of $G(r_0)$, $\Phi(R) = 0$ identically. $h - mG(r_0)$ is bounded above and so therefore is $\Phi(R)$.

Thus all the conditions of this theorem are satisfied. We remark that we could establish a similar theorem using closed intervals and figures if the dimension of the set E was greater than one (or in the general case, greater than $q - 1$).

Application to a particular type of set. Theorem I can be used to evaluate the dimension of a number of different types of set. For example, we can prove the following theorem on a generalization of Cantor's ternary set.

Theorem II. *If E is a closed bounded set and $E = E_1 + \ldots + E_n$, where the E_i are closed non-overlapping sets similar geometrically to E but reduced in the ratios t_i then the dimension of E is p_0 and $0 < p_0 - mE < \infty$, where p_0 is the root of the equation $t_1^p + \ldots + t_n^p = 1$.* In particular, if $t_1 = \ldots = t_n = t$, we have $p_0 = \log n / \log 1/t$ in agreement with known results. Theorem II can be proved simply from Theorem I, but it is better to regard it as a special case of the following stronger theorem.

Theorem III. *Suppose E is a closed bounded set, defined in the following way. Let O_1 be an open bounded set and O_2^i $(i = 1, \ldots, n)$ be n non-overlapping open sets contained in O_1 and similar to it, but reduced in the ratios t_i. Similarly, let O_3^{ij} $(i, j = 1, \ldots, n)$ be n^2 sets bearing the same relations to the O_2^i as the latter do to O_1. Let the sequence be prolonged indefinitely and let $P_1, P_2, P_3, \ldots$ be the closures of the sets $O_1, \sum O_2^i, \sum O_3^{ij}, \ldots$, etc. Now let E be the common part $P_1, P_2, P_3, \ldots$. Then*

$$0 < p_0 - mE < \infty.$$

Proof. We first prove that $p_0 - mE < \infty$. Let the diameter of O_1 be δ and let $\delta < d < \infty$. Then P_m can be covered by n^m open sets whose diameters are of the form $dt_i \ldots t_k$ and the sum of these raised to the p_0th power is

$$\sum d^{p_0} t_i^{p_0} \ldots t_k^{p_0} = d^{p_0}.$$

Since all the t_i are less than 1, the largest diameter of the covering sets can be made arbitrarily small and

$$p_0 - mE < d^{p_0};$$

that is

$$p_0 - mE \leq \delta^{p_0},$$

since d is any number greater than δ.

We now show that $p_0 - mE > 0$. We may suppose the diameter of O_1 to be less than or equal to 1. We shall define a suitable function $\Psi(A)$ of half–open figures A, in such a way that

$$\Psi(A) \leq \text{constant}\,(\text{diameter } A)^{p_0}.$$

We define $\varphi(O_m^{ij\ldots k})$ to be a number associated with $O_m^{ij\ldots k}$ and equal to $(t_i t_j \ldots t_k)^{p_0}$. Then for any half–open figure A we write

$$\Phi(A) = \lim_{m \to \infty} \sum \varphi(O_m^{ij\ldots k}),$$

where the sum is taken over all sets of order m whose closure lies entirely in A. Let A_η be the set of points (x, y) such that there is a point $(x + h, y + k)$ in A, where $0 \leq h \leq \eta, 0 \leq k \leq \eta$. A_η is again half–open and

$$\Psi(A) = \lim_{\eta \to +0} \Phi(A_q)$$

is an additive function of half–open intervals. We must now show that it satisfies the conditions of Theorem I, that is, if d is the diameter of A, we want

$$\Psi(A) \leq \text{constant } d_0^p.$$

Suppose $t_1 \geq t_2 \geq \ldots \geq t_n$. Then the various sets $O_2^i, O_3^{ij}, \ldots$ are similar to O_1 but reduced in the ratios $t_i, t_{ij}, \ldots$, etc. Let these ratios be arranged in decreasing order and denoted by $T_1 \geq T_2 \geq \ldots$. Given any A, let d be its diameter and suppose r such that

$$T_r \geq d \geq T_{r+1}.$$

Let C be a circle whose centre is some point of A and whose radius is $2d$. Consider all the sets O in C, with reduction ratios lying between T_{r+1} and $T_{r+1} t_n$. If some of these are contained in others we count only those with largest reduction ratios. Let these be $\{P_i\}$ $(i = 1, \ldots, N)$. Then

$$\Psi(A) \leq \sum_1^N \varphi(P_i).$$

Each $\varphi(P_i)$ is less than or equal to $T_r^{p_0}$ and P_i has an area greater than or equal to $(T_{r+1} t_n)^2 |O_1|$. Therefore

$$N \leq \frac{4\pi d^2}{(T_{r+1} t_n)^2 |O_1|},$$

and so

$$\Psi(A) \leq \frac{4\pi d^2 T_r^{p_0}}{(T_{r+1} t_n)^2 |O_1|} \leq 4\pi d^{p_0} \frac{d^{2-p_0}}{T_r^{2-p_0}} \frac{T_r^2}{T_{r+1}^2} \frac{1}{t_n^2 |O_1|} \leq \frac{4\pi}{t_n^4 |O_1|} d^{p_0},$$

and this is the required result. The conditions of Theorem I now being satisfied we conclude that

$$p_0 - mE > 0.$$

The interest of this result is that it establishes the dimension of all sets constructed in a manner similar to Cantor's ternary set, whether they are in a line, plane or space of q dimensions, and whether or not the intervals or domains used to construct them have common boundaries.

Proof of Theorem II. Let O_1 be the open set consisting of all points whose distance from E is less than $\frac{1}{3}\eta$, where η is the greatest lower bound of the distances between the sets $E_1, \ldots, E_n$. We then construct O_2^i to bear the same relation to E_i that O_1 does to E. We then continue the process similarly. These sets define E in the fashion of Theorem III and it follows that $0 < p_0 - mE < \infty$.

Theorems on Product Sets

We now apply Theorem I to the problem of product sets. For the sake of simplicity we confine ourselves to the case where $h(t)$ is a power. We shall prove the following:

Theorem IV. *If A is any measurable p–dimensional set ($0 < p \leq 1$) on the OX axis such that $0 < p - mA < \infty$, and B is any measurable linear set on the OY axis such that $0 < |B| < \infty$, where $|B|$ is the linear measure of B, then there exist constants K_1, K_2, such that*

$$0 < K_1 \leq \frac{(1+p)\text{–measure}\,(A \times B)}{(p - mA) \times |B|} \leq K_2 < \infty,$$

where $(A \times B)$ is the product of A and B. One might imagine that it would be possible to prove a more general theorem on these lines in which B is a set of dimension q ($q < 1$). This in fact is not true and it has been shown recently that it is possible to construct bounded sets A, B of zero dimension on the OX and OY axes such that their product has positive linear measure (Besicovitch and Moran (3)). The above theorem shows that if A is of dimension p and B of dimension q, then $(A \times B)$ has a dimension at most equal to $1 + \min(p, q)$. This is the best possible result.

We begin by proving the right-hand inequality of Theorem IV. Write $C = (A \times B)$. Let $\{I_i\}$ be a set of intervals on the OY axis covering B such that $\sum |I_i| < |B| + \varepsilon$ and such that each interval is of length less than δ, where ε and δ are small numbers. For any particular interval I_k consider the set $(A \times I_k)$. Let n be a large integer and cover A by intervals $\{I_i'\}$ of length less than $|I_k|/n$ and such that

$$\sum |I_i'|^p < (p - mA) + \varepsilon.$$

The product of these intervals with I_k consists of a series of rectangles. Each of these rectangles can be covered by not more than

$$\{[|I_k|/|I_i'|] + 1\}$$

squares of side $|I_i'|$, where $[|I_k|/|I_i'|]$ is the integral part of

$$|I_k|/|I_i'|.$$

Then we have covered the set $(A \times I_k)$ by squares of diagonals d_i such that

$$\sum d_i^{1+p} \leq \sum_i \{[|I_k|/|I_i'|] + 1\}|I_i'|^{1+p}(\sqrt{2})^{1+p}$$

$$\leq \left\{ \sum_i |I_k| |I_i'|^p + \sum_i |I_i'|^{1+p} \right\} (\sqrt{2})^{1+p}$$

$$\leq |I_k| \{(p - mA) + \varepsilon\} (1 + (1/n)) (\sqrt{2})^{1+p}.$$

Carrying out the same process for the other intervals similar to I_k, we obtain a covering of $(A \times B)$ by squares such that the sum of their diagonals raised to the $(1 + p)$th power is

$$\sum d_i^{1+p} \leq (1 + (1/n)) (\sqrt{2})^{1+p} \{(p - mA) + \varepsilon\} (|B| + \varepsilon).$$

Since δ, ε and $1/n$ are arbitrarily small, we have

$$(1 + p)\text{-measure } (A \times B) \leq (\sqrt{2})^{1+p}(p - mA) |B|.$$

Suppose now that A is a measurable set such that $0 < p - mA < \infty$, and that B is a measurable set such that $0 < |B| < \infty$. We show first that we can take A and B, and consequently $C = (A \times B)$ as closed.

To do this we need to prove two lemmas.

Lemma I. *If A is measurable, $0 < p - mA < \infty$, and given any positive ε, there exists a closed set A_1 such that*

$$p - m(A - A_1) + p - m(A_1 - A) < 3\varepsilon \,.$$

Proof. We choose an open set O_1 containing A and consisting of an enumerable sequence $\{C_1^i\}$ of open convex sets such that

$$\sum_i \{\delta(C_1^i)\}^p < p - mA + \varepsilon \,.$$

and so that all the sets C_1^i have diameters less than η, where η is small. Choose n_1 so that[*]

$$p - m \left(A \cdot \sum_{n_1+1}^{\infty} C_1^i \right) < \varepsilon/2 \,.$$

Let D_1 be the closure of $\sum_1^{n_1} C_1^i$ and write $E_1 = D_1 \cdot A$. Now suppose that E_1 is covered by an open set O_2 consisting of an enumerable sequence $\{C_2^i\}$ of open convex sets such that

$$\sum_i \{\delta(C_2^i)\}^p < p - mA + \varepsilon \,,$$

and such that the diameters of all C_2^i are less than $\eta/2$. Now choose n_2 so that

$$p - m \left(A \cdot \sum_{n_2+1}^{\infty} C_2^i \right) < \varepsilon/2^2 \,.$$

Let D_2 be the closure of $\left(\sum_1^{n_2} C_2^i \cdot D_1 \right)$ and $E_2 = D_2 \cdot E_1$. Continue the process with E_2. Now write

$$A_1 = \prod_i^{\infty} D_i \,.$$

---

[*]This can be done by using Lemma II of Theorem 1 and choosing n_1 so that all the C_1^i ($i = n_1 + 1, \ldots$) are less than a d_0 which will make the inequality (1) of that lemma hold with $\varepsilon/2$.

Then A_1 exists and is closed. Moreover, since it is contained in the closure of each of the finite sequences $\{C_r^i\}$ in which the convex sets have diameters less than $\eta/2^{r-1}$, we have

$$p - mA_1 \leq p - mA + \varepsilon$$

and

$$p - m(A - A_1) \leq \varepsilon/2 + \varepsilon/2^2 + \ldots \leq \varepsilon .$$

Also

$$p - m(A \cdot A_1) \geq p - mA - \varepsilon$$

so

$$p - m(A_1 - A) \leq (p - mA + \varepsilon) - (p - mA - \varepsilon) \leq 2\varepsilon .$$

Lemma II. *We now show that it is sufficient to prove the theorem when A and B are closed. For suppose they were not. Then there exists a closed set A_1 satisfying Lemma 1. Moreover,*

$$(1 + p) - m(A_1 \times B) \leq (1 + p) - m\{(A + \overline{A_1 - A}) \times B\}$$

$$\leq (1 + p) - m(A \times B) + (1 + p) - m\{\overline{A_1 - A} \times B\}$$

$$\leq (1 + p) - m(A \times B) + 3\varepsilon K_2 |B| ,$$

and ε can be chosen as small as we please.

It is therefore sufficient to prove the result for A_1 and B.

We can therefore take A as closed. By arguing similarly on B we can also take B as closed. $C = (A \times B)$ is then closed and we now construct a function of rectangles, $\Phi(R)$ satisfying Theorem I.

Now it has been shown by Besicovitch that the right–hand upper density is less than unity p.p. in A. Let ε be small and take A_η as the part of A in which

$$\frac{p - m(A \cdot I)}{|I|^p} \leq 1 + \varepsilon$$

whenever $|I| < \eta$, where I is an interval to the right from a point P of A. A_η is closed. By decreasing η we can make $p - mA_\eta$ as near $p - mA$ as we wish. Assume η_1 to be such that $p - mA_{\eta_1} \geq (p - mA)(1 - \varepsilon)$. Making a similar argument for B we obtain a number η_2 such that

$$|B_{\eta_2}| \geq |B| m(1 - \varepsilon) .$$

Let δ_0 be the smaller of η_1, η_2. Then, given any half–open rectangle R with sides AC, AB parallel to the axes, we define

$$\Phi(R) = [p - m(A_\eta . \overline{AC})] [|(B_\eta . \overline{AB})|].$$

Then, if (diameter R) $< \delta_0$, we shall show

$$\Phi(R) \le (1 + \varepsilon)^2 x^p y,$$

where x is the length of the side of the rectangle parallel to the OX axis and similarly for y.

For let $\overline{AC} = I_x$ be this side. Let K be the left–hand limiting point of points of A_η in I_x. Then

$$\frac{p - m(A_\eta . \overline{KC})}{(\overline{KC})^p} \le 1 + \varepsilon,$$

and the left–hand side of this is not less than

$$\frac{p - m(A_\eta . \overline{AC})}{(\overline{AC})^p}.$$

Arguing similarly on the OY axis, it follows that

$$\Phi(R) \le (1 + \varepsilon)^2 x^p y \le (1 + \varepsilon)^2 (\text{diameter } R)^{1+p}.$$

Applying Theorem I we get

$$(1 + p) - m(A \times B) \ge (p - mA) (|B_\eta|),$$

that is

$$(1 + p) - m(A \times B) \ge (p - mA_\eta) (|B|).$$

This completes the proof.

[*Note added 26 June 1945.* The methods used above can be applied with only slight modifications to product and cylinder sets in three dimensions. If A is a set on the (x, y) plane and B a set on the z–axis, there exist finite non–zero constants satisfying inequalities of the kind given in Theorem IV, in two cases: (i) when A is a 2–dimensional set and B is s–dimensional ($0 < s \le 1$), (ii) when A is an s–dimensional set ($0 < s \le 2$) and B is 1–dimensional. For more general conditions the theorem does not hold in general and in fact it is not difficult to construct three sets A, B, C, on the x, y, and z axes which are of zero dimension, and yet their product $(A \times B \times C)$ has positive 2–dimensional measure.]

St John's College
Cambridge

Here we see again (as in Selection 5 by Hausdorff) a generalization of the Hausdorff measure to include Hausdorff functions $h(r)$ other than a power function r^p. This added generality is useful in some cases, even when the primary interest is in the power–function Hausdorff dimension. For example [8, Chapter 16] [11], in the study of stochastic processes, it turns out in many cases that the natural function h for the Hausdorff measure is of the form $h(r) = r^p (1/\log(1/r))^{p_1}$, for appropriate p and p_1. [See the remark on the Brownian zero–set following Selection 15.] Then if the h–measure of a set E is positive and finite, the Hausdorff dimension of E is (of course) p. But if $p_1 > 0$, then we may think of the dimension as "infinitesimally greater" than p, while if $p_1 < 0$, "infinitesimally smaller" than p. The generality added by considering Hausdorff functions $h(r)$ other than power functions is also useful in the study of certain infinite–dimensional spaces, such as function spaces–this may involve functions like $h(r) = \exp(-C/r^a)$ or $h(r) = \exp(-C(\log(1/r))^a)$, which are asymptotically smaller than all functions r^p as $r \to 0$. [See the notes following Selection 17 (Kolmogorov and Tihomirov) for additional comments on this possibility.]

The set E described in Theorem II is **self–similar** in more recent terminology (it is the **attractor** of an **iterated function system** of similarities). But compare this to the more general construction in Theorem III. The set E is made up of parts $E_1, \ldots, E_n$ as before, and each of the parts E_i is constructed in the same way as E itself (but shrunk by a factor of t_i). However, there are many choices for placement of the subsets within the open set at each stage. The parts E_i may not be similar to E. The open sets O_3^{ij} are similar to O_1, but the similarity transformation F_{ij} need not be a composition $F_i \circ F_j$ of the transformations used to place the sets O_2^i. So, for example, Theorem III covers the "statistically self–similar" sets as in Selection 19 (Mandelbrot) as long as the open sets are non–overlapping.

Note that the set–function φ in Theorem I is not required to be countably additive (just finitely additive), and is defined only on half–open figures. Since the set E under discussion is compact, this is enough. If we begin with an arbitrary Borel set $E \subseteq \mathbf{R}^q$, we must postulate a countably additive measure μ satisfying the inequalities $\mu(R) \leq kh(\text{diameter } R)$. There will then be a compact subset $E_1 \subseteq E$ with $\mu(E_1) > 0$. (See [6, Prop. 8.1.10] or [9, Ex. 10, p. 40].) Then Theorem I may be applied to conclude that the Hausdorff h–measure of E_1 is positive, and therefore also the h–measure of E itself.

The theorem may be considered a **density** result. Let μ be a finite measure on $\mathbf{R}^n$, concentrated on the compact set K. The diameter of the ball $B_\varepsilon(x)$ of radius ε about the point x is 2ε. Let the **upper-p-density** of μ at the point $x \in \mathbf{R}^n$ be defined as

$$\bar{D}_p(\mu, x) = \limsup_{\varepsilon \downarrow 0} \frac{\mu(B_\varepsilon(x))}{(2\varepsilon)^p}.$$

If $\bar{D}_p(\mu, x) \leq C$ for all $x \in K$, for some constant C, then $\dim K \geq p$. If $\bar{D}_p(\mu, x) \geq c$ for all $x \in K$, for some constant $c > 0$, then $\dim K \leq p$.

Similarly, consider the **packing dimension** Dim, and the **lower-p-density**

$$\underline{D}_p(\mu, x) = \liminf_{\varepsilon \downarrow 0} \frac{\mu(B_\varepsilon(x))}{(2\varepsilon)^p}.$$

If $\underline{D}_p(\mu, x) \leq C$ for all $x \in K$, for some constant C, then $\operatorname{Dim} K \geq p$. If $\underline{D}_p(\mu, x) \geq c$ for all $x \in K$, for some constant $c > 0$, then $\operatorname{Dim} K \leq p$. For packing dimension, see [14, §3.3], [13], [12], [7, p. 180], [8, p. 47].

Another note. This selection (dated 1946) is the first one of our selections with its references collected together at the end. The earlier papers typically had references scattered throughout as footnotes. (I collected them and made reference lists for this volume.) Later papers typically follow the custom of a reference list at the end of the paper, classified alphabetically. –Ed.

Bibliography

[1] J. Gillis, *Note on a Theorem of Myrberg*, Proc. Cambridge Phi. Soc. **33** (1937), 419–424.

[2] A. S. Besicovitch, *On the fundamental geometrical properties of linearly measurable plane sets of points (II)*, Math. Annalen **115** (1938), 296–329.

[3] A. S. Besicovitch and P. A. P. Moran, *The measure of product and cylinder sets*, J. London Math. Soc. **20** (1945), 110–120.

[4] R. L. Jeffrey, *Sets of k–extent in n–dimensional space*, Trans. Amer. Math. Soc. **35** (1933), 629–647.

[5] A. S. Besicovitch, *On linear sets of points of fractional dimensions*, Math. Annalen **101** (1929), 161–193.

[6] D. L. Cohn, *Measure Theory*, Birkhäuser, Boston, 1980.

[7] G. A. Edgar, *Measure, Topology, and Fractal Geometry*, Springer-Verlag, New York, 1990.

[8] K. Falconer, *Fractal Geometry: Mathematical Foundations and Applications*, Wiley, 1990.

[9] P. R. Halmos, *Measure Theory*, Van Nostrand, 1950.

[10] E. J. Hannon, *Obituary: Some memories of Pat Moran*, J. Applied Probability **26** (1989), 215–218.

[11] S. J. Taylor, *The measure theory of random fractals*, Math. Proc. Cambridge Phil. Soc. **100** (1986), 383–406.

[12] S. J. Taylor and C. Tricot, *Packing measure, and its evaluation for a Brownian path*, Trans. Amer. Math. Soc. **228** (1985), 679–699.

[13] C. Tricot, *Two definitions of fractional dimension*, Math. Proc. Cambridge Phil. Soc. **91** (1982), 57–74.

[14] H. Wegmann, *Die Hausdorff-Dimension von kartesischen Produkten metrischen Räume*, J. Reine Angew. Math. **246** (1971), 46–75.

Received 8 October 1953

The natural conjecture is that the Hausdorff dimension of a Cartesian product $A \times B$ is the sum of the dimension of A and the dimension of B. In general that is false ([2]; a more elementary example is in [8, Theorem 5.11]).The correct result for Hausdorff dimension of general Cartesian products is just an inequality $\dim A \times B \geq \dim A + \dim B$. The original proofs of this relied on density results ([2], [5], Selection 13), and do not apply directly to the case of sets of infinite measure. In this selection, J. M. Marstrand proposed another method to prove these product theorems, and the requisite density results.

The s-dimensional Hausdorff measure of a set X is written $\Lambda^s X$. The length (or Lebesgue measure) of an interval I is written $|I|$.

FOURTEEN

The Dimension of Cartesian Product Sets

J. M. Marstrand

Given a plane set E, we denote by E_x the set of its points whose abscissae are equal to x.

Throughout this paper we use the letters A and B to denote subsets of the x–axis and y–axis respectively, and we denote by $A \times B$ their Cartesian product set. We use the letters s and t to denote positive numbers; we denote by $\Lambda^s E$ the outer Hausdorff s–dimensional mesure of the set E.

We shall prove the following

Theorem. *Suppose that E is a plane set and that p is a positive number such that for every point x of a given set A we have $\Lambda^t E_x > p$. Then*

$$\Lambda^{s+t} E \geq kp\Lambda^s A,$$

where k is a positive absolute constant.

Two corollaries follow imediately:

Corollary 1. *If $\Lambda^s A$ and $\Lambda^t B$ are positive then*

$$\Lambda^{s+t}(A \times B) \geq k\Lambda^s A . \Lambda^t B .$$

Corollary 2. *Given arbitrary sets A and B the dimension of $A \times B$ is not less than the sum of their dimensions.*

Corollary 1 has been proved by Besicovitch and Moran [2] in the case where A is an s–set (that is, a set which is measurable with respect to Λ^s–measure and satisfies $0 < \Lambda^s A < \infty$) and B is a t–set.

Moran [6] obtained some results similar to the theorem of this paper, but he imposed the same conditions upon his linear sets.

The question has been raised by Eggleston [5] as to whether theorems of this type can be proved without restricting the sets in any way; difficulties arise if the sets have infinite measure. In a previous paper [4] he had given a proof of Corollary 2 as stated, but it was pointed out to him by Mr R.O. Davies that his proof was in error. He acknowledged this correction in [5], and also gave a new proof, but this time restricted the set A by supposing that it contained an s-set.

Our method is quite different from previous ones which have been used, as it does not depend upon the idea of density; we make use of the special L^s-measure introduced by Besicovitch [1] and also used by Davies [3]. We now define this measure:

Notation and definitions. We shall call a closed linear interval a binary interval if it is of the form $[m2^{-n}, (m+1)2^{-n}]$, where m and n are integers. We shall call a closed square a binary square if its vertices are of the form

$$(m_1 2^{-n}, m_2 2^{-n}), \quad ((m_1+1)2^{-n}, m_2 2^{-n}), \quad ((m_1+1)2^{-n}, (m_2+1)2^{-n})$$

and

$$(m_1 2^{-n}, (m_2+1)2^{-n}).$$

Suppose that X is a linear set and that q is a positive number. Let $\sum_{j=1}^{\infty} I_j$ denote any set of binary intervals which covers X, and such that the length of each interval is less than q. Then by

$$L_q^s X,$$

we denote the lower bound of $\sum_{j=1}^{\infty} |I_j|^s$, taken over all such sets of intervals.

We extend this L_q^s-measure to plane sets of points by replacing, in the definition, binary intervals by binary squares, and the lengths of the intervals by the lengths of the sides of the squares.

We write

$$L^s X = \lim_{q \to 0} L_q^s X.$$

Lemma 1. *Given any linear set X,*

$$\Lambda^s X \leq L^s X \leq 2\Lambda^s X, \tag{1}$$

and given any plane set E,

$$2^{-\frac{1}{2}s}\Lambda^s E \le L^s E \le 2^{s+2}\Lambda^s E. \tag{2}$$

The inequality (1) is proved in Besicovitch (1), and (2) is left to the reader.

Lemma 2. *Given any increasing sequence of linear sets X_n ($n = 1, 2, \ldots$), we have*

$$\lim_{n\to\infty} L_q^s X_n = L_q^s \lim_{n\to\infty} X_n \tag{3}$$

and

$$\lim_{n\to\infty} L^s X_n = L^s \lim_{n\to\infty} X_n. \tag{4}$$

The equality (3) is proved in Besicovitch (1). We deduce (4) from the fact that L^s–measures is regular (see Saks (7), §6, p. 50); in fact, any set X is contained in a G_δ set of the same measure.

Lemma 3. *If a linear set X is contained in a finite set of binary intervals, we may select from these a non–overlapping set which contains X.*

We may select all those intervals of our set which are not contained in any other.

Lemma 4. *Suppose that a linear set X is contained in a finite set $\sum_{j=1}^n I_j$ of binary intervals each of length less than q, and that with each interval is associated a positive number $f(j)$. Suppose further that there exists a positive number p such that for every point x of X,*

$$\sum_{x\in I_j} f(j) > p.$$

Then

$$\sum_{j=1}^n f(j) |I_j|^s > p L_q^s X.$$

Proof. Since $f(j)$ takes only a finite number of values we may suppose without loss of generality that all these values, and the value of p, are rational. We may express all these rational numbers as fractions with a common denominator. Hence we may suppose that the values are all integers. Finally, by modifying our set $\sum_{j=1}^n I_j$ by repeating each interval I_j, $f(j)$ times in the sum, we may simplify the lemma to the case where $f(j) = 1$ ($j = 1, 2, \ldots, n$).

Thus each point X is contained in I_j for at least $p + 1$ values of j, and we are required to prove that

$$\sum_{j=1}^{n} |I_j|^s > pL_q^s X . \tag{5}$$

We may suppose that X does not contain any end–points of binary intervals, since the removal of such points does not affect the value of $L_q^s X$.

Since $\sum_{j=1}^{n} I_j \supset X$, it follows from Lemma 3 that we can select from this set of intervals a set of non–overlapping intervals

$$\sum_{j=1}^{n(1)} I_{1,j} \supset X.$$

Then each point of X is contained in at least p intervals of the set

$$\sum_{j=1}^{n} I_j - \sum_{j=1}^{n(1)} I_{1,j}.$$

From this latter set we can select a set of non--overlapping intervals

$$\sum_{j=1}^{n(2)} I_{2,j} \supset X.$$

Then each point of X is contained in at least $p - 1$ intervals of the set

$$\sum_{j=1}^{n} I_j - \sum_{j=1}^{n(1)} I_{1,j} - \sum_{j=1}^{n(2)} I_{2,j}.$$

We proceed in this way $p + 1$ times, and from $\sum_{j=1}^{n} I_j$ select $p + 1$ mutually exclusive sets of intervals

$$\sum_{j=1}^{n(\nu)} I_{\nu,j} \supset X \quad (\nu = 1, 2, \ldots, p + 1).$$

Since $|I_{\nu,j}| < q$, it follows that for each value of ν,

$$\sum_{j=1}^{n(\nu)} |I_{\nu,j}|^s \geq L_q X ,$$

whence

$$\sum_{j=1}^{n} |I_j|^s \geq \sum_{v=1}^{p+1} \sum_{j=1}^{n(v)} |I_{v,j}|^s \geq (p+1) L_q^s X,$$

and (5) follows immediately. Thus the lemma is established.

Notation. By $[S]$ we denote the length of the sides of the square S. By proj S we denote the orthogonal projection of S on to the x-axis.

Proof of the theorem. By virtue of (4) of Lemma 2, we may suppose without loss of generality that A is bounded. Hence for any value of q, $L_q^s A$ is finite.

Let φ be an arbitrarily small number. To every point x of A there corresponds a positive number $q(x) < \varphi$ such that

$$L_q^t E_x > p \quad \text{whenever} \quad q < q(x).$$

By (3) of Lemma 2, given $\varepsilon > 0$ we can find a set $A_1 \subset A$ and a positive number

$$q < \varphi, \tag{6}$$

such that

$$L_\varphi^s A_1 \geq (1 - \varepsilon) L_\varphi^s A, \tag{7}$$

and for every point x of A_1,

$$L_q^t E_x > p. \tag{8}$$

Choose a set $\sum_{j=1}^{\infty} S_j$ of binary squares such that

$$\sum_{j=1}^{\infty} S_j \supset E, \tag{9}$$

$$[S_j] < q \quad (j = 1, 2, \ldots), \tag{10}$$

and

$$\sum_{j=1}^{\infty} [S_j]^{s+t} \leq L_q^{s+t} E + \varepsilon. \tag{11}$$

From (9), for every point x on the x-axis we have

$$\sum_{j=1}^{\infty} (S_j)_x \supset E_x.$$

Consequently, from (8) and (10), for every point x of A_1,

$$\sum_{j=1}^{\infty} |(S_j)_x|^t > p,$$

and hence there exists a positive integer $n(x)$ such that

$$\sum_{j=1}^{n(x)} |(S_j)_x|^t > p.$$

From (3) of Lemma 2 it follows that we can find a set $A_2 \subset A_1$ and a positive integer n such that

$$L_q^s A_2 \geq (1 - \varepsilon) L_q^s A_1, \tag{12}$$

and for every point x of A_2,

$$\sum_{j=1}^{n} |(S_j)_x|^t > p. \tag{13}$$

Let us temporarily use the notation

$$X = A_2; \quad I_j = \text{proj } S_j, \quad f(j) = [S_j]^t \quad (j = 1, 2, \ldots, n).$$

From (10) and (13), we see that the hypotheses of Lemma 4 are satisfied by X, $\sum_{j=1}^{n} I_j$, $f(j)$, p, and q. Using the we deduce that

$$\sum_{j=1}^{n} [S_j]^{s+t} = \sum_{j=1}^{n} [S_j]^t |\text{proj } S_j|^s > p L_q^s A_2.$$

Hence, from (11),(12),(6) and (7),

$$L_q^{s+t} E + \varepsilon \geq p(1 - \varepsilon) L_q^s A_1 \geq p(1 - \varepsilon) L_\varphi^s A \geq p(1 - \varepsilon)^2 L_\varphi^s A.$$

This holds for arbitrarily small values of ε, q, and φ, and therefore

$$L^{s+t} E \geq p L^s A.$$

The inequality of the theorem now follows by using (1) and (2) of Lemma 1.

I wish to thank Mr R. O. Davies for his useful advice and criticism.

Another consequence of the main theorem is this [8, Exercise 5.2]: Let $E \subseteq \mathbb{R}^2$ be a plane set, and write

$$E_y = \{x \in \mathbb{R} : (x, y) \in E\}$$

for each y. If s, t, k are as in the Theorem, then

$$\Lambda^{s+t}(E) \geq k \int \Lambda^s(E_y)d\Lambda^t(y).$$

Note that the theorem is proved for sets in $\mathbb{R} \times \mathbb{R}$. The same method can be adapted to sets in $\mathbb{R}^p \times \mathbb{R}^q$. The case of general metric spaces, proved using a different method, can be found in [11]. For conditions under which equality $\dim(A \times B) = \dim A + \dim B$ holds see [11], [10].

Questions. Consider the Hausdorff f-measure for Hausdorff functions f other than power functions (as in Selection 13). Are there inequalities like

$$\Lambda^h(A \times B) \geq \Lambda^f(A) \Lambda^g(B)?$$

Given Hausdorff functions f and g, the appropriate "product" Hausdorff function to consider is $h(t) = f(t)g(t)$, or perhaps $h(t) = f(t/\sqrt{2})g(t/\sqrt{2})$ [9, p. 131], [11]. —Ed.

Bibliography

[1] A. S. Besicovitch, *On existence of subsets of finite measure of sets of infinite measure*, Indag. Math. **14** (1952), 339–344.

[2] A. S. Besicovitch and P. A. P. Moran, *The measure of product and cylinder sets*, J. London Math. Soc. **20** (1945), 110–120.

[3] R. O. Davies, *Subsets of finite measure in analytic sets*, Indag. Math. **14** (1952), 488–489.

[4] H. G. Eggleston, *The Besicovitch dimension of a cartesian product sets*, Proc. Cambridge Phil. Soc. **46** (1950), 383–386.

[5] H. G. Eggleston, *A correction to a paper on the dimension of cartesian product sets*, Proc. Cambridge Phil. Soc. **49** (1953), 437–440.

[6] P. A. P Moran, *On plane sets of fractional dimensions*, Proc. London Math. Soc. **51** (1949), 415–423.

[7] S. Saks, *Theory of the Integral*, Warsaw 1937.

[8] K. J. Falconer, *The Geometry of Fractal Sets,* Cambridge University Press, 1985.

[9] C. A. Rogers, *Hausdorff Measures,* Cambridge University Press 1970.

[10] C. Tricot, *Two definitions of fractional dimension,* Math. Proc. Cambridge Phil. Soc. **91** (1982), 57–74.

[11] H. Wegmann, *Die Hausdorff-Dimension von kartesischen Produkten metrischen Räume,* J. Reine Angew. Math. **246** (1971), 46–75.

Received 19 March, 1954;
read 22 April, 1954

In this selection, the fractal dimension of a compact subset of the line is investigated by considering the lengths of the intervals that constitute the complement of the set.

The β-dimensional Hausdorff measure of a set E is denoted $\Lambda^{\beta}(E)$. The Hausdorff dimension of E is denoted dim E.

Samuel James Taylor was born in Northern Ireland in 1929. A. S. Besicovitch was Rouse Ball Professor of Mathematics at Cambridge University from 1950 to 1958; S. James Taylor was his research student, and this paper forms part of the dissertation for which Taylor was awarded a Ph.D. by Cambridge University in 1954. Taylor has taught at the universities of Birmingham, London, Liverpool, British Columbia, and Virginia. His work deals primarily with probability and measure theory. [7]

FIFTEEN

On the Complementary Intervals of a Linear Closed Set of Zero Lebesgue Measure

A. S. Besicovitch & S. J. Taylor

§1. *Introduction.* Suppose E is a bounded linear closed set. We will assume that it is contained in the unit interval $[0,1]$. Then, If I is the open interval $(0,1)$, $I - E$ is an open linear set and therefore consists of an enumerable sequence of open intervals. The lengths of these intervals may be arranged as a non–increasing sequence

$$a_n \quad (n = 1, 2, \ldots)$$

of positive numbers. If E has zero Lebesgue measure, then

$$\sum_{n=1}^{\infty} a_n = 1. \tag{1}$$

Given any non–increasing sequence $\psi = \{a_n\}$ satisfying (1), there are many possible linear closed sets E such that the complementary intervals $(I - E)$ have lengths given by the elements of ψ. Such sets E will be said to *belong to the sequence* ψ. It is the object of this note to establish relations between some properties of sequences ψ and the α–dimensional $(0 < \alpha \leq 1)$ measure* of sets E belonging to ψ. In §2 we obtain upper bounds for the measure of E in terms of ψ, and in §3

---

*For definitions and notation relevant to the measure of linear sets see, for example [5].

we show the existence of sets E belonging to a given ψ with measure preassigned, but lying within certain limits. Finally, in §4, the techniques established are used to examine the dimensional number of a linear set arising in the study of Brownian paths.

§2. For a sequence ψ whose sum converges, let

$$r_n = \sum_{s=n}^{\infty} a_s \quad (n = 1, 2, \ldots). \tag{2}$$

Then r_n decreases to zero. We first see how the measure of sets E belonging to ψ depends on the rate of this decrease.

Lemma I. *Suppose ψ is a non–increasing sequence of positive numbers satisfying (1), and E is a closed linear set belonging to ψ. Then for $0 < \beta \le 1$, and r_n defined by (2),*

$$\Lambda^{\beta}(E) \le \liminf_{n \to \infty} \left[n, \left(\frac{r_n}{n} \right)^{\beta} \right].$$

Define $l(\beta, \psi)$ by

$$l(\beta, \psi) = \liminf_{n \to \infty} \left[n, \left(\frac{r_n}{n} \right)^{\beta} \right]. \tag{3}$$

We may assume that $l(\beta, \psi)$ is finite, as otherwise there is nothing to prove. Suppose E is a fixed set belonging to ψ and

$$I - E = \bigcup_{n=1}^{\infty} Q_n,$$

where Q_n $(n = 1, 2, \ldots)$ are disjoint open intervals of lengths a_n. Let $E_1 = [0, 1]$ and

$$E_n = E_1 - \bigcup_{s=1}^{n-1} Q_s \quad (n = 2, 3, \ldots).$$

Then E_n consists of n closed intervals (an isolated point is considered a degenerate interval) of total length r_n; and $E_n \supset E$.

Given $\varepsilon > 0$, $\eta > 0$, by (2) and (3), there exists an integer m such that

$$r_m < \eta,$$

and

$$m \left(\frac{r_m}{m}\right)^\beta < l(\beta, \psi) + \varepsilon.$$

Thus E_m is a $U(E, \eta)^*$ containing m intervals of total length r_m, and it is easy to see that

$$\sum_{E_m} l^\beta \leq m \left(\frac{r_m}{m}\right)^\beta < l(\beta, \psi) + \varepsilon.$$

Since η is arbitrary we have

$$\Lambda^\beta E \leq l(\beta, \psi) + \varepsilon;$$

and the lemma is proved, since ε is also arbitrary.

Suppose as before that, for a sequence ψ, r_n is defined by (2). Define numbers α_n by

$$n \left(\frac{r_n}{n}\right)^{\alpha_n} = 1 \quad (n = 1, 2, \ldots),$$

and let

$$\alpha(\psi) = \liminf_{n \to \infty} \alpha_n. \tag{4}$$

Clearly if ψ satisfies (1), we have $\alpha(\psi) \leq 1$. We can now state a result which is less exact than Lemma 1.

Corollary 1. *Suppose E is a closed linear set belonging to a sequence ψ and $\alpha(\psi)$ defined by (4). Then*

$$0 \leq \dim E \leq \alpha(\psi).$$

If $\alpha(\psi) = 1$, there is nothing to prove. If $\alpha(\psi) < 1$, and β satisfies $1 \geq \beta > \alpha(\psi)$, we have $l(\beta, \psi) = 0$; and $\Lambda^\beta E = 0$, by Lemma 1. Hence $\dim E \leq \alpha(\psi)$.

There are other ways of measuring the rate of decrease of sequences ψ satisfying (1). Define $\delta(\psi)$ as the exact lower bound of values of β for which $\sum a_n^\beta$ converges. Then $\delta(\psi)$ is known as the *exponent of convergence* of the sequence ψ. It is clear that $0 \leq \delta(\psi) \leq 1$ for sequences satisfying (1).

Lemma 2. *If E is a closed linear set belonging to a sequence ψ, and $\sum_{n=1}^\infty a_n^\beta$ converges $(0 < \beta \leq 1)$, then*

$$\Lambda^\beta E = 0.$$

---

*$U(E, \eta)$ is a set of intervals each of length less than η which contains the set E.

If $\beta = 1$, there is nothing to prove. Assume $\beta < 1$; then

$$r_n = \sum_{s=n}^{\infty} a_s = \sum_{s=n}^{\infty} a_s^{1-\beta} a_s^{\beta} < a_n^{1-\beta} \sum_{s=n}^{\infty} a_s^{\beta}.$$

Hence

$$n \left(\frac{r_n}{n}\right)^{\beta} < (a_n^{1-\beta})^{\beta} n^{1-\beta} \left[\sum_n^{\infty} a_s^{\beta}\right]^{\beta} = [n a_n^{\beta}]^{1-\beta} \left[\sum_n^{\infty} a_s^{\beta}\right]^{\beta}.$$

But $\sum a_n^{\beta}$ converges by hypothesis and therefore each of the terms $[n a_n^{\beta}]$ and $\left[\sum_n^{\infty} a_s^{\beta}\right]$ tends to zero as n increases. Hence $l(\beta, \psi) = 0$ and the lemma follows immediately using Lemma 1.

Corollary 2. *If E is a closed linear set belonging to ψ and $\delta(\psi)$ is the exponent of convergence, then*

$$0 \leq \dim E \leq \delta(\psi).$$

This follows from Lemma 2 in the same way as Corollary 1 followed from Lemma 1.

§3. In Corollaries 1 and 2 we have obtained two upper bounds $\alpha(\psi)$, and $\delta(\psi)$ for the dimension of linear sets E belonging to sequences ψ.

In some cases we will have $\alpha(\psi) = \delta(\psi)$: for example, if $\beta > 1$, and $\psi = \{a_n\}$ is defined by $a_n = k n^{-\beta}$ $(n = 1, 2, \ldots)$, where k is a constant, then $\alpha(\psi) = \delta(\psi) = \beta^{-1}$. Further, the argument of Lemma 2 shows that, for any ψ, $\alpha(\psi) \leq \delta(\psi)$.

Lemma 3. *Given α, δ satisfying $0 \leq \alpha \leq \delta \leq 1$, there exists a sequence ψ satisfying (1) such that*

$$\alpha(\psi) = \alpha, \quad \delta(\psi) = \delta.$$

Suppose first $0 < \alpha \leq \delta < 1$. Let n_i be a rapidly increasing sequence of integers, say defined by $n_0 = 1$, $n_1 = 10$, $n_{s+1} = n_s^{n_s}$ $(s = 1, 2, \ldots)$.

Put $\beta_2 = 1/\alpha$, $\beta_1 = 1/\delta$; then

$$\infty > \beta_2 \geq \beta_1 > 1.$$

Define

$$a_s = \mu s^{-\beta_2} \quad (n_{2i} \leq s < n_{2i+1}),$$

$$(i = 0, 1, 2, \ldots),$$

$$a_s = \mu n_{2i+1}^{\beta_1-\beta_2} s^{-\beta_1} \quad (n_{2i+1} \leq s < n_{2i+2}),$$

where μ is a positive constant chosen so that $\psi = \{a_n\}$ satisfies (1). The reader will easily verify that ψ is a non-increasing sequence and

$$\alpha(\psi) = \alpha, \quad \delta(\psi) = \delta.$$

If $\alpha = 0$, or $\delta = 1$, or both, the modifications required to the above example are obvious.

Because of Lemma 3, we see that there are sequences ψ satisfying (1) for which sets E belonging to ψ cannot attain the dimensional number $\delta(\psi)$.

§3. We now consider the bounds given by Lemma 1 and its corollary for the fractional measures of sets E belonging to ψ. Now for any ψ satisfying (1) there are sets E belonging to ψ which have zero dimension: in particular if we remove from $[0, 1]$ open intervals of lengths given by ψ placed end to end, we are left with an enumerable set E belonging to ψ. Thus the lower bound for dim E given by Corollary 1 is certainly best possible. That the upper bound given by Lemma 1 is best possible in general is seen by considering the Cantor ternary set for which

$$\Lambda^\beta E = l(\beta, \psi) = 1 \quad \text{when } \beta = \log 2 / \log 3.$$

However this is a very special sequence ψ. For quite general sequences ψ satisfying (1), we will prove the existence of sets E belonging to ψ such that dim $E = \alpha(\psi)$, and examine the possible for $\Lambda^\beta E$ when $0 < \beta \leq \alpha(\psi)$.

Consider the enumerable sequence of points given by

$$\eta_i = \frac{2i+1}{2^{r(i)}} - 1 \quad (i = 1, 2, \ldots), \tag{5}$$

with $r(i)$ the unique integer satisfying

$$2^{r(i)-1} \leq i < 2^{r(i)}.$$

The sequence $\{\eta_i\}$ contains all rational points of the interval $(0,1)$ whose denominator is a power of 2. If ψ is any sequence of positive numbers a_n whose sum converges, we define $E(\psi)$ as follows:

Let I_r be an open interval of length a_r with left–hand end–point given by

$$\delta_r = \sum_{s \in Q_r} a_s,$$

where Q_r is the set of those integers i for which $\eta_i < \eta_r$ with $\{\eta_i\}$ defined by (5). Then if

$$\delta_0 = \sum_{s=1}^{\infty} a_s,$$

put

$$E(\psi) = [0, \delta_0] - \bigcup_{r=1}^{\infty} I_r.$$

Thus if ψ is a sequence satisfying (1), $E(\psi)$ is a closed linear set belonging to ψ such that

$$(0,1) - E(\psi) = Q$$

consists of a sequence of open intervals

$$I_r \quad (r = 1, 2, \ldots)$$

and

(i) I_r has length a_r,

(ii) with the obvious ordering in $(0,1)$, I_r comes before I_s $(r \neq s)$ if, and only if $\eta_r < \eta_s$.

Suppose

$$T_N = \bigcup_{r=1}^{2^N - 1} I_r$$

and

$$E_N = [0, \delta_0] - T_N.$$

Then E_N consists of 2^N closed intervals, and

$$E(\psi) = \bigcap_{N=1}^{\infty} E_N.$$

In the next lemma we restrict the sequences ψ rather severely.

Lemma 4. *Suppose $\psi = \{a_n\}$ is a non–increasing sequence of positive numbers whose sum converges, and*

$$a_i = k_r \text{ for } 2^{r-1} \le i < 2^r \quad (r = 1, 2, \ldots),$$

where $\{k_r\}$ is a decreasing sequence. Then, if $0 < \beta \le 1$, and $l(\beta, \psi)$ is defined by (3), we have

$$\Lambda^\beta E(\psi) \ge \frac{1}{2} l(\beta, \psi).$$

Case (i). $l(\beta, \psi) = 0$. There is nothing to prove.

Case (ii). $0 < l(\beta, \psi) < \infty$.

E_N $(N = 1, 2, \ldots)$ consists of 2^N closed intervals of equal length, say l_N. Given $\varepsilon > 0$, there exists N_0 such that

$$2^N l_{N^\beta} > (1 - \varepsilon) l(\beta, \psi) \quad \text{for } N \ge N_0. \tag{6}$$

Let η be any number satisfying

$$0 < \eta < a_r, \quad r = 2^{N_0+1}. \tag{7}$$

Write E for $E(\psi)$ defined as above.

Suppose $R = N(E, \eta)$ is a covering of E by open intervals of length less than η. Since E is closed we may assume that R contains a finite number of open intervals. There exists an integer N_1 such that

$$R \supset E_{N_1},$$

as otherwise $E_N - R$ $(N = 1, 2, ldots)$ would be a decreasing sequence of non–void closed sets with a void intersection $E - R$. We may further assume that the intervals of R are disjoint. Now replace each interval W of R by the smallest possible single closed interval $V \supset W \cap E_{N_1}$. In this way a set C consisting of a finite number of closed intervals covering E_{N_1} is obtained.

If the interval V contains τ_ν closed intervals of E_{N_1},

$$\sum_C \tau_\nu = 2^{N_1}, \tag{8}$$

since each of the intervals of E_{N_1} must be contained in exactly one interval of C. Now define s_ν to be the unique integer such that

$$2^{s_\nu} \le \tau_\nu < 2^{s_\nu+1}. \tag{9}$$

Now any interval I which contains 2^{s_ν} intervals of E_{N_1} has length not less than $l_{N_1-s_\nu}$, since the sequence $\{k_n\}$ is non–increasing. Hence d_ν, the length of the interval V, must satisfy

$$\eta > d_\nu \geq l_{N_1-s_\nu}.$$

By (7), $N_1 - s_\nu > N_0$, and therefore by (6),

$$d_\nu^\beta > (1 - \varepsilon)l/2^{N_1-s_\nu}.$$

Applying (8) and (9) we have

$$\sum_C d_\nu^\beta > \frac{1}{2}(1 - \varepsilon)l(\beta, \psi).$$

Thus for any $R = U(E, \eta)$ we have

$$\sum_R l^\beta \geq \sum_C d_\nu^\beta > \frac{1}{2}(1 - \varepsilon)l(\beta, \psi).$$

Since ε is arbitrary,

$$\Lambda^\beta E(\psi) \geq \frac{1}{2}l(\beta, \psi).$$

Case (iii). $l(\beta, \psi) = +\infty$.
Obvious modifications of the argument used in Case (ii) show that
$$\Lambda^\beta E(\psi) = +\infty.$$

Lemma 5. *Suppose* $\psi = \{a_n\}$ *is a non–increasing seqeunce whose sum converges, and* $\psi' = \{a_n\}$ *is defined by*

$$a'_s = a_q \quad (q = 2^r - 1, \ 2^{r-1} \leq s < 2^r, \ r = 1, 2, \ldots).$$

Then, if $0 < \beta \leq 1$, *and* $l(\beta, \psi)$ *is defined by* (3),

$$l(\beta, \psi) \geq l(\beta, \psi') \geq \frac{1}{2}l(\beta, \psi).$$

As in (2), let

$$\tau'_n = \sum_{s=n}^\infty a'_s.$$

Now in the definition of ψ', $q < 2s$. hence, for any s,

$$a'_s \geq a_{2s} \geq a_{2s+1}.$$

and

$$r'_n \geq \frac{1}{2} \left[\sum_{s=n}^{\infty} a_{2s} + \sum_{s=n}^{\infty} a_{2s+1} \right] = \frac{1}{2} r_{2n}.$$

Thus

$$n \left(\frac{r_n}{n} \right)^{\beta} \geq n \left(\frac{r'_n}{n} \right)^{\beta} \geq \frac{1}{2} 2n \left(\frac{r_{2n}}{2n} \right)^{\beta}.$$

Taking lower limits gives the required result, because

$$\liminf_{n \to \infty} 2n \left(\frac{r_{2n}}{2n} \right)^{\beta} \geq \liminf_{n \to \infty} n \left(\frac{r_n}{n} \right)^{\beta}.$$

Lemma 6. *If ψ is any non–increasing sequence satisfying (1), $0 < \beta \leq 1$, and $l(\beta, \psi)$ is defined by (3); then*

$$l(\beta, \psi) \geq \Lambda^{\beta} E(\psi) \geq \frac{1}{4} l(\beta, \psi).$$

Since $E(\psi)$ is a closed linear set belonging to ψ we have, by Lemma 1,

$\Lambda^{\beta} E(\psi) \leq l(\beta, \psi)$. From ψ, form the sequence ψ' by the process of Lemma 5. Since $a'_n \leq a_n$ ($n = 1, 2, \ldots$), if we set up the obvious correspondence between complementary intervals I'_r of $E' = E(\psi')$ and I_r of $E = E(\psi)$, we see that the closed intervals of E'_N are not greater in length than the corresponding closed intervals of E_N. Hence if R is a $U(E, \eta)$, there is a covering $R' = U(E', \eta)$ of intervals of the same lengths as those of R. Thus, for any $\eta > 0$, we have

$$\sum_{U(E,\eta)} l^{\beta} = \sum_{U(E',\eta)} l^{\beta},$$

and therefore

$$\Lambda^{\beta} E \geq \Lambda^{\beta} E'.$$

Since ψ' satisfies the conditions of Lemma 4,

$$\Lambda^{\beta} E(\psi') \geq \frac{1}{2} l(\beta, \psi') \geq \frac{1}{4} l(\beta, \psi),$$

by Lemma 5. Thus we have $\Lambda^{\beta} E(\psi) \geq \frac{1}{4} l(\beta, \psi)$, as required.

Remark. There is no significance in the constant $\frac{1}{4}$ appearing in Lemma 6, and no doubt it could be improved upon. We have contented ourselves with showing that the β–measure of $E(\psi)$ is zero, finite and positive, or infinite, according as $l(\beta, \psi)$ is zero, finite and positive, or infinite.

To obtain other sets E belonging to ψ, we need

Lemma 7. *Suppose E is a closed linear set of dimension δ $(0 < \delta \leq 1)$. Then*

(i) *if $0 < \beta < \delta$ and $\gamma \geq 0$ (γ may be infinite), there is a closed subset E_1 of E such that*

$$\dim E_1 = \beta, \quad \Lambda^\beta E_1 = \gamma ,$$

(ii) *if ζ satisfies $0 \leq \zeta \leq \Lambda^\delta E$, there is a closed subset E_2 of E such that*

$$\dim E_2 = \delta, \quad \Lambda^\delta E_2 = \zeta .$$

(i) E is a closed set of infinite β–measure. Hence, if $0 < \gamma < +\infty$, E_1 exists by the main theorem of **(1)**. This theorem can be shown to hold for a wider class of measure functions and in particular for

$$k(x) = x^\beta \log \frac{1}{x} \quad \text{and} \quad h(x) = x^\beta \left/ \log \frac{1}{x} \right. .$$

If $\gamma = \infty$, let E_1 be a closed subset such that $h - m(E_1) = 1$. This set will have the required dimensional number and measure. Similarly, if $\gamma = 0$, we define E_1 as a closed subset such that

$$k - m(E_1) = 1.$$

(ii) If $\Lambda^s E = \infty$, this is effectively the same as (*i*). If $\Lambda^\delta E = 0$, there is nothing to prove. If $0 < \Lambda^\delta E = \infty$, and $\zeta = 0$; then $k - m(E) = \infty$, with $k(x) = x^\delta \log(1/x)$. A closed subset E_1 such that $k - m(E_1) = 1$ will satisfy the conditions. If $0 < \Lambda^\delta E < \infty$ and $0 < \zeta \leq \Lambda^\delta E$, the result follows from Lemmas 2 and 3 of **(2)**.

We will now see that the measure of linear sets E belonging to ψ can have almost any preassigned value between the limits prescribed in Lemma 1.

Theorem 1. *Suppose ψ is a non–increasing sequence of positive numbers satisfying (1), and $l(\beta, \psi)$, $\alpha(\psi)$ are defined by (3) and (4). Then*

(i) *for any β satisfying* $0 \leq \beta \leq \alpha(\psi)$, *there are linear closed sets of dimension β which belong to ψ;*

(ii) *if* $\alpha(\psi) > 0, 0 < \beta \leq \alpha(\psi)$ *and γ is any number satisfying*

$$0 \leq \gamma \leq \frac{1}{4} l(\beta, \psi) \quad (\gamma \text{ may be infinite}),$$

there is a linear set E belonging to ψ such that

$$\dim E = \beta, \quad \Lambda^{\delta} E = \gamma .$$

Proof. (i) If $\alpha(\psi) = 0$, every set E, and in particular $E(\psi)$, will have zero dimension by Corollary 1. If $0 = \beta < \alpha(\psi)$, an enumerable set E belonging to ψ will satisfy the conditions. Finally, if $0 < \beta \leq \alpha(\psi)$, the result is included in (ii).

(ii) By Lemma 6, $E(\psi)$ is a closed set belonging to ψ such that

$$\dim E(\psi) = \alpha(\psi)$$

and

$$\Lambda^{\alpha} E(\psi) \geq \frac{1}{4} l(\alpha, \psi) \quad (\alpha = \alpha(\psi)), .$$

Now apply Lemma 7 to $E(\psi)$: let F be a closed subset of $E(\psi)$ such that

$$\dim F = \beta, \quad \Lambda^{\beta} F = \gamma .$$

The set F obtained in this way need not belong to ψ, but we will·see that it differs from a set E belonging to ψ by only an enumerable set. Since F is closed, $(0, 1) - F$ is an enumerable sequence of open intervals, say

$$I_r \quad (r = 1, 2, \ldots) .$$

Suppose the complement, $(0, 1) - E(\psi)$, of $E(\psi)$ consists of the intervals Q_s $(s = 1, 2, \ldots)$. Then end–points of an interval I_r are in F, and therefore in $E(\psi)$ and not in any Q_s. Hence for any integer r, I_r contains a finite or enumerable subset of the intervals Q_s. The total length of this subset must be the length of I_r. Rearrange the intervals Q_s on I_r into a decreasing sequence of abutting intervals and denote by P_r the set of end–points of the rearranged intervals.

Put

$$E = F \cup \bigcup_{r=1}^{\infty} P_r .$$

Since P_r $(r = 1, 2, \ldots)$ is a countable set

$$\Lambda^\beta E = \Lambda^\beta F = \gamma, \quad \dim E = \dim F = \beta.$$

Also the only limit points of $\cup_{r=1}^{\infty} P_r$ are contained in F, and so $E = F \cup \bigcup P_r$ is closed. It is clear that E belongs to ψ, and the theorem is proved.

§4. If Ω is the space of Brownian paths ω in one dimension,[*] it is known[†] that for almost all ω of Ω, the values of t for which

$$x(t, \omega) = 0, \quad 0 \le t \le 1,$$

form a perfect linear set which we denote by $\mathcal{E}(\omega)$. We need

Lemma 8. *Suppose $\mathcal{E}(\omega)$ belongs to a non–increasing sequence $\psi = \{a_n\}$: then, for almost all ω of Ω, $\lim_{n \to \infty} n a_n^{\frac{1}{2}}$ and $\lim_{n \to \infty} r_n a_n^{-\frac{1}{2}}$ exist and satisfy*

$$0 < \lim_{n \to \infty} n a_n^{\frac{1}{2}} = \lim_{n \to \infty} r_n a_n^{-\frac{1}{2}} < \infty.$$

This is just a restatement in our notation of Theorem 47.2 of (3), noting that the stochastic variables $s_0 = S_0(t)$ of Lévy is finite and non–zero for $t = 1$, with probability 1.

This allows us to prove

Theorem 2. *Suppose $\mathcal{E}(\omega)$ belongs to a non–increasing sequence $\psi = \{a_n\}$: then, for almost all ω of Ω,*

(i) $\Lambda^{\frac{1}{2}} \mathcal{E}(\omega)$ *is finite.*

(ii) *the exponent of convergence of ψ is $\frac{1}{2}$, and $\sum a_n^{\frac{1}{2}}$ diverges.*

By Lemma 8, with probability 1, $0 < l\left(\frac{1}{2}, \psi\right) < \infty$. (i) follows immediately using Lemma 1. Further the proof of Lemma 2 shows that $\sum a_n^{\frac{1}{2}}$ cannot converge, for this would imply $l\left(\frac{1}{2}, \psi\right) = 0$. Thus $\delta(\psi) \ge \frac{1}{2}$.

Now let $\beta > \frac{1}{2}$. Define numbers δ_s and integers n_s by

$$\delta_s = 2^{-s}, \tag{10}$$

$$(s = 1, 2, \ldots)$$

$$a_{n_s} \ge \delta_s > a_{n_s+1}. \tag{11}$$

---

[*]For definitions, and notation, the reader is referred to [6].
[†]See, for example, section 7 of [3].

Then

$$\sum_{r=1}^{\infty} a_r^{\beta} \leq \sum_{r=1}^{n_1} a_r^{\beta} + \sum_{s=1}^{\infty} (n_{s+1} - n_s)\delta_s^{\beta}$$

$$= k_1 + k_2 \sum_{s=2}^{\infty} n_s \delta_s^{\beta},$$

where k_1 and k_2 are constants.

By Lemma 8, there exists a constant k_3 and an integer s_0 such that, for $n > n_{s_0}$,

$$na_n^{\frac{1}{2}} < k_3. \tag{12}$$

By (11),

$$n_s \delta_s^{\beta} = n_s \delta_s^{\frac{1}{2}} \delta_s^{\beta - \frac{1}{2}} < n_s a_{n_s}^{\frac{1}{2}} \delta_s^{\beta - \frac{1}{2}};$$

and so by (12), for $s > s_0$,

$$n_s \delta_s^{\beta} < k_3 \delta_s^{\beta - \frac{1}{2}}.$$

By (10), $\sum \delta_s^{\beta - \frac{1}{2}}$ converges: this implies that $\sum a_n^{\beta}$ converges. Since β was any number greater than $\frac{1}{2}$,

$$\delta(\psi) = \frac{1}{2}.$$

The methods of the present note have not been successful in obtaining a lower bound for dim $\mathcal{E}$. All we have proved is that $0 \leq \dim \mathcal{E} \leq \frac{1}{2}$. By quite a different approach we have been able to obtain dim $\mathcal{E} > \frac{1}{5}$, but this seems a bad lower estimate for the dimension: in fact there are strong reasons for conjecturing that dim $\mathcal{E}(\omega) = \frac{1}{2}$ for almost all ω. However, we have proved a result slightly stronger than Theorem 2(i), namely:

For almost all ω of Ω, $\Lambda^{\frac{1}{2}} \mathcal{E}(\omega) = 0$.

In the selection, a fractal dimension $\delta(\psi)$ is defined, and the inequality dim $E \leq \delta(\psi)$ is proved. In general, they are not equal. This dimension $\delta(\psi)$ is in fact equal to the dimensions defined in Selection 7 (Bouligand), Selection 8 (Pontrjagin and Schnirelmann), and Selection 17 (Kolmogorov and Tihomirov). Twelve equivalent formulations (for subsets of $\mathbb{R}$) are discussed in [10].

The result of this selection may be applied in many ways. One application is to the evaluation of the Hausdorff dimension of the zero set Z of one–dimensional

Brownian motion. Paul Lévy determined in [4] the behavior of the complementary intervals; use of the Besicovitch–Taylor estimate proves dim $Z \leq 1/2$. Taylor [8] proved dim $Z \geq 1/2$. But $\Lambda^{1/2}(Z) = 0$; we might say that the Hausdorff dimension is "infinitesimally smaller" than $1/2$. The "exact" Hausdorff dimension function for the set Z is not $x^{1/2}$, but

$$h(x) = (x \log \log 1/x)^{1/2} .$$

That is, when the Hausdorff measure is computed using this function of the diameter, we get

$$0 < \Lambda^h(Z) < \infty .$$

This story (and much more) can be found in [9].

Historical note. This selection (date 1954) is the first one of our selections to use the "cup" $\cup$ and "cap" $\cap$ for union and intersection, rather than addition and multiplication signs. *–Ed.*

Bibliography

[1] A. S. Besicovitch, *On existence of subsets of finite measure of sets of infinite measure*, Indag. Math. **14** (1952), 339–344.

[2] A. S. Besicovitch and P.A.P. Moran, *The measure of product and cylinder sets*, J. London Math. Soc. **20** (1945), 110–120.

[3] P. Lévy, *Sur certains processus stochastiques homogène*, Compos. Math. **7** (1940), 283–339.

[4] P. Lévy, *Processus Stochastiques et Mouvement Brownien*, Paris, 1948.

[5] S. J. Taylor, *On Cartesian product sets*, J. London Math. Soc. **27** (1952), 295–304.

[6] S. J. Taylor, *The Hausdorff α–dimensional measure of Brownian paths in n–space*, Proc. Cambridge Phil. Soc. **49** (1953), 31–30.

[7] A. Debus, editor, *Marquis Who's Who*, A.N. Marquis, Chicago, 1990.

[8] S. J. Taylor, *The α–dimensional measure of the graph and set of zeros of a Brownian path*, Proc. Cambridge Phil. Soc. **51** (1955), 265–274.

[9] S. J. Taylor, *The measure theory of random fractals*, Math. Proc. Cambridge Phil. Soc. **100** (1986), 383–406.

[10] C. Tricot, *Douze définitions de la densité logarithmique*, Compte Rendus Acad. Sci. Paris, Série I **293** (1981), 549–552.

In this selection, certain self-affine curves are considered, related
to the curves of Paul Lévy (Selection 12). Another self-affine curve
is in Selection 16 (Rosenwasser).

A notation for a continued fraction:

$$[a_0, a_1, a_2, \ldots] = a_0 + \cfrac{1}{a_1 + \cfrac{1}{a_2 + \cfrac{1}{a_n}}}$$

Georges Cham (1905?–1960) was born in Switzerland and
taught at his doctorate in Paris. He held positions at Lausanne
and Geneva. He was interested in all aspects of geometry and
topology. He is best known for a theorem connecting the cohomology of a simplex and the properties of information on that
manifold. This ... in turn is over a half-dozen papers he wrote on
curves and ... and representational equations. [5], [6]

In this selection, certain **self-affine** curves are considered, related to the curves of Paul Lévy (Selection 12). Another self-affine graph is in Selection 18 (Kiesswetter).

A notation for a continued fraction:

$$[a_0, a_1, a_2, \cdots, a_n] = a_0 + \cfrac{1}{a_1 + \cfrac{1}{a_2 + \cfrac{1}{\cdots \cfrac{1}{a_n}}}}$$

Georges de Rham (1903[1]–1990) was born in Switzerland, and completed his doctorate in Paris. He held positions at Lausanne and Geneva. He was interested in all aspects of geometry and topology. He is best known for a theorem connecting the cohomology of a manifold with the properties of integration on that manifold. This selection is one of a half-dozen papers he wrote on curves and functions defined by functional equations. [15], [17]

---

[1]1903 according [17], but 1901 according to [15].

SIXTEEN

On Some Curves Defined by Functional Equations

Georges de Rham

The curves that are the subject of this talk are related to the following problem: *given two transformations F_0 and F_1 of the plane into itself, find a map $t \to M(t)$ of the interval $0 \leq t \leq 1$ to the plane satisfying the functional equations*

[1]
$$M\left(\frac{t}{2}\right) = F_0 M(t), \qquad M\left(\frac{1+t}{2}\right) = F_1 M(t).$$

I will first show that if F_0 and F_1 satisfy certain conditions, then this problem has a bounded solution, which is unique and continuous. The continuous map $t \to M(t)$ thus parametrizes a curve, the one defined by the equations [1].

I will further study certain specific examples by fixing the transformations F_0 and F_1. By substituting complex numbers for points in the plane we also consider $M(t)$ as a complex valued function. In certain cases F_0 and F_1 will take the real axis to itself and so $M(t)$ will be a real valued function.

1. We say that a transformation F of the pane to itself is *contracting* if the ratio of distances of images under F of arbitrary points and the distance between the points themselves never surpasses a fixed number less than 1, this number being called the *ratio of contraction of F*. Such a transformation cannot have two fixed points since the distance between the points would not be decreased. On the other hand,

the sequence of images $F^n P$, $(n = 0, 1, 2, \ldots)$ of a point P by the iterates of such a transformation F converges since the distance between $F^n P$ and $F^{n-1} P$ decreases with n faster than the terms of a geometric progression with multiplier equal to the ratio of contraction. The limit point of this sequence is clearly a fixed point of F so a contracting transformation always has a unique fixed point.

Given this, I will show that *if the transformations F_0 and F_1 are contracting and if the image under F_0 of a fixed point of F_1 coincides with the image under F_1 of a fixed point of F_0 then equations* [1] *have a unique bounded solution and this solution is continuous.*

We first remark that if one lets F_0 and F_1 be contracting then the condition relating to the fixed points is necessary for equations [1] to admit a uniform solution. For if $M(t)$ is a solution, $M(0)$ the fixed point of F_0 and $M(1)$ the fixed point of F_1 then, by replacing t by 1 in the first equation and t by 0 in the second one gets

[2]
$$F_0 M(1) = F_1 M(0) = M\left(\frac{1}{2}\right).$$

In the proof we denote by T_a the substitution $T_a t = \frac{a+t}{2}$. Our equations can thus be written as

[1']
$$M(T_a t) = F_a M(t), \qquad (a = 0, 1).$$

It easily follows by induction that

[3]
$$M(T_{a_1} T_{a_2} \cdots T_{a_n} t) = F_{a_1} F_{a_2} \cdots F_{a_n} M(t),$$

for any sequence $a_1 a_2 \ldots a_n$ of digits equal to 0 or 1.

Expanding t in base 2:

$$t = \sum_1^\infty a_k 2^{-k},$$

and letting

$$\tau_n = \sum_1^n a_k 2^{-k}, \qquad t_n = \sum_1^\infty a_{k+n} 2^{-k},$$

one has

$$t = \tau_n + 2^{-n} t_n,$$

and this can be written, as is easily checked,

$$t = T_{a_1} T_{a_2} \cdots T_{a_n} t_n.$$

It follows from this, by taking (3)^2 into account, that if $M(t)$ satisfies [1] then

[4] $$M(t) = F_{a_1} F_{a_2} \ldots F_{a_n} M(t_n).$$

To complete the proof we will use the following lemma that will be proved later:

Lemma. The image of an arbitrary bounded set under $F_{a_1} F_{a_2} \ldots F_{a_n}$ tends, as $n \to \infty$, to a point that depends only on t.

If $M(t)$ is a bounded solution of equations [1] then, because of (4), $M(t)$ is exactly the point determined by the lemma since, in that case, $M(t_n)$ stays in a bounded set. Moreover, this point can be represented as

[5] $$M(t) = \lim_{n=\infty} F_{a_1} F_{a_2} \ldots F_{a_n} P,$$

where P is an arbitrary point independent of t and n and the function thus defined actually satisfies equations [1] or equation [1']: in fact, this follows because the sequence of digits of the binary expansion of $\frac{a+t}{2}$ can be found from the binary digits of t by prepending the digits of a. Existence and uniqueness of the bounded solution is established given the truth of the lemma.

To prove the lemma denote by M_a the fixed point of F_a ($a = 0, 1$), by P the center point of the segment $M_0 M_1$, by $2e$ the length of this segment, by k ($0 < k < 1$) a ratio of contraction of F_0 and F_1, by $C(M_a, r)$ the circle with center M_a and radius r, and by $C = C(P, R)$ the circle with center P and radius R.

The transformation F_a takes $C(M_a, r)$ into a part of $C(M_a, kr)$ and if

$$R > \frac{1+k}{1-k} e$$

which implies $kR + ke < R - e$, it takes $C(M_a, R + e)$ to a part of $C(M_a, R - e)$. Since

$$C(M_a, R + e) \supset C \supset C(M_a, R_e),$$

it follows that F_a takes C to a part of C.

It follows immediately that each of the sets $C_n = F_{a_1} F_{a_2} \ldots F_{a_n} C$ ($n = 0, 1, 2, \ldots$) contains the next one, and since the diameter of C_n

---

^{2}Formulas are refered to with parentheses like (3) or [3] interchangeably. –Ed.

tends to 0, as k^n is a ratio of contraction of $F_{a_1} F_{a_2} \ldots F_{a_n}$, this clearly implies that C_n tends to a point, this point being given by [5] and independent of the sequence $a_1 a_2 \ldots a_n \ldots$.

In general, to each value of t there corresponds a unique sequence $a_1 a_2 \ldots$, the only exceptions being binary fractions for which there correspond two sequences of the form

$$a_1 a_2 \ldots a_p 1000 \ldots \quad \text{and} \quad a_1 a_2 \ldots a_p 0111 \ldots .$$

But these two sequences give rise to the same limit point since $F_a^k P$ tends to M_a as $k \to \infty$ since, by virtue of [2], one has

$$\lim_{k=\infty} F_1 F_0^k P = F_1 M_0 = F_0 M_1 = \lim_{k=\infty} F_0 F_1^k P .$$

This completes the proof of the lemma.

We still have to establish the continuity of the above solution. This will follow from the following remark. If $a_1, a_2, \ldots$ and a_n stay fixed then t_n varies between 0 and 1 and t varies between τ_n and $\tau_n + 2^{-n}$. Since $M(t_n)$ stays in a circle C, $M(t)$ will stay in the set $C_n = F_{a_1} F_{a_2} \ldots F_{a_n} C$ which, as we have seen, has diameter as small as desired for n sufficiently large.

2. We consider the simplest case of our equations by taking F_0 and F_1 to be linear transformations in one variable. By supposing that the fixed points are 0 and 1 and taking [2] into account we see that these are of the form

$$F_0 z = az \quad \text{and} \quad F_1 z = a + (1 - a)z ,$$

where a is arbitrary except for satisfying the conditions

$$|a| < 1 \quad \text{and} \quad |1 - a| < 1 ,$$

which express the fact that F_0 and F_1 are contracting.

The solution of equations [1] is thus a continuous function $M(t)$. For $a = 1/2$ one has $M(t) = t$. I will show that *if $a \neq 1/2$ then the derivative $M'(t)$ cannot exist without vanishing*: for any value of t either the derivative exists and is zero or it does not exist.

If this derivative exists then it is the limit, as $n \to \infty$, of the average slope of $M(t)$ in the interval $(\tau_n, \tau_n + 2^{-n})$. But, from the fact that the derivative of the linear function

$$F_{a_1} F_{a_2} \ldots F_{a_n} \quad \text{is equal to} \quad D_n = a^{n-s_n} (1 - a)^{s_n} ,$$

where $s_n = a_1 + a_2 + \cdots a_n$, and taking [4] into account, it follows that the increase in $M(t)$ in this interval is equal to D_n. Its average slope is thus $2^n D_n$ and if $M'(t)$ exists one has

$$M'(t) = \lim_{n=\infty} 2^n D_n .$$

If $M'(t) \neq 0$ then the ratio $2^{n+1} D_{n+1} : 2^n D_n$ would tend to 1 and $D_{n+1} : D_n$ to $1/2$ which is impossible since $D_{n+1} : D_n$ can only take on the values a and $1 - a$.

Let us suppose that $r = \inf\{|a|, |1 - a|\} > 1/2$ which clearly necessitates that a be complex (for example, this holds when $a = \frac{1+i}{2}$), then $|D_n| \geq r^n$ and $|2^n D_n| \to \infty$. It follows that $M'(t)$ cannot exist and $M(t)$ is *a continuous function that is nowhere differentiable*. It is a complex valued function but its real and imaginary parts are also nowhere differentiable: this holds because neither the real nor the imaginary part of $2^n D_n$ is bounded, as is easily verified, and so neither can have a limit as $n \to \infty$. This example of a continuous non differentiable function was already discovered by **Cesaro** using another method: **Cesaro** defined this function directly by an expansion (deducible from [5]) without recourse to the functional equations [1].

If a is real and $\neq 1/2$, then the fact that D_n is always positive shows that the function $M(t)$ is strictly increasing. By a theorem of **Lebesgue** it is differentiable almost everywhere. The derivative is thus zero almost everywhere and the function $M(t)$ is *a singular function*. This same function was also defined in another way and studied by **Cesaro, Faber**, and **Salem**. One can show that, for values of t for which s_n/n tends to a limit l as $n \to \infty$ and letting

$$l_0 = \frac{\log 2a}{\log a - \log(1 - a)} ,$$

there is a derivative $M'(t) = 0$ when $(l - l_0)(a - 1/2) > 0$.

One easily checks that, as a increases from 0 to 1, l_0 decreases from 1 to 0 taking the value $1/2$ when $a = 1/2$.

It follows that $(1/2l_0)(a - 1/2) > 0$ so that $M'(t) = 0$ if s_n/n tends to $1/2$. Since the set of values of t for which s_n/n does not tend to $1/2$ has measure zero, this shows (without appealing to the theorem of **Lebesgue**) that $M'(t) = 0$ almost everywhere.

One can also show that, for values of t for which $s_n - nl_0$ stays bounded in absolute value (it is easy to find such values of t), $M(t)$ has a finite set of limiting derivative values, these being positive and not all equal: the ratio of the largest to the smallest is at least equal to the maximum of $2a$ and $2 - 2a$.

I came to study the functional equations discussed here by searching for the simplest equations of the form [1] after a particular problem to be presented below led me to study other less simple equations of the same form. One is led to the same equations with a real by considering the following problem from the calculus of probabilities where they have a very simple interpretation.

Suppose that a number ξ in the interval $(0, 1)$ is determined by choosing at random and independently the digits of its binary expansion with the values 0 and 1 always having the probabilities a and $1-a$, respectively. The problem at hand is to find the distribution function of ξ, i.e., the probability $f(t)$ that $\xi \leq t$. If one sets $\xi = \frac{a + \xi_1}{2}$, where a is the first digit in the expansion of ξ, then it is clear that ξ_1 has the same distribution function $f(t)$ as ξ and the two random variables a and ξ_1 are independent. The event $\xi \leq t/2$ thus consists of two events $a = 0$ and $\xi_1 \leq t$ and its probability $f(t/2)$ is equal, by the rule for joint probabilities, to $a f(t)$ one gets the equation $f(t/2) = a f(t)$. By noting that the event $\xi \leq \frac{1+t}{2}$ consists of the event $a = 0$ or of the joint event $a = 1$ and $\xi_1 \leq t$, one obtains the equation $f\left(\frac{1+t}{2}\right) = a + (1 - a) f(t)$. These are exactly the equations studied above and the distribution function $f(t)$ is exactly the function $M(t)$ previously discussed.

3.   A case that is almost as simple is that of the transformations

$$F_0 z = a \bar{z}, \qquad F_1 z = a + (1 - a) \bar{z},$$

where $\bar{z}$ denotes the complex conjugate of z and a is a complex number again satisfying $|a| < 1$ and $|1 - a| < 1$. The triangle Δ with vertices $(0, a, 1)$ is mapped by F_0 and F_1 to two triangles similar to Δ and contained in Δ. The curve defined by the equations [1] is therefore contained in Δ and, more generally, is contained in the union of 2^n triangles that are images of Δ by the transformations $F_{a_1} F_{a_2} \dots F_{a_n}$. For $a = 1/2 + i\sqrt{3}/6$ this is the celebrated curve of **von Koch**. If $|a - 1/2| = 1/2$ one obtains Peano curves, already considered by **Cesaro** and **Polya**, filling the triangle Δ.

4.   We now take F_0 and F_1 to be the homographic transformations

$$F_0 m = \frac{m}{m + 1}, \qquad F_1 M = 1 + m.$$

These transformations are not contracting and F_1 does not have a fixed point at finite distance. However, by considering $M = \infty$ as the fixed point of F_1, condition [2] is verified and the method used in paragraph 1 allows one to show that equations [1], where M is replaced by m, have a unique *positive* solution $m(t)$ which is continuous and increasing for $0 \le t \le 1$.

Using continued fraction notation one has

$$F_a m = a + \cfrac{1}{1 - a + \cfrac{1}{m}} = [a, 1 - a, m], \quad (a = 0, 1).$$

Formula [4] then becomes

$$m(t) = [a_1, 1 - a_1, a_2, 1 - a_2, \ldots, a_n, 1 - a_n, m(t_n)],$$

and if one supposes that $m(t_n)$ stays positive, this expression converges, as $n \to \infty$, to the canonical continued fraction of **Denjoy**

$$m(t) = [a_1, 1 - a_1, a_2, 1 - a_2, \ldots, a_n, 1 - a_n, \ldots].$$

One can also expand $m(t)$ using the regular continued continued fraction. Denote by S the transformation $Sm = 1/m$ so that $F_0 = SF_1S$, $F_0^h = SF_1^h S$. Now suppose that the sequence $a_1 a_2 \ldots$ starts with k_1 digits "1" ($k_1 \ge 0$), then k_2 digits "0", then k_3 digits "1", then k_4 digits "0", etc., one has

$$F_{a_1} F_{a_2} \ldots = F_1^{k_1} F_0^{k_2} F_1^{k_3} \ldots = F_1^{k_1} SF_1^{k_2} SF_1^{k_3} S \ldots$$

and since $F_1^h Sm = k + 1/m$ it follows immediately from [4] that

$$m(t) = [k_1, k_2, k_3, \ldots].$$

This function is essentially the inverse function of the function that **Minkowski** denoted by the symbol ? (more precisely, $t = 2?m$). A deep study of it was made by **Denjoy** who, in particular, showed that this function is singular.

5. Here now is the curve that led me to study equations of the form in [1].

Consider the sequence of polygonal lines P_n ($n = 0, 1, 2, \ldots$) constructed in the following way. $P_0 = ABC$ is a line with two sides and P_{n+1} has as its vertices the points that divide into three equal parts the sides of P_n taken in order of traversal of P_n. These polygonal lines are convex and, as $n \to \infty$, P_n tends to a limit curve C joining the center of AB to the center of BC.

The middle third of each side of P_n is a side of P_{n+1} and it follows that the middle points of the sides of P_n are still middle points of the sides of all the polygons P_{n+k} $(k = 1, 2, \ldots)$, and so belong to the curve C. We number the sides of these polygons by always taking the same direction from A to B. The number of sides of P_n is $2^n + 1$ and the middle point of the $(h+1)$st side of P_n coincides with the middle point of the $(2h + 1)$st side of P_{n+1} and, more generally, with the middle point of the $(2^k h + 1)$st side of P_{n+k}. We denote this point by $M(h \, 2^{-n})$, which is justified since it depends only on the binary number $h \, 2^{-n}$. Thus, to each binary fraction t $(0 \le t \le 1)$ there corresponds a point $M(t)$ of the curve. This correspondence extends, by continuity, to the whole interval $0 \le t \le 1$ and gives a parametric representation of the curve C.

Denote by $A'B'C'D'$ the vertices of P_1, then A' and B' divide AB into three equal parts and C' and D' divide BC into three equal parts. Let F_0 be the linear (or affine) transformation of the plane into itself taking ABC to $A'B'C'$. It takes P_n to a polygon constructed from $A'B'C'$ just as P_n was constructed from ABC, and this is none other than the polygon formed by the first $2^n + 1$ sides of P_{n+1}. It follows that F_0 takes $M(h \, 2^{-n})$ to $M(h \, 2^{-n-1})$. By continuity one finds that $F_0 M(t) = M(t/2)$ for all t $(0 \le t \le 1)$.

Similarly, if F_1 denotes the linear transformation taking ABC to $B'C'D'$, one sees that $F_1 M(t) = M\left(\frac{1+t}{2}\right)$.

So the curve C can be defined by functional equation of the form (1), where F_0 and F_1 are linear transformations. It is easy to write equations for these transformations. We introduce Cartesian coordinates in such a way that $A = (-1, 0)$, $B = (1, 0)$, $C = (1, 2)$. The endpoints of C are the points $M(0) = (0, 0)$, $M(1) = (1, 1)$, and the formulas below give the coordinates (x', y') of the image of the point (x, y) and the slope m' of the image of a line with slope m for each of the two transformations:

$$F_0 \begin{cases} x' = \dfrac{x + y}{3} \\ y' = \dfrac{y}{3} \\ m' = \dfrac{m}{m + 1} \end{cases} \qquad F_1 \begin{cases} x' = \dfrac{x + 2}{3} \\ y' = \dfrac{x + y + 1}{3} \\ m' - m + 1 . \end{cases}$$

In particular, one sees that m transforms like the transformations in paragraph 4. As a consequence, the tangent to C at the point $M(t)$ has slope equal to $m(t)$ and one sees that the function of **Minkowski** plays a natural role in the study of this curve.

I will confine myself to mentioning a number of properties of this curve (which I have studied elsewhere).

Like the function $m(t)$, the coordinates $x(t)$ and $y(t)$ of $M(t)$ are strictly increasing functions of t whose derivatives are zero whenever they exist, i.e., almost everywhere. The same is true for the function $m(x) = dx/dy$ that gives the slope of the tangent as a function of the abscissa. It follows that *the curvature of C is zero almost everywhere*. But at points where the parameter t is equal to a binary fraction the curvature is infinite.

Finally, this curve also enjoys some arithmetic properties. At a point of C for which t is a binary fraction with denominator 2^k, the coordinates x and y are ternary fractions with common denominator 3^k, and the slope m of the tangent there is a rational number which has the property that the sum of the partial quotients of its regular continued fraction expansion is equal to k. Conversely, if m is rational t is a binary fraction and x and y are ternary fractions. At a point of C for which t is a rational number whose denominator is not a power of 2 then x and y are rational number whose common denominator is not a power of 3 and m is quadratic irrational. Conversely, if m is quadratic irrational then x, y, and t are rational.

6. We modify the construction of the polygons P_n considered in paragraph 5 in the following way: we take as vertices of P_{n+1} the points which divide the sides of P_n into three parts, the outer ones being equal and the ratio of the middle one to the others being a fixed number γ. One obtains a curve C_γ which reduces to C for $\gamma = 1$ and which once again satisfies equations of the form (1), where F_0 and F_1 are linear transformations.

For $\gamma < 1$, this curve has an infinite number of angular points corresponding to values of t which are binary fractions. The curvature to the left and the curvature to the right at these points is zero if $\gamma < \sqrt{2} - 1$ and is infinite if $\sqrt{2} - 1 < \gamma < 1$. For $\gamma > 1$, the curve C_γ has a tangent everywhere that varies in a continuous way and, at points for which t is a binary fraction, the curvature is infinite if $1 \leq \gamma < 1$ and zero if $\gamma > 2$. It follows that, for $\gamma \neq 2$, the curve C_γ contains no arc with all points having a curvature varying in a continuous way.

For $\gamma = 2$ this curve is an arc of a parabola. In fact, this follows from the well known property that all the polygons are then circumscribed by the parabola tangent to each of the sides of P_0 at its middle point.

Therefore, among the curves and functions defined by the equations [1] by taking F_0 and F_1 as suitable linear transformations, one obtains continuous functions that are nowhere differentiable singular functions, the curve of **von Koch**, Peano curves of **Cesaro** and **Polya**, and the curves C_γ. These are only examples. To my knowledge a systematic study of all the types of curve obtainable in this way remains to be done.

Bibliographical Note

The functions appearing as solutions to the functional equation of paragraph 2 were discovered in different ways by a number of authors. **Cesaro** (*Fonctions continues sans dérivée*, Archiv der Mathematik und Physik **10** (1906), p. 57–63) noted the fact that the derivative cannot exist without vanishing as well as the non existence of the derivative for certain complex values of a. Then **G. Faber** (*Ueber stetige Functionen II*, Mathematische Annalen **69** (1910), p. 372–443, especially p. 395–400) and R. Salem (*On some Singular Monotonie Functions which are strictly increasing*, Transactions of the American Mathematical Society **53** (1943), p. 427–439) defined the function for real a and proved its singular character as well as other properties. One can find much bibliographical information in: **P. Hartmann** and **R. Kershner**, *The Structure of Monotone Functions*, American Journal of Mathematics, **59** (1937), p. 809–822. The study of functional equations in paragraph 2 above is treated in the same way as in my article *Sur certaines équations fonctionelles*, Ecole Polytechnique de l'Université de Lausanne, published on the occasion of its centenary, (1953), p. 95–97.

There is a large literature on the **von Koch** curve and the curves of **Cesaro**. I will confine myself to citing: **H. von Koch**, *Sur une courbe continue sans tangente obtenue par construction géometrique élémentaire*, Arkiv för Math., Astronomi och Fysik **1** (1904), p. 681–702.; **E. Cesaro**, *Remarques sur la courbe de von Koch*, Atti della R. Accad. Sc. Fis. Mat. (Napoli), Series 2, **12** (1905); **G. Polya**, *Ueber eine Peanosche Kurve*, Bull. intern. Acad. Sc. Cracovie, Classe des Sciences mathématiques et naturelles, Series A (1913), p. 305–313.; **P. Lévy**, *Les courbes planes ou gauches et les*

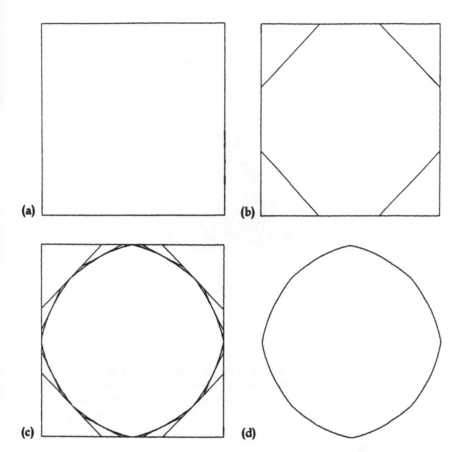

Figure 1. Cutting corners: (a) a polygon; (b) with the next stage; (c) the first five stages; (d) the limit.

surfaces composées de parties semblables au tout, Journal de l'Ecole Polytechnique (1939), p. 227–292.

Concerning the function of Minkowski: H. **Minkowski**, *Zur Geometrie der Zhalen,* Gesamelte Abhandlungen, vol. 2, p. 50–51; A. **Denjoy,** *Notice sur les travaux scientiphiques de M. Arnaud Denjoy,* Hermann, Paris 1934; and, by the same author: *Sur une fonction réele de Minkowski,* J. Math. pures et appl. (1938), p. 105–151, as well as a number of notes in the Comptes rendus de l'Acad. Sc. mentioned in the Notice.

I studied the curves C and C_γ in the article *Sur une courbe plane,* J. Math. pures et appl. **35** (1956), p. 25–42, where a proof of the properties stated here can be found. This is also done in a slightly different fashion in *Un peu de mathématiques à propos d'une courbe plane,* Elemente der Matematik II (1947), p. 73–76 and 89–97.

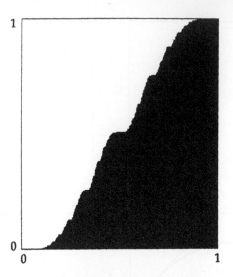

Figure 2.   Minkowski's '?' function.

The curve described here is related to many others that may be obtained by "cutting corners"; Figure 1. Begin with a polygon, trisect each of the edges, and "cut off" the corners by joining the last trisection point of each segment with the first trisection point of the next segment. Repeat, and observe convergence to a limiting curve. When we begin with a square, the result is *not* a circle. In fact, the limit curve (d) has tangent everywhere, but curvature zero almost everywhere. See [16].

Minkowski's '?' function is singular (has derivative zero almost everywhere) but strictly increasing. See Figure 2. –Ed.

Bibliography

[1] E. Cesàro, *Fonctions continues sans dérivée*, Archiv der Math. und Phys. **10** (1906), 57–63.

[2] G. Farber, *Über stetige Funktionen II*, Math. Annalen **69** (1910), 372–443.

[3] R. Salem, *On some singular monotonic functions which are strictly increasing*, Trans. Amer. Math. Soc. **53** (1943), 427–439.

[4] P. Hartmann and R. Kershner, *The structure of monotone functions*, Amer. J. Math. **59** (1937), 809–822.

[5] G. de Rham, *Sur certaine équations fonctionnelle*, École Polytechnique de l'Université de Lausanne; Ouvrage Publié à l'Occasion de son Centenaire, 1853–1953, Lausanne, 1953, pp. 95–97.

[6] H. von Koch, *Sur une courbe continue sans tangentes obtenus par une construction géométrique élémentaire*, Arkiv för Mat., Astron. och Fys. **1** (1904), 681–702.

[7] E. Cesàro, *Remarques sur la courbe de von Koch*, Atti della R. Accad. Sc. Fis. Mat. Napoli **12** (1905), 1–12.

[8] G. Polya, *Über eine Peanosche Kurve*, Bull. Intern. Acad. Sc. Cracovie (1913), 305–313.

[9] P. Lévy, *Les curbes planes ou gauches et les surfaces composées de parties semblables au tout*, J. l'École Poly. **81** (1939), 227–292.

[10] H. Minkowski, *Zur Geometrie der Zahlen*, Gesammelte Abhandlungen, vol. II, pp. 50–51.

[11] A. Denjoy, *Notice sur les travaux scientiphiques de M. Arnaud Denjoy*, Herman, Paris, 1934.

[12] A. Denjoy, *Sur une fonction réele de Minkowski*, J. Math. Pures et Appl. (1938), 105–151.

[13] G. de Rham, *Sur une courbe plan*, J. Math. Pures et Appl. **35** (1956), 25–42.

[14] G. de Rham, *Un peu des mathématiques à propos d'une courbe plane*, Elemente der Mathematik **2** (1947), 73–76, 89–97.

[15] R. Bott, *Georges de Rham, 1901–1990*, Notices Amer. Math. Soc. **38** (1991), 114–115.

[16] C. de Boor, *Cutting corners always works*, Computer Aided Geometric Design **4** (1987), 125–131.

[17] Georges de Rham, *Oeuvres Mathématiques*, L'Enseignement Mathématique, Geneva, 1981.

[18] S.C. Altheon, K.E. Schilling, and M.F. Wyneken, *Cutting Corners: A four-gon conclusion*, College Mathematics Journal **25** (1994), 266-279.

Here we will see a discussion of fractal dimension and its use for spaces other than subsets of Euclidean space. This selection is just the first 4 chapters of a longer paper. The remaining chapters describe the ε-entropy and ε-capacity for certain infinite-dimensional sets, namely spaces of functions.

Andrei Nikolaevich Kolmogorov (1903–1987) was one of the leading mathematicians of the century. He was born at Tambov, and graduated from Moscow University in 1925, a student of Luzin. He became famous at the age of 19 by constructing a summable function with everywhere divergent Fourier series. Kolmogorov established a mathematical formulation of probability theory in terms of measure theory (this is a solution to half of Hilbert's sixth problem—the other half asks for a mathematical formulation of mechanics). Besides probability, his work includes mechanics (KAM theory), turbulence, genetics, Fourier series, logic, entropy, cohomology, and complexity of algorithms. In his later years, he was interested in mathematics education. [30]

ε-Entropy and ε-Capacity of Sets in Functional Spaces (Excerpt)

A. N. Kolmogorov & V. M. Tihomirov

Introduction

This article is principally devoted to the systematic exposition of results published in the years 1954–1958 by K. I. Babenko [1], A. G.
Vituškin [2], [3], V. D. Erohin [4], A. N. Kolmogorov [5], [6], and V. M.
Tihomirov [7]. It is natural that certain new theorems were proved
while organizing the material and that certain examples were worked
through in more detail. Some results included in this survey that have
not been previously published, and which go beyond a mere system-
atization, belong to V. I. Arnol'd (§6) and V. M. Tihomirov (§4, 7, and 8).

One can acquaint one's self with the principal line of the investigations set forth here by reading §1, looking through the examples
of §2 and reading §3, which serves as an introduction to the following part of the article. The idea of the possibility of characterizing
the "massiveness" of sets in metric spaces by the help of the order
of growth of the number of elements of their most economical coverings as ε → 0 was developed in the work of L. S. Pontryagin and
L. G. Šnirel'man[1] [8] (appendix to the translation). Among the early
related (but not immediately adaptable for our ends) constructions,
one should mention the definition of measures of fractional orders
in the well-known work of F. Hausdorff [9]. Interest in this circle of

---

[1] *Transliterations here are those used by the American Mathematical Society in 1961. This*
Šnirel'man (Шнирельман) *is the same as the Schnirelmann on page 132. Nowadays the Am-*
erican Mathematical Society uses a different system, and this would be Shnirel'man.

ideas arose anew in Moscow when A. G. Vituškin [2] obtained a lower bound (in the notation of our §1) for the function $M_\varepsilon(A)$ for classes of functions of n variables with bounded partial derivatives up to a given order p, and applied this lower bound to the proof of the theorem of the inescapable decrease of smoothness in representing an arbitrary function of n variables by superposition of functions of $m < n$ variables. A. N. Kolmogorov [5] proved that the second part of the proof of the theorem of A. G. Vituškin, carried out by the author with the help of the theory of many-dimensional variation, can be carried out very simply with the help of an upper estimate of $M_\varepsilon(A)$ (see Appendix I of our article). Proceeding from this work, and using the currently popular general ideas of the theory of information, A. N. Kolmogorov [6] formulated a general program of investigation of ε-entropy and ε-capacity, interesting from the point of view of compacta in functional spaces. The connection between the investigations of all of the type set forth here with the probabilistic theory of information is at the present time only one of analogy and parallelism (see Appendix II). For example, the results of V. M. Tihomirov set forth in §8 were inspired by the "theorem of Kotel'nikov" from the theory of information.

There is no doubt that the results obtained here can be of interest in the non-probabilistic theory of information in the study of the necessary size of memory and the number of operations in computational algorithms (see the works of N. S. Bahvalov [10], [11] and A. G. Vituškin [12]). However, to obtain practically useful results the presently known estimates of the functions C_ε and $\mathcal{H}_\varepsilon$ would have to be significantly improved. Our §2 contains some as yet highly imperfect efforts in this direction. In the work of L. S. Pontryagin and L. S. Šnirel'man cited above, it was shown that the asymptotic behavior of M_ε and N_ε can lead also to topological invariants (that is, to the definition of topological dimension). In the note of A. N. Kolmogorov [13] it is shown how to obtain by an analogous route a certain definition of the linear dimension of topological vector spaces.

It seems to us that the connections of the investigations of the direction described with various parts of mathematics are both interesting and varied. We remark in particular that V.D. Erohin [14], in solving the problem of sharpening the estimates of the function C_ε and $\mathcal{H}_\varepsilon$ for certain classes of analytic functions, has discovered a new method of approximating analytic functions by linear forms $c_1\varphi_1(z) + \cdots + c_n\varphi_n(z)$, which has interesting properties, capable of being stated without the concepts of ε-entropy and ε-capacity.

§1. Definition and basic properties of the functions $\mathcal{H}_\varepsilon(A)$ and $\mathcal{C}_\varepsilon(A)$.

Let A be a nonvoid set in a metric space R. We introduce the following definitions:

Definition 1. A system γ of sets $U \subset R$ is called an ε-*covering* of the set A if the diameter $d(U)$ of an arbitrary $U \in \gamma$ does not exceed 2ε and

$$A \subseteq \bigcup_{U \in \gamma} U .$$

Definition 2. A set $U \subseteq R$ is called an ε-*net* for the set A if every point of the set A is at a distance not exceeding ε from some point of U.

Definition 3. A set $U \subset R$ is called ε-*separated* if every pair of distinct points of U are at a distance greater than ε from each other.

It will be clear later on why it was convenient in definition 1 to use 2ε instead of the usual ε. The number ε, unless the contrary is specifically stated, will be taken to be positive.

We shall always be concerned with *totally bounded sets*. It is useful to keep in mind the three equivalent definitions of such sets, expressed in the following theorem:

Theorem I. *The following three properties of the set A are equivalent and depend upon the metric of A itself (that is, they hold or do not hold when A with its given metric is imbedded in any superspace R):*

α) *For every ε there exists a finite ε-covering of the set A.*

β) *For every ε, there exists a finite ε-net for the set A.*

γ) *For every ε, every ε-separated set is finite.*

The proof of Theorem I can be left to the reader (see [15], pp. 312–320, where β) is taken as the definition). As is known, all *compacta* are totally bounded, and, under the assumption that the space is complete, a set A is totally bounded if and only if the closure of A in R is compact (see [15], p. 315).

For totally bounded set A, it is natural to introduce the following three functions, which in a certain sense characterizes the "massiveness" of the set A^*):

---

*If we wish to extend the definitions to the case when A is not totally bounded, then we must alter them somewhat. (See [6].) It is natural that for non-totally bounded sets A, all of the functions $\mathcal{N}_\varepsilon, \mathcal{N}_\varepsilon^R$, and $\mathcal{M}_\varepsilon$ assume infinite values for sufficiently small ε.

$\mathcal{N}_\varepsilon(A)$ is the minimal number of sets in an ε-covering of A;

$\mathcal{N}_\varepsilon^R(A)$ is the minimal number of points in an ε-net for the set A;

$\mathcal{M}_\varepsilon(A)$ is the maximal number of points in an ε-separated subset of the set A.

The functions $\mathcal{N}_\varepsilon(A)$ and $\mathcal{M}_\varepsilon(A)$, for a given metric on A, do not depend upon the choice of the containing space R (for $\mathcal{M}_\varepsilon$ it is evident from the definition itself, and for $\mathcal{N}_\varepsilon$ it is proved without difficulty). On the contrary, the function $\mathcal{N}_\varepsilon^R(A)$ in general depends upon R, a fact reflected in our notation.

We assign special notations for the logarithms to the base 2 of the functions define above:*

$\mathcal{H}_\varepsilon(A) = \log_2 \mathcal{N}_\varepsilon(A)$ is called the *minimal ε-entropy of the set A*, or simply the *ε-entropy* of the set A.

$\mathcal{H}_\varepsilon^R(A) = \log_2 \mathcal{N}_\varepsilon^R(A)$ is the *ε-entropy of the set A relative to R*.

$\mathcal{C}_\varepsilon(A) = \log_2 \mathcal{M}_\varepsilon(A)$ is the *ε-capacity of A*.**

These names are connected with designations from the theory of information. To explain them, the following approximation remarks suffice: a) in the theory of information, the unit of a "collection of information" is the amount of information in one binary sign (that is, designating whether it is 0 or 1); b) the "entropy" of a collection of possible "communications", undergoing preservation or transmission with a specified accuracy, is defined as the number of binary signs necessary to transmit an arbitrary one of these communications with the given accuracy (that is, an h such that to every communication x, one can assign a sequence $s(x)$ of h binary signs from which the communication x can be reconstructed with the needed accuracy); c) the "capacity" of a transmitting or remembering apparatus is defined as the number of binary signs that it can reliably transmit or remember.

If we regard A as a set of possible communications and suppose that by finding x' with the distance $\rho(x, x') \le \varepsilon$ permits us to reconstruct the communication x with sufficient accuracy, then it is evident that it suffices to demand $\mathcal{N}_\varepsilon^R(A)$ different signals to transmit any of the communications $x \in A$; as these signals, we may take sequences of binary signs of length not greater than $\mathcal{H}_\varepsilon^R(A) + 1$. To succeed with sequences of length less than $\mathcal{H}_\varepsilon^R(A)$ is obviously impossible.

On the other hand, if we consider A as a set of possible signals and suppose that two signals $x_1 \in A$ and $x_2 \in A$ are reliably different

---

*Since A is by hypothesis nonvoid, we always have $\mathcal{N}_\varepsilon(A) \ge 1$, $\mathcal{N}_\varepsilon^R(A) \ge 1$, $\mathcal{M}_\varepsilon(A) \ge 1$, and hence $\mathcal{H}_\varepsilon(A) > 0$, $\mathcal{H}_\varepsilon^R(A) \ge 0$, and $\mathcal{C}_\varepsilon(A) \ge 0$.

**$\log N$ will always denote the logarithm of the number N to the base 2.

in the case $\rho(x_1, x_2) > \varepsilon$, then it is evident that one can choose in A $\mathcal{M}_\varepsilon(A)$ reliably different signals and with their help fix for storage or transmissions any binary sequence of length $\mathcal{C}_\varepsilon(A) - 1$. It is obviously impossible to find a system of reliably different signals in A for the transmission of arbitrary binary sequences of length $\mathcal{C}_\varepsilon(A) + 1$.

The role of the function $\mathcal{H}_\varepsilon(A)$ is as follows: 1) in view of the inequality $\mathcal{H}_\varepsilon(A) \leq \mathcal{H}_\varepsilon^R(A)$ (see also Theorem IV), it serves to give a lower bound for $\mathcal{H}_\varepsilon^R(A)$; 2) in case the space is *centered* (see further definition 4), $\mathcal{H}_\varepsilon^R(A) = \mathcal{H}_\varepsilon(A)$, and $\mathcal{H}_\varepsilon(A)$ receives the direct meaning of entropy. Every totally bounded metric space A can be imbedded in a centered space R. Hence, according to an idea of A. G. Vituškin, one can consider $\mathcal{H}_\varepsilon(A)$ in a known sense of the word as the preferred or "absolute" entropy of A. The practical meaning of this assertion, however, is not totally clear, except for those cases when the actually demanded accuracy of reproduction of $x \in A$ is in fact comprised in the designation of x' with $\rho(x, x') \leq \varepsilon$ from a certain centered space R, as happens in the case of problems about the approximate specification of *real* functions $f(t)$ of an arbitrary argument t with definite uniform precision ε. For all this, see the sequel, in connection with Theorems VI, VII.

We shall state several simple theorems, giving basic general properties of the functions defined above.

Theorem II. *All six of the functions* $\mathcal{N}_\varepsilon(A)$, $\mathcal{N}_\varepsilon^R(A)$, $\mathcal{M}_\varepsilon(A)$, $\mathcal{H}_\varepsilon(A)$, $\mathcal{H}_\varepsilon^R(A)$, $\mathcal{C}_\varepsilon(A)$, *as functions of the set A are semiadditive, that is, if*

$$A \subseteq \bigcup_k A_k,$$

then

$$F(A) \leq \sum_k F(A_k).$$

The proof can be left to the reader. From the semiadditivity and nonnegativity of our functions, it follows that if $A' \subseteq A$, then for each of them the inequality $F(A') \leq F(A)$ holds.

Theorem III. *All six of the functions* $\mathcal{N}_\varepsilon(A)$, $\mathcal{N}_\varepsilon^R(A)$, $\mathcal{M}_\varepsilon(A)$, $\mathcal{H}_\varepsilon(A)$, $\mathcal{H}_\varepsilon^R(A)$, $\mathcal{C}_\varepsilon(A)$, *as functions of ε, are nonincreasing (for increasing ε), that is, they are nondecreasing as ε decreases. The functions* $\mathcal{N}_\varepsilon(A)$, $\mathcal{M}_\varepsilon(A)$, $\mathcal{H}_\varepsilon(A)$, $\mathcal{C}_\varepsilon(A)$ *are continuous on the right.*

The first half of the theorem follows at once from the definitions, so that carrying out the proof can be left to the reader.

It suffices to prove the second part of the theorem for the functions $M_\varepsilon(A)$. Let $M_\varepsilon(A) = n$ and let $x_1, x_2, \ldots, x_n$ be a corresponding ε_0-separated set. Then

$$\varepsilon_1 = \min_{i \neq j} \rho(x_i, x_j) > \varepsilon_0,$$

and for all ε within the limits $\varepsilon_0 \leq \varepsilon < \varepsilon_1$, we must have $M_\varepsilon(A) \geq n$; in view of monotonicity, this implies that $M_\varepsilon(A) = n$.

For $N_\varepsilon(A)$, the proof is somewhat more complicated and is carried out by contradiction. If we assume that there exists a sequence

$$\varepsilon_1 > \varepsilon_2 > \cdots > \varepsilon_k > \cdots \to \varepsilon_0,$$

for which $N_{\varepsilon_k}(A) = m < N_{\varepsilon_0}(A) = n$, then, for an arbitrary $k \geq 1$ there must exist an ε_k covering of A sets $A_{k1}, A_{k2}, \ldots, A_{km}$. The closure $\bar{A}$ of the set A in the completion R^* of the space R is compact.

In the metric "distance" $\alpha(F, F')$, the closed subsets of the compactum $\bar{A}$ form a compactum (see [16], 548–550).

We consider the set

$$F_{ki} = \bar{A}_{ki} \cap \bar{A}.$$

In view of what has been said, it is possible to choose a sequence $k_s \to \infty$ such that

$$F_{k_s i} \to F_i \quad (i = 1, 2, \cdots, m).$$

It is easy to verify that $F_1, \cdots, F_m$ form an ε_0-covering of the set A, that is, the assumption $n \geq m$ is contradicted.

Theorem IV. *For every totally bounded set A in the metric space R, the following inequalities hold:*

$$M_{2\varepsilon}(A) \leq N_\varepsilon(A) \leq N_\varepsilon^R(A) \leq N_\varepsilon^A(A) \leq M_\varepsilon(A), \tag{1}$$

and consequently

$$C_{2\varepsilon}(A) \leq H_\varepsilon(A) \leq H_\varepsilon^R(A) \leq H_\varepsilon^A(A) \leq C_\varepsilon(A). \tag{2}$$

We shall prove the inequalities (1) from right to left. Let $x_1, \ldots, x_{M_\varepsilon}$ be a maximal ε-separated set in A. It is then evidently an ε-net, since in the contrary case, there would be a point $x' \in A$ such that $\rho(x', x_i) > \varepsilon$; this last would contradict the maximality of $x_1, \ldots, x_{M_\varepsilon}$. In view of the fact that $x_i \in A$ by definition, we obtain

$$N_\varepsilon^A(A) \leq M_\varepsilon(A).$$

It is evident that every ε-net consisting of points $\in A$ is also an ε-net consisting of points of $R \supset A$, that is

$$\mathcal{N}_\varepsilon^R(A) \leq \mathcal{N}_\varepsilon^A(A).$$

Let $y_1, \cdots, y_n$ be an ε-net with respect to A in R. We consider the sets $S_\varepsilon(y_i) \cap A = U_i$, where $S_\varepsilon(y_i)$ denotes the sphere of radius ε and center at the point y_i. $\{U_i\}$ is an ε-covering of A, that is, every ε-net generates an ε-covering, for which we have

$$\mathcal{N}_\varepsilon(A) \leq \mathcal{N}_\varepsilon^R(A).$$

And, finally, for every ε-covering and every 2ε-separated set, the number of points in the latter does not exceed the number of sets in the former, since in the contrary case, two points at distance $> 2\varepsilon$ would lie in a single set of diameter $\leq 2\varepsilon$; from this we have

$$M_{2\varepsilon}(A) \leq \mathcal{N}_\varepsilon(A).$$

As has already been mentioned, the relations between the functions defined above becomes simpler in the case of a centered space.

Definition 4. A space R is said to be *centered*, if for every subset U of R of diameter $d = 2r$, there exists a point x_0 in R from which every point of U is at a distance not greater than r.

Theorem V. *In a centered space R, for every totally bounded set A, we have*

$$\mathcal{N}_\varepsilon^R(A) = \mathcal{N}_\varepsilon(A),$$

$$\mathcal{H}_\varepsilon^R(A) = \mathcal{H}_\varepsilon(A).$$

In fact, the inequality

$$\mathcal{N}_\varepsilon(A) \leq \mathcal{N}_\varepsilon^R(A)$$

is already stated in Theorem IV. In a centered space, to every ε-covering A by sets

$$U_1, \ldots, U_N$$

there exists an ε-net of the same number of points

$$x_0^{(1)}, \ldots, x_0^{(N)},$$

and hence

$$\mathcal{N}_\varepsilon^R(A) \leq \mathcal{N}_\varepsilon(A).$$

Theorem VI. *The space D^X of real functions on an arbitrary set X with the metric*

$$\rho(f, g) = \sup_{x \in X} |f(x) - g(x)|$$

is centered.

Let $U \subset D^X$ be a set of diameter

$$d = \sup_{f, g \in U} \sup_{x \in X} |f(x) - g(x)|.$$

Setting

$$\bar{f}(x) = \sup_{f \in U} f(x), \quad \underline{f}(x) = \inf_{x \in U} f(x),$$

one can see that

$$d = \sup_{x \in X} (\bar{f}(x) - \underline{f}(x)).$$

It is now easy to see that for the function

$$f_0(x) = \frac{1}{2}(\bar{f}(x) + \underline{f}(x))$$

and for an arbitrary function $f \in U$, we have

$$|f(x) - f_0(x)| \leq \frac{d}{2},$$

for all $x \in X$. This means that

$$\rho(f, f_0) \leq \frac{d}{2}, \quad f \in U,$$

as we wished to show.

Theorem VII. (A. G. Vituškin [3]). *Any totally bounded space A can be imbedded in a centered space R.*

In view of the theorem of Mazur–Banach ([17], ch. IX), an arbitrary A (we are concerned only with totally bounded, that is, in any event separable A) can be isometrically imbedded in the space C, and consequently in the space D^I which contains C, where I is the unit interval $[0, 1]$. This proves the theorem.

By Theorem VI, the space D^I is centered, and by Theorem V, for any A imbedded in it, we have

$$\mathcal{N}_\varepsilon^{D^I}(A) = \mathcal{N}_\varepsilon(A).$$

In view of the inequality

$$\mathcal{N}_\varepsilon(A) \le \mathcal{N}_\varepsilon^R(A)$$

(Theorem IV), and Theorem VII, we have

$$\mathcal{N}_\varepsilon(A) = \min_{A \subseteq R} \mathcal{N}_\varepsilon^R(A), \quad \mathcal{H}_\varepsilon(A) = \min_{A \subseteq R} \mathcal{H}_\varepsilon^R(A),$$

which justifies the designation of $\mathcal{H}_\varepsilon(A)$ as the *minimal ε-entropy*.

§2. Examples of exact computation of the functions $\mathcal{H}_\varepsilon(A)$ and $\mathcal{C}_\varepsilon(A)$ and estimates of them in certain simple cases

1. A is the interval Δ: $\{a \le x \le b\}$ of length $|\Delta| = b - a$ on the line D with metric $\rho(x, x') = |x - x'|$.
   In this case we have:*

$$\mathcal{N}_\varepsilon(A) = \mathcal{N}_\varepsilon^D(A) = \mathcal{M}_{2\varepsilon}(A) = \left\{ \begin{array}{ll} \frac{|\Delta|}{2\varepsilon} & \text{for } \frac{|\Delta|}{2\varepsilon} \text{ integral} \\ \left[\frac{|\Delta|}{2\varepsilon}\right] + 1 & \text{for } \frac{|\Delta|}{2\varepsilon} \text{ nonintegral,} \end{array} \right\} \quad (3)$$

that is,

$$\left. \begin{array}{ll} \mathcal{N}_\varepsilon(A) = \dfrac{|\Delta|}{2\varepsilon} + O(1) & \mathcal{M}_\varepsilon(A) = \dfrac{|\Delta|}{\varepsilon} + O(1), \\[2mm] \mathcal{H}_\varepsilon(A) = \log \dfrac{|\Delta|}{2\varepsilon} + O(\varepsilon) & \mathcal{C}_\varepsilon(A) = \log \dfrac{|\Delta|}{\varepsilon} + O(\varepsilon). \end{array} \right\} \quad (4)$$

In fact, it is easy to see that Δ is *centered* in the sense of §1, and hence, in accordance with Theorem V, we have

$$\mathcal{N}_\varepsilon^D(\Delta) = \mathcal{N}_\varepsilon(\Delta). \quad (5)$$

The number of sets of an ε-covering of Δ cannot be less than $\left[\frac{|\Delta|}{2\varepsilon}\right]$ or equals $\left[\frac{|\Delta|}{2\varepsilon}\right]$, in case $\frac{|\Delta|}{2\varepsilon}$ is not an integer, since in this case the union of these sets would have measure not larger than $2\varepsilon \left[\frac{|\Delta|}{2\varepsilon}\right] < \Delta$. However, the reader will easily convince himself that one can always construct an ε-covering of $\left[\frac{|\Delta|}{2\varepsilon}\right] + 1$ sets, where, if $\frac{|\Delta|}{2\varepsilon}$ is an integer, $\frac{|\Delta|}{2\varepsilon}$ sets will suffice (fig. 1).

---

*The sign $[A]$ means the integral part of the number A.

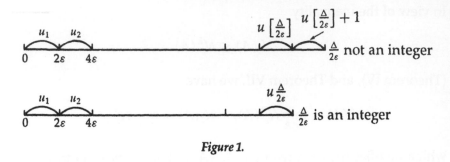

Figure 1.

On the other hand, if $\frac{|\Delta|}{2\varepsilon}$ is not an integer, then, if we divide Δ into $\left[\frac{|\Delta|}{2\varepsilon}\right]$ equal parts by the points $a = x_1, x_2, \ldots, x_{\left[\frac{|\Delta|}{2\varepsilon}\right]+1} = b$, we obtain $\left[\frac{|\Delta|}{2\varepsilon}\right] + 1$ points, which form a 2ε-separated set. If $\frac{|\Delta|}{2\varepsilon}$ is an integer, we divide Δ in a similar way into $\frac{|\Delta|}{2\varepsilon} - 1$ equal parts and obtain $\frac{|\Delta|}{2\varepsilon}$ points that form a 2ε-separated set. From what we have said it follows that

$$\mathcal{M}_{2\varepsilon}(\Delta) \geq \mathcal{N}_\varepsilon(\Delta),$$

from which, upon using (5) and the inequality (1), we obtain (3).

2. A is the set $F_1^\Delta(L)$ of the function $f(x)$, defined on the interval $\Delta = [a, b]$, satisfying the Lipschitz condition

$$|f(x) - f(x')| \leq L\,|x - x'|$$

and vanishing at the point a.

We consider the given set in the space D^Δ, that is, with the metric

$$\rho(f(x), g(x)) = \sup_{x \in \Delta} |f(x) - g(x)|.$$

We shall show that

$$\mathcal{H}_\varepsilon(A) = \mathcal{C}_{2\varepsilon}(A) = \mathcal{H}_\varepsilon^{D^\Delta}(A) = \begin{cases} \dfrac{|\Delta|L}{\varepsilon} - 1 & \text{for } \dfrac{|\Delta|L}{\varepsilon} \text{ an integer,} \\[2mm] \left[\dfrac{|\Delta|L}{\varepsilon}\right] & \text{for } \dfrac{|\Delta|L}{\varepsilon} \text{ a noninteger,} \end{cases} \tag{6}$$

that is

$$\mathcal{H}_\varepsilon(A) = \frac{|\Delta|L}{\varepsilon} + O(1), \\[2mm] \mathcal{C}_\varepsilon(A) = \frac{2|\Delta|L}{\varepsilon} + O(1). \tag{7}$$

We note first of all that by the change of the independent variable $t = L(x - a)$, this set is isometrically and one to-one mapped onto the set $F_1^{\Delta'}(1)$, that is, onto the set of functions defined on the interval $[0, \Delta']$, where $\Delta' = |\Delta|L$, satisfying the Lipschitz condition with constant equal to one, and equal to zero at zero. It is clear that under this transformation, the quantities $\mathcal{H}_\varepsilon$ and $\mathcal{C}_\varepsilon$ do not change. In the last-named set, we shall compute the functions $\mathcal{H}_\varepsilon$ and $\mathcal{C}_\varepsilon$.

The idea of our constructions consists in constructing an ε-covering of $F_1^{\Delta'}(1)$ and a 2ε-separated set in $F_1^{\Delta'}(1)$, which consist of the same number of elements, K_ε. Then, in agreement with the definitions of the function $\mathcal{N}_\varepsilon$ and $\mathcal{M}_\varepsilon$, we obtain

$$\left.\begin{aligned} \mathcal{N}_\varepsilon(F_1^{\Delta'}(1)) \leq K_\varepsilon, \\ \mathcal{M}_{2\varepsilon}(F_1^{\Delta}(1)) \geq K_\varepsilon, \end{aligned}\right\} \tag{8}$$

for which, using the inequality (1), we obtain, just as in para. 1 of this section, the relations

$$\mathcal{M}_{2\varepsilon}(F_1^{\Delta'}(1)) = \mathcal{N}_\varepsilon(F_1^{\Delta'}(1)) = K_\varepsilon. \tag{9}$$

In addition to the equalities (9), we also have

$$\mathcal{N}_\varepsilon(F_1^{\Delta'}(1)) = \mathcal{N}_\varepsilon^{D^{\Delta'}}(F_1^{\Delta'}(1))$$

in view of the fact that the space $D^{\Delta'}$ is centered. Let $\varepsilon > 0$ and be such that $\frac{\Delta'}{\varepsilon}$ is not an integer. The number $\left[\frac{\Delta'}{\varepsilon}\right]$ will be denoted by n, and the quantities $k \cdot \varepsilon$ by t_k ($k = 1, 2, \ldots, n$). We divide the interval $[0, \Delta']$ into $n + 1$ intervals:

$$\Delta_k = [(k-1)\varepsilon, k\varepsilon] \quad (k = 1, 2, \ldots, n), \quad \Delta_{n+1} = \left[\left[\frac{\Delta'}{\varepsilon}\right]\varepsilon, \Delta'\right].$$

Let $\varphi(t)$ denote the function defined on the interval $[\varepsilon, \Delta']$, equal to ε at the point $t = \varepsilon$, and on the intervals Δ_k linear with *angular coefficient* equal to $+1$ or -1.

The set of points (t, u) of the plane such that

$$-t \leq u \leq t, \quad t \in \Delta_1,$$

$$\varphi(t) - 2\varepsilon \leq u \leq \varphi(t), \quad t \in [\varepsilon, \Delta'],$$

will be called an ε-*corridor* (fig. 2) and will be denoted by $K(\varphi)$.

Let K_ε be the number of all corridors. It is easy to see that under our choice of ε, K_ε is equal to $2^{\left[\frac{\Delta'}{\varepsilon}\right]}$.

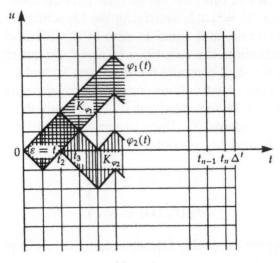

Figure 2.

We shall show that the collection of all ε-corridors forms an ε covering of the set $F_1^{\Delta'}(1)$. Furthermore, we say that a function $f(t)$ *belongs* to a certain corridor $K(\varphi)$ if

$$\varphi(t) - 2\varepsilon \leq f(t) \leq \varphi(t).$$

We obtain a proof of this assertion by induction.

In fact, on the interval $\Delta_1 = [0, \varepsilon]$, all functions of our set belong to *all* ε-corridors. Suppose now, as an inductive hypothesis, that for every function $f(t)$ in our set on the interval $[0, t_k]$ ($k \leq n$), there exists an ε-corridor $K(\varphi)$ such that

$$\varphi(t) - 2\varepsilon \leq f(t) \leq \varphi(t), \quad 0 \leq t \leq t_k.$$

Then, the Lipschitz condition implies (fig. 3) that $f(t_{k+1})$ belongs either to the interval

$$\delta_1 = [\varphi(t_k) - \varepsilon, \varphi(t_k) + \varepsilon],$$

or to the interval

$$\delta_2 = [\varphi(t_k) - 3\varepsilon, \varphi(t_k) - \varepsilon],$$

and on the interval $[0, t_{k+1}]$ the function $f(t)$ belongs to a corridor $K(\varphi)$, where φ is the function coinciding with $\varphi(t)$ on $[0, t_k]$ and on $\Delta_k = [t_k, t_{k+1}]$ linear with angular coefficients $+1$ if $f(t_{k+1}) \in \delta_1$ and angular coefficient -1 if $f(t_{k+1}) \in \delta_2$. The induction is finished. We

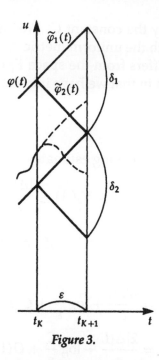

Figure 3.

must now construct a 2ε-separated set consisting of K_ε elements. We divide $[0, \Delta']$ into n equal intervals $\bar{\Delta}_1, \ldots, \bar{\Delta}_n$. The length of each interval $|\bar{\Delta}_i| = \frac{\Delta'}{n} > \varepsilon$ (we recall that $n = [\frac{\Delta'}{\varepsilon}]$). We consider the set M_n of functions that vanish at 0, and on the intervals $\bar{\Delta}_i$ are linear with angular coefficients equal to ± 1. Two distinct functions in M_n differ from each other at least by $2\frac{\Delta'}{n}$, that is, they are 2ε-separated, and their number is $2^n = K_\varepsilon$.

If $\frac{\Delta'}{\varepsilon}$ is an integer, then the quantity K_ε, equal to the number of all ε-corridors, turns out to be equal to $2^{\frac{\Delta'}{\varepsilon}-1}$. The set M_n, with $n = \frac{\Delta'}{\varepsilon} - 1$, analogous to that considered above, is, as is not hard to show a 2ε-separated set consisting of $2^n = K_\varepsilon$ functions.

Thus we have proved that

$$\mathcal{N}_\varepsilon(A) = \mathcal{M}_{2\varepsilon}(A) = \mathcal{N}_\varepsilon^{D^\Delta}(A) = K_\varepsilon = \begin{cases} 2^{\frac{\Delta'}{\varepsilon}-1} & \text{if } \frac{\Delta'}{\varepsilon} \text{ is an integer,} \\ 2^{\left[\frac{\Delta'}{\varepsilon}\right]} & \text{if } \frac{\Delta'}{\varepsilon} \text{ is a noninteger,} \end{cases}$$

from which (6) follows by taking logarithms, since $\Delta' = \Delta L$.

3. Now let A be the set $F_1^\Delta(C, L)$ of functions $f(t)$ defined on the interval $\Delta = [a, b]$, satisfying the Lipschitz condition with the constant

L, and bounded on Δ by the constant C. We shall also consider this set as a metric space with the uniform metric Δ.

The set $F_1^\Delta(C, L)$ differs from the space $F_1^\Delta(L)$ that we studied in the preceding paragraph in that we have introduced the condition

$$|f(t)| \leq C, \quad f \in \Delta,$$

instead of $f(0) = 0$.

We shall obtain the following estimates

$$\frac{|\Delta| L}{\varepsilon} + \log\frac{C}{\varepsilon} - 3 \leq C_{2\varepsilon}(A) \leq \mathcal{H}_\varepsilon(A) \leq \frac{|\Delta| L}{\varepsilon} + \log\frac{C}{\varepsilon} + 3$$

$$\text{if} \quad \varepsilon \leq \min\left(\frac{C}{4}, \frac{C^2}{16\Delta'}\right), \text{ (10)}$$

that is

$$\left. \begin{aligned} \mathcal{H}_\varepsilon(A) &= \frac{|\Delta| L}{\varepsilon} + \log\frac{C}{\varepsilon} + O(1), \\ C_\varepsilon(A) &= \frac{2|\Delta| L}{\varepsilon} + \log\frac{C}{\varepsilon} + O(1). \end{aligned} \right\} \tag{11}$$

We see that here for $\mathcal{H}_\varepsilon$ and C_ε we obtain upper and lower estimates, which for ε not too large, differ from each other only by a few units. Results of this sort can yet be regarded as totally satisfactory from the point of view of practical applications.

We proceed to the proof of the inequalities (10).

As in the preceding paragraph, we reduce at once to the space $F_1^{\Delta'}(C, 1)$, where $\Delta' = [0, |\Delta| L]$.

We obtain an upper estimated for $\mathcal{H}_\varepsilon$ thanks to the use of the results of the preceding paragraph. In fact, having given $\varepsilon > 0$, we consider the set of points of the plane (t, u) with co-ordinates $(-\varepsilon, 2k\varepsilon)|k|$ $\leq \left[\frac{C}{2\varepsilon}\right]$ (fig. 4). At each of these points, we construct a set of ε-corridors just as we did for zero in the previous paragraph. The set of ε-corridors constructed at the point $(-\varepsilon, 2k\varepsilon)$ is, in accordance with the foregoing, an ε-covering for the set of functions $f(t) \in F_1^{\Delta'}(C, 1)$ for which $2k\varepsilon - \varepsilon \leq f(0) \leq 2k\varepsilon + \varepsilon$, and consequently the set of all ε-corridors is an ε-covering for $F_1^\Delta(C, 1)$. In this way, we obtain

$$\mathcal{N}_\varepsilon(F_1^{\Delta'}(C, 1)) \leq \left(2\left[\frac{C}{2\varepsilon}\right] + 1\right) \mathcal{N}_\varepsilon(F_1^{\Delta'+\varepsilon}(1)),$$

from which, upon taking logarithms and using (6), we obtain

$$\mathcal{H}_\varepsilon(A) \leq \frac{\Delta'}{\varepsilon} + \log\left(\frac{C}{\varepsilon} + 1\right) + 1 \leq \frac{|\Delta'|}{\varepsilon} + \log C\varepsilon + 2, \tag{12}$$

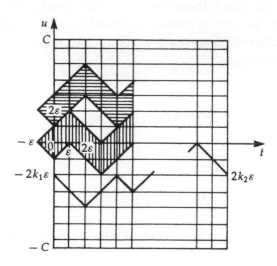

Figure 4.

since for $\varepsilon \le C$, $\log\left(\frac{C}{\varepsilon} + 1\right) \le \log C\varepsilon + 1$.

It is somewhat more complicated to obtain the upper estimate for $C_{2\varepsilon}$. Let $2r$ denote the maximal *even* integer such that $2r\varepsilon < \Delta'$. We divide $[0, \Delta']$ into $2r$ equal intervals $\Delta_1, \Delta_2, \ldots, \Delta_{2r}$. We construct the set $\bar{M}_{2r}$ of functions assuming one of the values $2k\varepsilon$, $|k| \le \left[\frac{C}{2\varepsilon}\right]$, at zero, and linear with angular coefficient ± 1 on the intervals Δ_i $(i = 1, 2, \ldots, 2r)$. The functions in $\bar{M}_{2r}$ are 2ε-separated and belong to $F_1^\Delta(1)$. It remains for us to estimate the number of these functions that belong to $F_1^{\Delta'}(C, 1)$, which in absolute value do not exceed C. Furthermore, we shall establish this estimate by elementary methods. From the point of view of the theory of probability, the problem reduces to the estimate of the number of paths (corresponding to the functions in $\bar{M}_{2r}$ constructed above) which do not pass beyond the given limits under the ordinary random walk with probability $\frac{1}{2}$ at each step to go up or go down, and we could obtain the needed estimate much more quickly, if we used inequalities known in the theory of probability.

We carry out the required estimation in several steps.

α) Let $N(k_1, k_2)$ be the number of functions $\varphi(t) \in \bar{M}_{2r}$ such that

$$\varphi(0) = 2k_1\varepsilon, \quad \varphi_2(\Delta') = 2k_2\varepsilon,$$

where k_1 and k_2 are any admissible integers; suppose for the sake of definiteness that $k_2 \ge k_1$; then $N(k_1, k_2)$ is equal to the number of functions $\varphi(t)$ for which there are n_1 angular coefficients equal to $+1$ (the number of "ascents"), minus the number n_2 of coefficients equal to -1 (the number of "descents"), equal to $2(k_2 - 1k_1)$.

From this we have

$$n_1 + n_2 = 2r,$$

$$n_1 - n_2 = 2(k_2 - k_1),$$

that is, $n_1 = r + k_2 - k_1$, and the number of functions with n_1 ascents is equal to $C_{2r}^{n_1}$. In this way,

$$N(k_1 k_2) = C_{2r}^{k_2 - k_1 + r}. \tag{13}$$

β) We consider the set $U_1(k)$ of functions $\bar{M}_{2r}$ that satisfy the conditions

$$\varphi(0) \le 2k\varepsilon \quad \varphi(\Delta') > 2k\varepsilon.$$

The number $N_1(k)$ of such functions satisfies the relation

$$N_1(k) = \sum_{\substack{k_1 \le k \\ k_2 > k}} N(k_1 k_2) \le \sum_{\substack{k_1 \le k \\ k_2 > k}} C_{2r}^{k_2 - k_1 + r}. \tag{14}$$

In (14), the term C_{2r}^{s+r} occurs s times, and thus

$$N_1(k) \le \sum_{0 \le s \le r} s\, C_{2r}^{s+r}. \tag{15}$$

γ) Let $U_2(k)$ denote the set of functions in $\bar{M}_{2r}$ such that

$$\varphi(0) < 2k\varepsilon$$

$$\max_{t \in [0, \Delta']} \varphi(t) > 2k\varepsilon.$$

Suppose that $\varphi(t) \in U_2(k)$ but does not belong to $U_1(k)$. We consider the last of the points s_ε at which $\varphi(s_\varepsilon) = (k+1)\varepsilon$. We construct a new function $\tilde{\varphi}(t)$ which coincides with $\varphi(t)$ up to the point s_ε and from there on is its mirror image in the line $u = k + 1$. It is obvious that $\tilde{\varphi}(t) \in U_1(k)$.

The functions $\varphi(t) \in U_1(k) \cap U_2(k)$ will be put in correspondence with themselves. We have obtained a mapping of the set $U_2(k)$ into $U_1(k)$ under which, as is easy to see, the inverse image of each function in $U_1(k)$ consists of not more than *two* functions, that is, the number of functions $N_2(k)$ in $U_2(k)$ is no larger than $2N_1(k)$.

It is easy to understand that the number of functions $\in \bar{M}_{2r}$ for which

$$|\varphi(0)| \leq 2k\varepsilon, \quad \max_{t \in [0,\Delta']} |\varphi(t)| > 2k\varepsilon, \tag{16}$$

does not exceed $2N_2(k) \leq 4N_1(k)$.

It remains only to remark that the set of functions $\in \bar{M}_{2r}$ but *not belonging* to (16) for $k = \left[\frac{C}{2\varepsilon}\right]$ is the set, the number of functions in which we originally had to estimate.

Thus

$$\mathcal{M}_{2\varepsilon}(F_1^{\Delta'}(C,1)) \geq \left(1\left[\frac{C}{2\varepsilon}\right] + 1\right) 2^{2r} - 4N_1(k) \geq$$

$$\geq \left(2\left[\frac{C}{\varepsilon}\right] + 1\right) 2^{2r} - 4\sum_{s=0}^{r} sC_{2r}^{s+r}. \tag{17}$$

We now make an estimate of the right side of (17).

We have

$$\sum_{s=0}^{r} sC_{2r}^{s+r} = \left(2rC_{2r-1}^r - \frac{rC_{2r}^r}{2}\right) = \frac{r}{2}C_{2r}^r. \tag{18}$$

The proof of the simple equality (18) is left to the reader.

We also use the elementary inequality

$$C_{2r}^r < \frac{1}{\sqrt{2r+1}} 2^{2r} \tag{19}$$

(which is found, for example, in [18] and is proved without difficulty).

Thus, from (17) (using (18) and (19)), we obtain:

$$\mathcal{M}_{2\varepsilon}(F_1^{\Delta'}(C,1)) \geq 2^{2r}\left(2\left[\frac{C}{2\varepsilon}\right] + 1 - \frac{2r}{\sqrt{2r+1}}\right) \geq 2^{2r}\left(\frac{C}{\varepsilon} - 1 - \sqrt{2r}\right). \tag{20}$$

Taking logarithms in (20) and using the fact that $\Delta' - 2\varepsilon \leq 2r\varepsilon \leq \Delta'$, we obtain

$$C_{2\varepsilon}(A) \geq \frac{\Delta'}{\varepsilon} - 2 + \log\frac{C}{\varepsilon} + \log\left(1 - \frac{\varepsilon}{C} - \sqrt{\frac{\varepsilon\Delta'}{C^2}}\right) \geq$$

$$\geq \frac{\Delta'}{\varepsilon} + \log\frac{C}{\varepsilon} - 3 \quad \text{for } \varepsilon \leq \min\left(\frac{C}{4}, \frac{C^2}{16\Delta'}\right),$$

which we wished to prove.

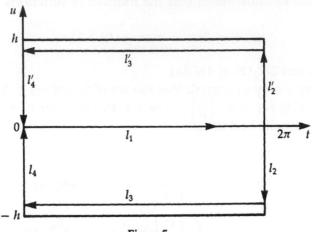

Figure 5.

4. Let A be the set $A_h(C)$ of functions $f(x)$, defined on D, periodic with period equal to 2π, and *analytic* in the strip of width h: $|\operatorname{Im} z| \leq h$, $z = x + iy$, and bounded there by the constant C. We consider the given set with the uniform metric on D. For this set, the following formulas hold:

$$\left.\begin{aligned} \mathcal{H}_\varepsilon(A) &= 2\frac{\left(\log\frac{1}{\varepsilon}\right)^2}{h\log e} + O\left(\log\frac{1}{\varepsilon}\log\log\frac{1}{\varepsilon}\right), \\ C_\varepsilon(A) &= 2\frac{\left(\log\frac{1}{\varepsilon}\right)^2}{h\log e} + O\left(\log\frac{1}{\varepsilon}\log\log\frac{1}{\varepsilon}\right). \end{aligned}\right\} \tag{21}$$

We obtain the relations (21) just as we did in the preceding paragraphs, making upper estimates of the number of sets forming an ε-covering of $A_h(C)$, and estimating below the number of 2ε-separated function in $A_h(C)$.

To obtain the upper estimate of the quantity $\mathcal{H}_\varepsilon(A)$, we expand the function $f(x) \in A_h(C)$ in its Fourier series:

$$f(z) = \sum_{h=-\infty}^{+\infty} c_k e^{ikx}. \tag{22}$$

We shall use the following inequalities for the coefficients in (22):

$$|c_k| \leq C e^{-|k|h}. \tag{23}$$

The inequalities (23) are obtained in the following way. Let $k > 0$; we integrate $f(z)e^{ikz}$ around the contour sketched in fig. 5. In view of the periodicity of $f(z)$, we have

$$\int_{l_2} f(\xi)e^{ik\xi}\,d\xi = -\int_{l_4} f(\xi)e^{ik\xi}\,d\xi ,$$

from which by Cauchy's theorem

$$|c_k| = \left| \frac{1}{2\pi} \int_0^{2\pi} f(x)e^{ikx}\,dx \right|$$

$$= \left| \frac{1}{2\pi} \int_0^{2\pi} f(x - ih')e^{ikx}e^{-kh'}\,dx \right| \leq Ce^{-|k|h'} \leq Ce^{-|k|h} ,$$

since h' is an arbitrary number less than h. The case $k < 0$ is analogous.

Using (23), we rewrite (22) in the form

$$\left.
\begin{aligned}
f(x) &= \sum_{|k|\leq n} c_k e^{ikx} + R_n(x), \\
|R_n(x)| &\leq \left| \sum_{|k|>n} c_k e^{ikx} \right| \leq 2Ce^{-nh}\frac{e^{-h}}{1-e^{-h}} .
\end{aligned}
\right\}
\tag{24}$$

Suppose that $\varepsilon > 0$. We choose the least n such that $|R_n(x)| < \frac{\varepsilon}{2}$. We obtain from (24) that

$$n = \frac{\log \frac{1}{\varepsilon}}{h \log e} + O(1) .
\tag{25}$$

We next approximate each coefficient $c_k = \alpha_k + i\beta_k$ by a quantity $c'_k = \alpha'_k + i\beta'_k$ to within $\frac{\varepsilon}{2(2n+1)}$ (absolute value), setting for this purpose

$$\left.
\begin{aligned}
\alpha'_k &= \left[\frac{2\sqrt{2}(2n+1)\alpha_k}{\varepsilon} \right] \frac{\varepsilon}{2\sqrt{2}(2n+1)} = m_k^1 \frac{\varepsilon}{2\sqrt{2}(2n+1)} , \\
\beta'_k &= \left[\frac{2\sqrt{2}(2n+1)\beta_k}{\varepsilon} \right] \frac{\varepsilon}{2\sqrt{2}(2n+1)} = m_k^2 \frac{\varepsilon}{2\sqrt{2}(2n+1)} .
\end{aligned}
\right\}
\tag{26}$$

In doing this, we obviously obtain that

$$\left| f(x) - \sum_{|k|\leq n} c'_k e^{ikx} \right| \leq \varepsilon .
\tag{27}$$

It is clear from (27) that if two functions f and g correspond to one and the same collection $\{c'_k\}$, then $\|f - g\| < 2\varepsilon$, that is, the set of elements in an ε-covering can be approximated by the number of all collections $\{c'_k\}$ "induced" by the space $A_h(C)$.

In agreement with (26), each set of the 2ε-covering is defined by the following matrix consisting of integers:

$$U = \left\| \begin{array}{ccccc} m^1_{-n} & \cdots & m^1_0 & \cdots & m^1_n \\ m^2_{-n} & \cdots & m^2_0 & \cdots & m^2_n \end{array} \right\| .$$

For this, using (23), we obtain inequalities for m^i_k:

$$|m^i_k| \leq \frac{2\sqrt{2}C(2n+1)e^{-|k|h}}{\varepsilon} = N_k \quad (i = 1, 2), \tag{28}$$

from which it follows that

$$\log N_k = \log \frac{1}{\varepsilon} - |k|h \log e + \log n + D, \tag{29}$$

where D is a bounded constant, not depending upon k. It follows from (28) that the number of all possible matrices U does not exceed

$$N = \prod_{k=-n}^{+n} N^2_k$$

and hence, recalling (25), we obtain

$$\mathcal{H}_\varepsilon(A) \leq \log N = 2 \sum_{|k| \leq n} \log N_k \leq 4n \log \frac{1}{\varepsilon} - 2n^2 h \log e + O(n \log n)$$

$$= 2\frac{(\log \frac{1}{\varepsilon})^2}{h \log e} + O\left(\log \frac{1}{\varepsilon} \log \log \frac{1}{\varepsilon} \right) .$$

We now derive our lower estimate for $C_{2\varepsilon}(A_h(C))$. For this we use two facts:

α) For every $h' > h$, the validity of the inequalities

$$|c_k| \leq C \frac{1 - e^{-(h'-h)}}{1 + e^{-(h'-h)}} e^{-|k|h'} \quad (k = 0, \pm 1, \pm 2, \ldots) \tag{30}$$

implies that the functions

$$f(z) = \sum_{k=-\infty}^{+\infty} c_k e^{ikz} \tag{31}$$

belongs to the space $A_h(C)$, since then the series (31) converges uniformly in the strip of width h and

$$|f(z)| \leq \sum_{k=-\infty}^{+\infty} |c_k| e^{|k|h} \leq C \frac{1 - e^{(h'-h)}}{1 + e^{-(h'-h)}} \sum_{k=-\infty}^{+\infty} e^{-|k|(h'-h)} = C,$$

$$|\operatorname{Im} z| < h.$$

$\beta)$

$$\|f\| = \max_{x \in D} |f(x)| \geq \max_k |c_k|, \tag{32}$$

where c_k are the coefficients of (31). The last fact follows from the fact that

$$|c_k| = \left| \frac{1}{2\pi} \int_0^{2\pi} f(x) e^{ikx} dx \right| \leq \max_{x \in D} |f(x)|.$$

Choosing $\varepsilon > 0$ and setting $h' = h \left(1 + \frac{1}{\log \frac{1}{\varepsilon}} \right)$, we choose the largest m such that

$$C \frac{1 - e^{-(h'-h)}}{1 + e^{-(h'-h)}} e^{-mh'} \geq 2\sqrt{2}\varepsilon,$$

from which we obtain that

$$m = \frac{\log \frac{1}{\varepsilon}}{h \log e} + O \left(\log \log \frac{1}{\varepsilon} \right). \tag{33}$$

We now construct a set $\{\Phi\}$ of polynomials of degree m:

$$\left. \begin{aligned} \varphi(z) &= \sum_{|k| \leq m} c_k e^{ikz}, \\ c_k &= (s_k^1 + i s_k^2) 2\varepsilon, \end{aligned} \right\} \tag{34}$$

where s_k^i ($i = 1, 2$) are arbitrary integers such that

$$|s_k^i| \leq \left[C \frac{1 - e^{-(h'-h)}}{1 + e^{-(h'-h)}} \cdot \frac{e^{-|k|h'}}{2\sqrt{2}\varepsilon} \right] = M_k. \tag{35}$$

The inequalities (35) guarantee the validity of the inequalities (30), so that $\{\Phi\} \subset A_h(C)$. Furthermore, two different polynomials φ_1 and φ_2 from $\{\Phi\}$ have the property that for at least one k, $|k| \leq m$,

$$|c_k^1 - c_k^2| \geq 2\varepsilon,$$

that is, all of the polynomials in $\{\Phi\}$ are 2ε-separated.

In agreement with (34), every polynomial in $\{\Phi\}$ is defined by an integer matrix:

$$
U = \left\| \begin{array}{ccccc} s'_{-m} & \cdots & s'_0 & \cdots & s'_m \\ s^2_{-m} & \cdots & s^2_0 & \cdots & s^2_m \end{array} \right\|,
$$

and the number of all matrices U is not less than the number

$$
M = \prod_{|k| \le m} M_k^2 .
$$

It also follows from (35) that

$$
\log M_k = \log \frac{1}{\varepsilon} - h|k| \log e + O\left(\log\log \frac{1}{\varepsilon} \right)
$$

uniformly in k, from which we obtain, using (33), that

$$
C_{2\varepsilon}(A) \ge \log M \ge 2\log \sum_{|k| \le m} \log M_k \ge 4m \log \frac{1}{\varepsilon} - 2m^2 h \log e +
$$

$$
+ O\left(\log \frac{1}{\varepsilon} \log\log \frac{1}{\varepsilon} \right) = 2 \frac{\left(\log \frac{1}{\varepsilon} \right)^2}{h \log e} + O\left(\log \frac{1}{\varepsilon} \log\log \frac{1}{\varepsilon} \right) .
$$

This completes the proof of the formulas (21).

§3. Typical orders of growth of the function $\mathcal{H}_\varepsilon$ and C_ε

The functions to be considered from now on, $f(\varepsilon)$, will usually be positive and defined for all ε in the limits $0 < \varepsilon < \varepsilon_0$. Since we shall constantly be studying their limit behavior as $\varepsilon \to 0$, we shall often omit the sign $\varepsilon \to 0$. To describe the limit behavior of functions

$f(\varepsilon)$, besides the usual signs O and o, we shall also use the following notation:

$$f \sim g, \quad \text{if } \lim \frac{f}{g} = 1,$$

$$\left.\begin{array}{l} f \gtrsim g \\ g \lesssim f \end{array}\right\}, \quad \text{if } \overline{\lim} \ \frac{f}{g} \leq 1,$$

$$f \asymp g, \quad \text{if } f = O(g) \text{ and } g = O(f),$$

$$\left.\begin{array}{l} f \succeq g \\ g \preceq f \end{array}\right\}, \quad \text{if } f = O(g),$$

$$\left.\begin{array}{l} f \gg g \\ g \ll f \end{array}\right\}, \quad \text{if } f = o(g).$$

The relation $f \sim g$ is called *strong equivalence*, and the relation $f \asymp g$ is called *weak equivalence* of the functions f and g.

The examples considered in §2 give us models of three typical orders of growth of the functions $\mathcal{H}_\varepsilon$ and $\mathcal{C}_\varepsilon$ as $\varepsilon \to 0$.

1. *If A is a bounded set in n-dimensional Euclidean space D^n having interior points, then*

$$M_\varepsilon(A) \asymp N_\varepsilon(A) \asymp \left(\frac{1}{\varepsilon}\right)^n. \tag{36}$$

The formula (36) is valid in any *n-dimensional Banach space* (see §4). From it we infer that

$$\mathcal{C}_\varepsilon(A) \sim \mathcal{H}_\varepsilon(A) \sim h \log \frac{1}{\varepsilon}. \tag{37}$$

Since, in view of (37), the behavior of $\mathcal{H}_\varepsilon$ and $\mathcal{C}_\varepsilon$ is defined principally by the dimension n of the space D, the idea naturally arises to define the *lower* and the *upper metric dimension* for an arbitrary totally bounded set A by the formulas

$$\underline{\mathrm{dm}}\,(A) = \underline{\lim} \frac{\mathcal{H}_\varepsilon(A)}{\log \frac{1}{\varepsilon}}, \tag{38}$$

$$\overline{\mathrm{dm}}\,(F) = \overline{\lim} \frac{\mathcal{H}_\varepsilon(A)}{\log \frac{1}{\varepsilon}}, \tag{39}$$

One infers without difficulty from Theorem IV of §1 that the formulas

$$\underline{dm}(A) = \varliminf \frac{C_\varepsilon(A)}{\log \frac{1}{\varepsilon}},$$

$$\overline{dm}(A) = \varlimsup \frac{C_\varepsilon(A)}{\log \frac{1}{\varepsilon}},$$

define just the same quantities $\underline{dm}(A)$ and $\overline{dm}(A)$. If

$$\underline{dm}(A) = \overline{dm}(A) = dm(A),$$

then $dm(A)$ is called simply the *metric dimension* of the set A. It is clear that in our case of a totally bounded set A in D^n, having interior points, we have

$$dm(A) = n.$$

It is easy to prove (see Theorem XII in §4) that for a convex, infinite dimensional set A lying in a Banach space, the metric dimension is always equal to $+\infty$, that is, the order of growth of the functions $\mathcal{H}_\varepsilon$ and C_ε exceeds $\log \frac{1}{\varepsilon}$:

$$\mathcal{H}_\varepsilon(A) \gg \log \frac{1}{\varepsilon}, \quad C_\varepsilon(A) \gg \log \frac{1}{\varepsilon}.$$

Metric dimension is useless for distinguishing the massiveness of sets of this sort.

II. Among infinite dimensional sets of least massiveness that are important for analysis are sets of totally bounded *analytic functions*, such as the set $A_h(C)$ considered in example 3 of § 2. Here a typical order of growth of the functions $\mathcal{H}_\varepsilon$ can C_ε is their growth according to the law

$$C_\varepsilon \asymp \mathcal{H}_\varepsilon \asymp \left(\log \frac{1}{\varepsilon} \right)^s \tag{40}$$

with a certain finite constant $r > 1$, or interval orders of growth of such a form as

$$\frac{\left(\log \frac{1}{\varepsilon} \right)^2}{\log \log \frac{1}{\varepsilon}} \tag{41}$$

(the last function grows more slowly than $\left[\log \frac{1}{\varepsilon}\right]^2$, but more quickly than $\left[\log \frac{1}{\varepsilon}\right]^{2-\delta}$ for an arbitrary $\delta > 0$). The simplest numerical characteristic of such orders of growth is*

$$\mathrm{df} = \lim \frac{\log \mathcal{H}_s}{\log \log \frac{1}{\varepsilon}}.$$

In §7 we shall see that in many important and fairly general cases of sets of analytic functions, $\mathrm{df}(A)$ coincides with the number of independent variables, or, geometrically, for classes of functions $f(P)$ of the point P of a complex manifold with the (complex) dimension of this manifold. Somewhat provisionally, we call $\mathrm{df}(A)$ the *functional dimension* of the set A.

III. The examples of para. 2 and 3 of §2 show that the growth of $\mathcal{H}_\varepsilon$ and C_ε for functions subject only to a Lipschitz condition is significantly faster than the orders of growth set down above for classes of analytic functions. It turns out that the orders of growth

$$C_\varepsilon \asymp \mathcal{H}_\varepsilon \asymp \left(\frac{1}{\varepsilon}\right)^q \tag{42}$$

with finite positive q are typical for classes of functions *differentiable a certain number of times*, say p, *and having derivatives of highest order p that satisfy Hölder's condition of order α.* Here, for functions of n variables we obtain the formula

$$C_\varepsilon \asymp \mathcal{H}_\varepsilon \asymp \left(\frac{1}{\varepsilon}\right)^{\frac{n}{p+\alpha}} \tag{43}$$

(see §5). Crudely speaking, formula (43) can be explained as follows: The massiveness of the set of p times differentiable functions of n variables, for which the p-th derivatives satisfy Hölder's condition α, is defined by the ratio

$$q = \frac{n}{p + \alpha} \tag{44}$$

of the number of independent variables n to the "exponent of smoothness" $q + \alpha$. For the derivation of the important relations (43), (44), see Appendix I.

---

*The reader will himself without difficulty write down the definition of the lower functional dimension $\underline{\mathrm{df}}(A)$ and upper functional dimension $\overline{\mathrm{df}}(A)$. The use of the function C_ε instead of $\mathcal{H}_\varepsilon$ leads to just the same df, $\underline{\mathrm{df}}$, and $\overline{\mathrm{df}}$, in view of Theorem IV of §1.

As a general definition of the *metric order* of a completely bounded set A, we use the formula*

$$q(A) = \lim \frac{\log \mathcal{H}_\varepsilon(A)}{\log \frac{1}{\varepsilon}} . \tag{45}$$

IV. We obtain essentially larger orders of growth of $\mathcal{H}_\varepsilon$ and $\mathcal{C}_\varepsilon$ upon considering functionals of various degrees of smoothness, defined on infinite dimensional totally bounded sets (see §9).

§4. ε-entropy and ε-capacity in finite dimensional spaces

We shall consider n-dimensional affine space as realized in the form of co-ordinate n-dimensional space D^n with points

$$x = (x_1, \ldots, x_n),$$

whose co-ordinates $x_1, \ldots, x_n$ are real numbers, although the relations of interest to us are for the most part not connected with the choice of co-ordinates. A certain amount of auxiliary apparatus is connected with the choice of a co-ordinate system, for example, the special norm

$$\|x\|_0 = \max_{1 \le k \le n} |x_k| .$$

The unit sphere in this norm, that is the set S_0 of points $x \in D^n$ with

$$\|x\|_0 \le 1,$$

is called the *unit cube*. Upon multiplying the unit cube by $b > 0$ and translating by a, we obtain the cube

$$S_0(a, b) = bS_0 - a ;$$

$2b$ is the *diameter* of the cube $S_0(a, b)$.

The set of points $x^{(i)}$ whose co-ordinates have the form $d(m_1^i, \ldots, m_n^i)$, where m_k^i ($k = 1, 2, \ldots, n$) run through all possible integers, will be called the *cubical lattice* of diameter d.

---

*Here it is also natural to introduce, in a completely understood fashion, the concept of lower metric order $\underline{q}(A)$ and of upper metric order $\overline{q}(A)$. The use of the function $\mathcal{C}_\varepsilon$ instead of $\mathcal{H}_\varepsilon$ here does not change the values of $q(A)$, $\underline{q}(A)$, $\overline{q}(A)$.

As is known, a point a is called an *interior point* of a set $A \subseteq D^n$, if A contains a certain cube $S_0(a, b)$ with center a. Sets consisting solely of interior points are called *open*, and define the usual topology in D^n.

A set $A \subseteq D^n$ is called convex if $x \in A$, $y \in A$, $\alpha + \beta = 1$, $\alpha \geq 0$, $\beta \geq 0$, imply that $\alpha x + \beta y \in A$. A closed convex set containing at least one interior point will be called a *convex body*.

Any convex body symmetric with respect to the origin of co-ordinates can be chosen as the unit sphere. The norm $\|x\|_S$ corresponding to the convex symmetric body S is defined by the formula

$$\|x\| = \inf_{\substack{x \\ b \in X}} b .$$

On the other hand, with respect to the norm $\|x\|$ the unit sphere which generates it is defined by the formula

$$\|x\| \leq 1 .$$

Thus we define in D^n an arbitrary *Banach metric*. The space D^n, with the metric defined by a symmetric convex body S will be denoted by D_S^n. In the cases

$$\|x\| = \left(\sum_h |x_k|^p \right)^{\frac{1}{p}} , \quad p \geq 1 ,$$

the unit spheres will be denoted by S_p^n, and the corresponding spaces D_p^n. The space D_2^n is ordinary *Euclidean* (or "Cartesian") co-ordinate space.

It is generally known that an arbitrary Banach metric defines one and the same topology in D^n. Also the concepts of *content* and *measure* do not essentially depend upon the choice of metric. In the standard theory of Lebesgue measure $\mu^n(A)$, the measure of $\frac{1}{2}S_0$ is taken equal to 1. If we take the content of $\frac{1}{2}S$ in an arbitrary other metric as the unit of measure, then the usual way of constructing Lebesgue measure leads to the measure

$$\mu_S^n(A) = \frac{\mu^n(A)}{\mu^n\left(\frac{1}{2}S\right)} ,$$

which differs from μ^n only by a constant factor.

A set A is Jordan measurable if it Lesbegue measurable and its boundary has measure zero. We shall call the measure $\mu^n(A)$ of such a set its "content" (if the measure of the boundary is not equal to zero, then the set does not have a content).

It is not hard to see that the question of finding minimal ε-nets in the space D_S^n is based on the problem of finding "maximally economical" coverings of subsets of the space D^n by translates of εS, and the problem of finding maximal 2ε-separated sets leads to the problem of most successful placement of non-intersecting translates of εS with centers on the given set.

There is a wide literature on the question under consideration. The reader will find a fairly complete bibliography in the book of Toth [19].*

In the present section, we shall touch on only the most general and simple results concerning estimates of the functions $\mathcal{H}_\varepsilon(A)$ and $C_\varepsilon(A)$ for $A \subset D_S^n$.

Theorem VIII. *For an arbitrary bounded set $A \subset D_S^n$ there exists a constant α such that for $\varepsilon \leq \varepsilon_1$, we have*

$$\mathcal{N}_\varepsilon(A) \leq \alpha \left(\frac{1}{\varepsilon}\right)^n. \tag{46}$$

If A contains an interior point in D^n, then there exists a constant β such that for $\varepsilon \leq \varepsilon_2$, we have

$$M_{2\varepsilon}(A) \geq 3 \left(\frac{1}{\delta}\right)^n. \tag{47}$$

In the notation of §3, Theorem VIII means that for a bounded set in D_S^n with an interior point, we have

$$\mathcal{N}_\varepsilon(A) \asymp M_\varepsilon(A) \asymp \left(\frac{1}{\varepsilon}\right)^n. \tag{48}$$

We have mentioned this theorem in §3, see (36).

Theorem IX. *There exist constants $\theta(n, S)$ and $r = r(n, S)$ for D_S^n such that for every set A with content $\mu^n(A) > 0$, we have*

$$\left.\begin{array}{l} \mathcal{N}_\varepsilon(A) \sim \theta \dfrac{\mu^n(A)}{\mu^n(S)} \left(\dfrac{1}{\varepsilon}\right)^n, \\[4mm] M_{2\varepsilon}(A) \sim \tau \dfrac{\mu^n(A)}{\mu^n(S)} \left(\dfrac{1}{\varepsilon}\right)^n. \end{array}\right\} \tag{49}$$

It is natural to call the constant θ the *"minimal density of covering D^n by the body S",* and the constant r the *maximal density of placements* of the body S in D^n.

---

*The reader will find much of value in the remarks of the editor, collected by I.M. Yaglom.

For θ and r these inequalities are evident:

$$r \le 1 \le \theta.$$

Proof of Theorem VIII. We choose constant b and B such that simultaneously we have

$$bS_0 \subset S \subset BS_0, \quad A \subset BS_0.$$

Let $\varepsilon > 0$. For a covering of the cube BS_0, we need no more than

$$\left(\frac{B}{sb} + 2\right)^n \tag{50}$$

cubes $\varepsilon b S_0 + a^{(i)} = S_0(a^{(i)}, \varepsilon b)$, where $a^{(i)}$ has co-ordinates $2b\varepsilon(m_1^{(i)}, m_2^{(i)}, \ldots, m_n^{(i)})$, and the $m_j^{(i)}$ are integers of absolute value less than $\left[\frac{B}{2\varepsilon b}\right]$ + 1. Describing the bodies $\varepsilon S + a^{(i)}$ around the cubes $\varepsilon b S_0 + a^{(i)}$, we obtain an ε-covering of A, and thus

$$\mathcal{N}_\varepsilon(A) \le \left(\frac{B}{sb} + 2\right)^n,$$

from which (46) follows.

Suppose further that A has an interior point a. We choose a c such that the cube $S_0(a, c) \subset A$. The cube $S_0(a, c)$ contains more than $2\left[\frac{c}{2\varepsilon B}\right]^n$ cubes $S_0(a^{(i)}, \varepsilon B)$, where the $a^{(i)}$ are vectors $2\varepsilon B(m_1^{(i)}, \ldots, m_n^{(i)})$, and the $m_j^{(i)}$ are integers not exceeding $\left[\frac{c}{2\varepsilon B}\right]$ in absolute value. The set of all $a^{(i)}$ is a 2ε-separated, from which (47) follows.

Proof of Theorem IX. We denote*

$$\varlimsup_{\varepsilon \to 0} \mathcal{N}_\varepsilon(S_0)\mu^n(S) \left(\frac{\varepsilon}{2}\right)^n = \varlimsup_{\varepsilon \to 0} \mathcal{N}_\varepsilon(S_0)\frac{\mu^n(S)}{\mu^n(S_0)} \varepsilon^n$$

by θ. We shall show that in fact, we have

$$\lim_{\varepsilon \to 0} \mathcal{N}_\varepsilon(S_0)\mu^n(S) \left(\frac{\varepsilon}{2}\right)^n = 0.$$

Let $\delta > 0$. We choose ε_1 such that for $\varepsilon \le \varepsilon_1$

$$\theta - \delta \le \mathcal{N}_\varepsilon(S_0)\mu^n(S) \left(\frac{\varepsilon}{2}\right)^n,$$

---

*We recall S_0 is the cube $\|x\|_0 \le 1$; its measure is equal to 2^n.

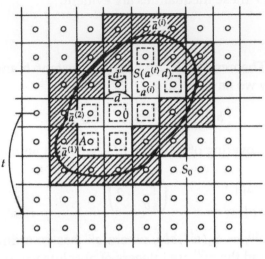

Figure 6.

while

$$\mathcal{N}_{\varepsilon_1}(S_0)\mu^n(S)\left(\frac{\varepsilon_1}{2}\right)^n \le \theta + \frac{\delta}{2}\,.$$

The cube S_0 can be covered by not more than

$$\left(\frac{1}{\lambda}+2\right)^n$$

cubes $S_0(a^{(i)},\lambda)$, for an arbitrary $0 < \lambda < 1$, see (50), and each of the latter is covered by $\mathcal{N}_{\varepsilon_1}$ bodies $\varepsilon_1\lambda S$.

Thus

$$\mathcal{N}_{\varepsilon_1\lambda}(S_0) \le \left(\frac{1}{\lambda}+2\right)^n \mathcal{N}_{\varepsilon_1}(S_0)\,,$$

that is

$$\mathcal{N}_{\varepsilon_1\lambda}(S_0)\mu^n(S)\left(\frac{\varepsilon_1}{2}\right)^n \cdot \lambda^n \le \lambda^n \left(\frac{1}{\lambda}+2\right)^n \mathcal{N}_{\varepsilon_1}(S_0)\left(\frac{\varepsilon_1}{2}\right)^n \mu^n(S)\,.$$

Choosing λ_1 such that for $\lambda \le \lambda_1$

$$\lambda^n \left(\frac{1}{\lambda}+2\right)^n \left(\theta+\frac{\delta}{2}\right) \le \theta + \delta\,,$$

we find that for all $\lambda \le \lambda_1$

$$\theta - \delta \le \mathcal{N}_{\varepsilon_1\lambda}(S_0)\mu^n(S)\left(\frac{\varepsilon_1\lambda}{2}\right)^n \le \theta+\delta\,, \tag{51}$$

that is,

$$\lim_{\varepsilon \to 0} \mathcal{N}_\varepsilon(S_0)\mu^n(S)\left(\frac{\varepsilon}{2}\right)^n = \theta.$$

Suppose further that A is an arbitrary body of content $\mu^n(A) > 0$, and that Π_{2d} is the cubical lattice of diameter $2d$. Being given $\delta' > 0$, we choose d so that the sum of the measures of the cubes $S_0(\bar{a}^{(i)}, d)$ $(i = 1, 2, \ldots, m)$, $\bar{a}^{(i)} \in \Pi_{2d}$, having points in common *with the boundary of A*, is less than δ'. Let $a^{(i)}$ $(i = 1, 2, \ldots, k)$ be the centers of the cubes $S(a^{(i)}, d)$ each of which *lies entirely within A*. Here we obviously have

$$k \geq \frac{\mu^n(A) - \delta'}{(2d)^n}. \tag{52}$$

Choosing an arbitrary $d' < d$, we consider the sets of cubes $S_0(a^{(i)}, d')$ (fig. 6).

In agreement with (51), for $\varepsilon \leq \varepsilon_1$, we have

$$\theta - \delta \leq \mathcal{N}_{\varepsilon d'}(d'S_0)\mu^n(S)\left(\frac{1}{2}\right)^n \leq \theta + \delta.$$

We choose ε_2 so small that $2\varepsilon d' < d - d'$ for $\varepsilon \leq \varepsilon_2$. Then evidently

$$\mathcal{N}_{\varepsilon d'}(A) \geq k\mathcal{N}_{sd'}(d'S_0),$$

since the coverings of contiguous cubes do not intersect, and for $\varepsilon \leq \varepsilon_2$, we have

$$\mathcal{N}_{\varepsilon d'}(A)\left(\frac{\varepsilon}{2}\right)^n \mu^n(S) \geq \frac{\mu^n(A) - \delta'}{(2d)^n}(\theta - \delta).$$

From this we have

$$\varliminf_{s \to 0} \mathcal{N}_\varepsilon(A)\varepsilon^n\mu^n(S) \geq (\mu^n(A) - \delta')(\theta - \delta) \geq \theta\mu^n(A)$$

in view of the arbitrariness of δ', δ, and $d' < d$. On the other hand, we also have

$$\mathcal{N}_{2d}(A) \leq (k + m)\mathcal{N}_{2d}(dS_0) \leq \frac{\mu^n(A) + \delta'}{(2d)^n}(\theta + \delta),$$

that is,

$$\varlimsup_{\varepsilon \to 0} \varepsilon^n\mu^n(S)\mathcal{N}_\varepsilon(A) \leq \theta\mu^n(A), \tag{53}$$

which we wished to prove.

We leave it to the reader to carry out for himself the proof of the theorem for $M_{2\varepsilon}$.

Besides Theorems VIII and IX, we point out the following extremely crude inequality, which gives upper and lower lower bounds on θ and r.

Theorem X. *For an arbitrary D_S^n, we have*

$$\frac{\theta}{r} \leq 2^n , \tag{54}$$

from which

$$\frac{1}{2^n} \leq r \leq 1 \leq \theta \leq 2^n . \tag{55}$$

Choosing $\delta > 0$, we then select ε_0 such that simultaneously, for a certain A of content $\mu^n(A) > 0$, the following inequalities hold for $\varepsilon \leq \varepsilon_0$:

$$\varepsilon^n \mu^n(S) \mathcal{N}_\varepsilon(A) \geq (\theta - \delta)\mu^n(A) ,$$

$$\varepsilon^n \mu^n(S) \mathcal{M}_{2\varepsilon}(A) \leq (\tau + \delta)\mu^n(A) ;$$

using also the inequality $\mathcal{N}_\varepsilon(A) \leq \mathcal{M}_\varepsilon(A)$ from Theorem IV of §1, we obtain

$$\theta - \delta \leq (r + \delta)2^n ,$$

from which Theorem X follows.

We point out, that while the fact of the *existence* of the constants θ and r (Theorem IX) is proved extremely simply, their calculation and even finding estimates for them for various concrete spaces not infrequently gives rise to very difficult problems.

A trivial case is the case of the space S_∞^n, since in this space the unit sphere is the unit cube S_0 and the entire space can be decomposed without intersections into unit cubes in such a way that the distance between centers of the cubes is equal to the diameter of the cubes (the space is "decomposable"). In every decomposable space, as is easy to see, we have

$$\theta(n, S) = r(n, S) = 1 . \tag{56}$$

We give two more examples of decomposable spaces; the space D_1^2 and the space D_S^2, where S is a regular hexagon. In the first space, the sphere is a square, and in the second a hexagon, by means of which the plane can be paved. (We note that we encountered a similar phenomenon of "decomposability" in an infinite dimensional space–the space $F_1^\wedge(L)$; see §1, para. 2).

The quantities θ and r have been computed in the case of the space D_2^2 [19]. It turns out that the best covering of the plane by disks is formed in the following way. The plane is covered by regular hexagons, and then, around each of these, a circle is described. The best placement of disks in the plane so that they do not intersect each other is obtained by inscribing a circle in each of the hexagons.

It is not hard to compute that this is the case

$$\tau = \frac{\pi}{\sqrt{12}} = 0,9069\ldots,$$

$$\theta = \frac{2\pi}{\sqrt{27}} = 1,2092\ldots.$$

Even in the case of three-dimensional space, the problem of finding the densest packing of sphere and the related problem of computing the constants r and θ for the space D^3 are not yet completely solved. Much more are the values of r_n and θ_n (as we denote θ and r for D_2^n) not known. Estimates for θ_n, the most precise up to the present time, were obtained by Davenport [20] and Watson [21].

Before stating their results, we make a definition. Let R be a lattice in n-dimensional space such that the spheres $S_i = S(a^{(i)}, 1)$, $a^{(i)} \in R$, cover all of D^n. Let $\sum_l$ denote summation over the values of i for which $a^{(i)} \in R \cap lS$. The quantity

$$\lim_{l \to \infty} \frac{\sum_l \mu^n(S_i)}{\mu^n(lS)}$$

under the assumption that the limit exists, is naturally called the density of the covering of the space D^n by the body S with centers on R. We write this quantity as $\theta_R(n, S)$. It is easy to see that the quantity $\theta_R(n, S)$ is an upper estimate for $\theta(n, S)$:

$$\theta(n, S) \le \theta_R(n, S). \tag{57}$$

Davenport and Watson constructed a lattice R in D^n such that

$$\theta_R(n, S_2^n) \le (1,07)^n, \tag{58}$$

and consequently

$$\theta_n \le (1,07)^n. \tag{59}$$

The theorems of Davenport and Watson are based on delicate constructions in n-dimensional space; we shall not reproduce them, referring the interested reader to the corresponding literature.

We can derive estimates of the quantity r_n form an older work of Blichfeldt [22].

The following simple lemma is basic for Blichfeldt's work.

Lemma I. (Blichfeldt). *Suppose that there are given m points in Euclidean space each at distance at least 2 from all the others. Then the sum of the squares of their distances from an arbitrary point of space is not less than $2(m-1)$.*

Proof. It is evident that it suffices to consider the case when the given point of space is zero. Choosing an orthonormal basis, we denote the co-ordinates of the i-th point by $(x_1^{(i)}, \ldots, x_n^{(i)})$. By hypothesis

$$\sum_{k=1}^{n} (x_k^{(i)} - x_k^{(j)})^2 \geq 4 . \tag{60}$$

Adding (60) for $i \neq j$, we obtain

$$m \sum_{i=1}^{m} ((x_1^{(i)})^2 + \cdots + (x_n^{(i)})^2) - (x_1^{(1)} + \cdots + x_1^{(m)})^2$$

$$- (x_n^{(1)} + \cdots + x_n^{(m)})^2 \geq 2m(m-1) ,$$

from which it follows that

$$\sum_{n=1}^{n} \sum_{i=1}^{m} (x_k^i)^2 \geq 2(m-1) ,$$

which we wised to show.

It follows at once from the lemma just proved that if $n+1$ points of Euclidean space have mutual distances at least 2, then the radius of a sphere containing all of these points is not less than $\sqrt{\frac{2n}{n+1}}$.

Suppose that we are given an arbitrary $\sqrt{\frac{2n}{n+1}} \cdot \varepsilon$-net. In every sphere, there are not more than $n+1$ points at distance 2ε.

Hence, for every body A, it is easy to obtain that

$$\mathcal{M}_{2\varepsilon}(A) \leq (n+1)\mathcal{N}_{\sqrt{\frac{2n}{n+1}}\varepsilon}(A) .$$

It follows quickly from (59) and (55) that

$$\left(\frac{1}{2}\right)^n \leq \tau_n \leq (n+1)\theta_n \left(\frac{n+1}{2n}\right)^{\frac{n}{2}}$$

$$\leq (n+1)\sqrt{e}\left(\frac{1,07}{\sqrt{2}}\right)^n \leq (n+1)\sqrt{e}\left(\frac{11}{14}\right)^n . \tag{61}$$

To conclude this section, we shall show that its main results do not admit useful generalizations to infinite dimensional Banach spaces. Bounded sets in infinite dimensional spaces do not need to be totally bounded (in contradistinction to the case of D^n). But even for totally bounded sets in an infinite dimensional space, it is impossible to obtain a single universal upper estimate for the growth of $\mathcal{H}_\varepsilon$ and $\mathcal{C}_\varepsilon$, which will be valid for all totally bounded sets in the given infinite dimensional E.(In D^n, as we saw, these orders are always $\preceq \left(\frac{1}{\varepsilon}\right)^n$.)This is evident from the following theorem.

Theorem XI. Let $\varphi(\varepsilon)$ be monotone decreasing as ε increases and have limit $+\infty$ as $\varepsilon \to 0$, and let E be an infinite dimensional Banach space. Then there is a compactum K in E for which

$$\mathcal{N}_\varepsilon(K) \geq \varphi(\varepsilon), \quad \mathcal{M}_\varepsilon(K) \geq \varphi(\varepsilon).$$

In view of Theorem IV of §1, it plainly suffices to prove Theorem XI for $\mathcal{M}_\varepsilon$. We find in E a sequence of n-dimensional linear subspaces

$$E_1 \subset E_2 \subset \cdots \subset E_n \subset \cdots .$$

We shall show that there exists a sequence of points

$$x_n \in E_n$$

with the properties

$$\|x_n\| = 1,$$

$$\rho(x_n, E_{n-1}) = 1.$$

For this we choose a point y_n in E_n that does not belong to E_{n-1}. In E_{n-1} we find (this is possible, see [23], p. 16) one of the "closest" points z_n to y_n, that is, a point z_n for which

$$\|y_n - z_n\| = \rho(y_n, E_{n-1}) = \rho_n,$$

and set

$$x_n = \frac{1}{\rho_n}(y_n - z_n).$$

It is easy to see that x_n has the required properties.

We now consider the function $\psi(n)$ inverse to $\varphi(\varepsilon)$. We form the set K from the point 0 and the points

$$\xi_n = 2\psi(n)x_n.$$

It is clear that K is a compactum, since

$$\|\xi_n\| = 2\psi(n) \to 0 \quad \text{as } n \to \infty.$$

For arbitrary ε and for

$$n = [\varphi(\varepsilon)]$$

we obtain

$$\psi(n) \geq \varepsilon,$$

from which it follows that the set of points

$$0, \xi_1, \ldots, \xi_n$$

is ε-separated in E. Since the number of its elements

$$n + 1 \geq \varphi(\varepsilon),$$

we find that

$$\mathcal{M}_\varepsilon(K) \succeq \varphi(\varepsilon).$$

A set A lying in a Banach space E is called n-dimensional if it lies in an n-dimensional linear subspace, but does not lie in a linear subspace of smaller dimension. In view of Theorem VIII, a convex n-dimensional bounded set in an arbitrary Banach space has metric dimension n, and furthermore satisfies the relations

$$\mathcal{N}_\varepsilon(A) \asymp \mathcal{M}_\varepsilon(A) \asymp \left(\frac{1}{\varepsilon}\right)^n.$$

A set A in a Banach space is called infinite dimensional if it lies in no finite dimensional linear subspace. The following theorem holds.

Theorem XII. *If a set A in a Banach space is infinite dimensional and convex, then*

$$\mathcal{N}_\varepsilon(A) \ggg \left(\frac{1}{\varepsilon}\right)^n, \quad \mathcal{M}_\varepsilon(A) \ggg \left(\frac{1}{\varepsilon}\right)^n,$$

for arbitrary n.

Proof. For arbitrary $n' > n$ and A, there exist n' linearly independent points. The simplex spanned by these points lies entirely in A, since A is convex, and for such a simplex, Theorem VIII shows that the function $\mathcal{N}_\varepsilon$ and $\mathcal{M}_\varepsilon$ have orders of growth $\left(\frac{1}{\varepsilon}\right)^{n'}$.

In equation (38) and (39) the selection defines the **upper** and **lower metric dimensions**. In fact, the upper dimension coincides with certain other fractal dimensions, at least in Euclidean space: for example the dimensions of Selection 7 (Bouligand), Selection 8 (Pontrjagin and Schnirelmann), and Selection 15 (Besicovitch and Taylor). Twelve equivalent formulations (for subsets of $\mathbb{R}$) are discussed in [32].

The Hausdorff dimension of a set A is $\leq$ both of the metric dimensions. Indeed, the set A can be covered by $\mathcal{N}_\varepsilon(A)$ sets of diameter $\leq 2\varepsilon$, so the s-dimensional Hausdorff measure Λ^s satisfies

$$\Lambda_{2\varepsilon}^s(A) \leq \mathcal{N}_\varepsilon(A)\,(2\varepsilon)^s.$$

If $s > \underline{\mathrm{dm}}\,(A)$, then for some arbitrarily small values of ε, we have

$$\frac{\mathcal{H}_\varepsilon(A)}{\log(1/\varepsilon)} < s,$$

so that $\mathcal{N}_\varepsilon(A) < 1/\varepsilon^s$. So for such ε we have

$$\Lambda_{2\varepsilon}^s(A) \leq 2^s.$$

When $\varepsilon \to 0$, we get $\Lambda^s(A) < \infty$, so $\dim A \leq s$. This is true for all $s > \underline{\mathrm{dm}}\,(A)$, so $\dim A \leq \underline{\mathrm{dm}}\,(A)$.

Hilbert's Thirteenth Problem. "Prove that the equation of the seventh degree $x^7 + ax^3 + bx^2 + cx + 1 = 0$ is not solvable with the help of any continuous functions of only two variables." A survey of progress related to this problem was given by G. G. Lorentz [31]. Kolmogorov (1957) solved the problem:

There exist fixed continuous functions $\varphi_{pq}(x)$, on $I = [0,1]$ so that each continuous function f on I^n can be written in the form

$$f(x_1, \cdots, x_n) = \sum_{q=1}^{2n+1} g_q\left(\sum_{p=1}^{n} \varphi_{pq}(x_p)\right),$$

where g_q are properly chosen continuous functions of one variable.

So, since the equation can be solved by a continuous function of the three variables a, b, c, in fact it can be solved by continuous functions of one variable, together with the single function $x + y$ of two variables. The summary: *generalized slide-rules suffice.*

The ε-entropy and ε-capacity can be used to prove results in the opposite direction. For example, *A continuously differentiable function of three variables need not be representable as a superposition of continuously differentiable functions of two variables.* Or: *A Lipschitz function of three variables need not be representable as a superposition of Lipschitz functions of two variables.* Or: *An analytic function of three variables need not be representable as a superposition of analytic functions of two variables.*

Results like these may be proved by showing that if the set F of functions is obtained by superposition from the sets $F_1, \cdots, F_m$ of functions, then the ε-entropy satisfies

$$\mathcal{H}_\varepsilon(F) \leq \sum_{i=1}^{m} \mathcal{H}_{\varepsilon/m}(F_i).$$

For example: let $F_1, F_2, \cdots, F_5$ be sets of functions of two variables. Let F be the set of all functions f of three variables of the form

$$f(x, y, z) = f_1(f_2(x, y), f_3(f_4(y, z), f_5(z, x))),$$

for some choices of $f_i \in F_i$. Then

$$\mathcal{H}_\varepsilon(F) \leq \mathcal{H}_{\varepsilon/5}(F_1) + \mathcal{H}_{\varepsilon/5}(F_2) + \mathcal{H}_{\varepsilon/5}(F_3) + \mathcal{H}_{\varepsilon/5}(F_4) + \mathcal{H}_{\varepsilon/5}(F_5).$$

Now recall in (7) we saw that a bounded set A of Lipschitz functions on an interval,

$$A = F_1^\Delta(C, L)$$

satisfies

$$\mathcal{H}_\varepsilon(A) \asymp \frac{1}{\varepsilon} \quad \text{or} \quad q(A) = 1,$$

with q as defined in (45). It is true similarly, for a bounded set A of Lipschitz functions of n variables, that

$$\mathcal{H}_\varepsilon(A) \asymp \left(\frac{1}{\varepsilon}\right)^n \quad \text{or} \quad q(A) = n.$$

But, for any m, when ε is small $(1/\varepsilon)^3$ is much larger than $m(1/\varepsilon)^2$. So the set of functions of three variables obtained by superposition of m Lipschitz functions of two variables is only a small part of the set of Lipschitz functions of three variables. "Most" Lipschitz functions of three variables cannot be written as superpositions of Lipschitz functions of two variables. More generally, "most" Lipschitz functions of n variables cannot be written as superpositions of Lipschitz functions of fewer than n variables.

Similar remarks apply to p times differentiable functions; or to analytic functions. Roughly speaking, a bounded set A of p times differentiable functions of n variables satisfies

$$\mathcal{H}_\varepsilon(A) \asymp \left(\frac{1}{\varepsilon}\right)^{n/p} \quad \text{or} \quad q(A) = \frac{n}{p};$$

a bounded set A of analytic functions of n complex variables satisfies

$$\mathcal{H}_\varepsilon(A) \asymp \left(\log \frac{1}{\varepsilon}\right)^{n+1} \quad \text{or} \quad \text{df}(A) = n + 1.$$

For reasons like those given before, "most" analytic functions of n complex variables cannot be written as superposition of finitely many analytic functions of fewer than n variables. —Ed.

Bibliography

[1] K. I. Babenko, *On the entropy of a class of analytic functions*, Nauč. Dokl. Vysš. Školy 1 (1958), no. 2.

[2] A. G. Vituškin, *On Hilbert's thirteenth problem*, Dokl. Akad. Nauk SSSR **95** (1954), 701–704. (Russian) [English translation: *American Mathematical Society Translations* Series 2, Volume 17, p. 365]

[3] A. G. Vituškin, *The absolute ε-entropy of metric spaces*, Dokl. Akad. Nauk SSSR **117** (1957), 745–747. (Russian)

[4] V. D. Erohin, *Asymptotic theory of the ε-entropy of analytic functions*, Dokl. Akad. Nauk SSSR **120** (1958), 949–952. (Russian)

[5] A. N. Kolmogorov, *Estimates of the minimal number of elements of an ε-net in different functional spaces and their application to the question of representability of functions of several variables by superposition of functions of a smaller number of variables*, Uspehi Math. Nauk **10** (1955), no. 1 (63), 192–193. (Russian)

[6] A. N. Kolmogorov, *On certain asymptotic characteristics of completely bounded metric spaces*, Dokl. Akad. Nauk. SSSR **108** (1956), 385–388.

[7] V. M. Tihomirov, *On the ε-entropy of certain classes of analytic functions*, Dokl. Akad. Nauk SSSR **117** (1957), 191–194. (Russian)

[8] W. Hurewicz and H. Wallman, *Dimension theory*, Princeton Univ. Press, Princeton, N.J., 1941; Russian translation, Izdat. Inost. Lit., Moscow, 1948.

[9] F. Hausdorff, *Dimension und äusseres Mass*, Math. Ann. **79** (1918/19), 157–179.

[10] N. S. Bahvalov, *On the number of arithmetic operations in solving Poisson's equation for a square by means of finite differences*, Dokl. Akad. Nauk SSSR **113** (1957), 252–254. (Russian)

[11] N. S. Bahvalov, *On setting up equations in finite differences for approximate solution of Laplace's equation*, Dokl. Akad. Nauk SSSR **114** (1957), 1146–1148. (Russian)

[12] A. G. Vituškin, *Some estimates from the tabulation theory*, Dokl. Akad. Nauk SSSR **114** (1957), 923–926. (Russian)

[13] A. N. Kolmogorov, *On the linear dimension of topological vector spaces*, Dokl. Akad. Nauk SSSR **120** (1958), 239–241. (Russian)

[14] V. D. Erohin, *On conformal transformations of rings and the fundamental basis of the space of functions analytic in an elementary neighbourhood of an arbitrary continuum*, Dokl. Akad. Nauk SSSR **120** (1958), 689–692. (Russian)

[15] P. S. Aleksandrov, *Introduction to the general theory of sets and functions*, Gosudarstv. Izdat. Tehn.–Teor. Lit., Moscow–Leningrad, 1948.

[16] P. S. Uryson, *Works on topology and other branches of mathematics*, Vol. 2, Gosudarstv. Izdat. Tehn.–Teor. Lit., Moscow–Leningrad, 1951.

[17] S. Banach, *Théorie des opérations linéaires*, Monogr. Math. Tom. 1, Warsaw, 1932.

[18] S. N. Bernštein, *Theory of probability*, Gosudarstv. Izdat. Tehn.–Teor. Lit., Moscow–Leningrad, 1946.

[19] L. Fejes Tóth, *Lagerungen in der Ebene, auf der Kugel und im Raum*, Springer, Berlin, 1953; Russian translation, Gosudarstv. Izdat. Tehn.–Teor. Lit., Moscow, 1958.

[20] H. Davenport, *The covering of space by spheres*, Rend. Circ. Mat. Palermo (2) **2** (1952), 92–107.

[21] G. L. Watson, *The covering of space by spheres*, Rend. Circ. Mat. Palermo (2) **5** (1956), 93–100.

[22] H. F. Blichfeldt, *The minimum value of quadratic forms, and the closest packing of spheres*, Math. Ann. **101** (1929), 605–608.

[23] N. I. Ahiezer, *Lectures on the theory of approximation*, OGIZ, Moscow–Leningrad, 1947.

[24] G. M. Fihtengol'c, *Course of differential and integral calculus*, Vol. 2, Vol. 3, Gosudarstv. Izdat. Tehn.–Teor. Lit., Moscow–Leningrad, 1951, 1949.

[25] A. I. Markuševič, *Theory of analytic functions*, Gosudarstv. Izdat. Tehn.–Teor. Lit., Moscow–Leningrad, 1950.

[26] V. A. Kotel'nikov, *Material for the first all–Union conference on questions of reconstruction of communication and development of low–current industry*, Izdat. Upr. RKKA, 1933.

[27] C. E. Shannon, *A mathematical theory of communication*, Bell System Tech. J. **27** (1948), 379–423, 623–656.

[28] B. Ya. Levin, *Distribution of zeros of entire functions*, Gosudarstv. Izdat. Tehn.–Teor. Lit., Moscow, 1956.

[29] M. A. Lavrent'ev and B.V. Šabat, *Methods of the theory of functions of a complex variable*, 2nd ed., revised, Gosudarstv. Izdat. Fiz.–Math. Lit., Moscow, 1958.

[30] D. Kendall, *Obituary: Andrei Nikolaevich Kolmogorov (1903–1987)*, Bull. London Math. Soc. **14** (1990), 31–100.

[31] G. G. Lorentz, *The 13-th problem of Hilbert*, Mathematical Developments Arising from Hilbert Problems (F.E. Browder, ed.), American Mathematical Society, 1976, pp. 419–430.

[32] C. Tricot, *Douze définitions de la densité logarithmique*, Compte Rendus Acad. Sci. Paris, Série I **293** (1981), 549–552.

This selection contains a simple example of continuous nowhere-differentiable function. An example quite similar to the one given here goes back to Bernard Bolzano [9]. Bolzano's manuscript was written about 1830 (well before the Weierstrass example), but published only a hundred years later (see [15], [10], pp. 30–32). See [19] for an interesting account of Bolzano's mathematical work—it was remarkably modern in outlook, but mostly unpublished, and therefore not influential. Bolzano's example is a self-affine function, slightly different from Kiesswetter's example. Bolzano proved only that his function is non-differentiable on a dense set; but in fact it, too, is nowhere differentiable.

Karl Kießwetter was born in 1930 in Sudetenland, and studied mathematics at the University of Köln. He was a gymnasium teacher, then held positions at the universities of Münster, Bielefeld, and Hamburg. He has been interested in mathematics education, and fostering creativity among gifted schoolchildren (the "Hamburg model" [16]).

EIGHTEEN

A Simple Example of a Function, which is Everywhere Continuous and Nowhere Differentiable

Karl Kiesswetter

We define $f(x)$ for all x in $0 \leq x \leq 1$ and formulate our theorems only for this closed interval. According to our definition $f(0) = 0$ and $f(1) = 1$, so by defining $g(x) = [x] + f(x - [x])$, one obtains a function with the desired characteristics for all real x.

Definition. Every number x from the closed interval $[0, 1]$ can be written in a base four expansion

$$x = \sum_{\nu=1}^{\infty} \frac{x_\nu}{4^\nu} \quad \text{where } x_\nu \text{ is } 0, 1, 2, \text{ or } 3.$$

The corresponding value of the function is

$$f(x) = \sum_{\nu=1}^{\infty} (-1)^{N_\nu} X_\nu / 2^\nu,$$

where $X_\nu = x_\nu - 2$ for $x_\nu > 0$ and $X_\nu = 0$ for $x_\nu = 0$ and N_ν is the number of x_k with $x_k = 0$ and $k < \nu$.

This sum is always absolutely convergent. Nevertheless there are for each quotient $V_n = p/4^n$ with $0 < V_n < 1$ two different base four representations. We must therefore still show that the value of the sum is independent of the representation chosen for V_n. Only then can we say that a function is given by our definition. Before we work through the proof, we should gain a notion of the given function through the calculation of several values for the function and through a sketch (Figure 1).

I. Let $x = 19/64$. Then $x_1 = 1$, $x_2 = 0$, and $x_3 = 3$, and therefore: $X_1 = -1$, $X_2 = 0$, $X_3 = 1$, and $N_1 = N_2 = 0$, $N_3 = 1$. From this it follows: $f(x) = -1/2 + (-1)(1/8) = -5/8$.

II. Let $x = 53/64$. Then $x_1 = 3$, $x_2 = 1$, and $x_3 = 1$, and therefore: $X_1 = 1$, $X_2 = -1$, $X_3 = -1$, and $N_1 = N_2 = N_3 = 0$. From this it follows: $f(x) = 1/2 - 1/4 - 1/8 = 1/8$.

III. Let $x = 1/3$. Then $x_\nu = 1$ for all ν and therefore: $X_\nu = -1$ for all ν and $N_\nu = 0$ for all ν. From this it follows: $f(x) = -1/2 - 1/4 - 1/8 - \cdots = -1$.

Now we want to work through the proof previously referred to. We show that the value of the sum is independent of the representation of V_n.

Let

$$a = \sum_{\nu=1}^{n} a_\nu/4^\nu = a^* + a_n/4^n \quad \text{with } a_n > 0,$$

$$b = \sum_{\nu=1}^{\infty} b_\nu/4^\nu \text{ with } b_\nu = a_\nu \text{ for } \nu < n, \ b_n = a_n - 1, \text{ and } b_\nu = 3 \text{ for } \nu > n.$$

Then $a = b$ and $b = a^* + (a_n - 1)/4^n + 3/4^{n+1} + 3/4^{n+2} + \cdots$. From this

case $a_n = 3$: $\quad f(a) = f(a^*) + (-1)^{N_n}/2^n$

$$f(b) = f(a^*) + \sum_{\nu=n+1}^{\infty} (-1)^{N_n}/2^\nu$$

case $a_n = 2$: $\quad f(a) = f(a^*)$

$$f(b) = f(a^*) - (-1)^{N_n}/2^n + \sum_{\nu=n+1}^{\infty} (-1)^{N_n}/2^\nu$$

case $a_n = 1$: $\quad f(a) = f(a^*) - (-1)^{N_n}/2^n$

$$f(b) = f(a^*) + \sum_{\nu=n+1}^{\infty} (-1)^{N_n+1}/2^\nu \quad \text{here } b_n = 0$$

In all three cases thus $f(a) = f(b)$.

Continuity and nondifferentiability of f may be shown now with relative simplicity.

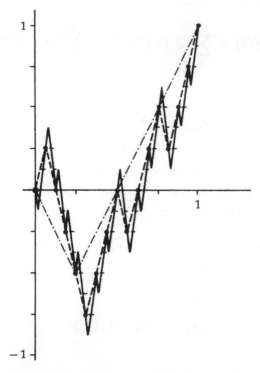

Figure 1. Here, for illustration, are drawn those points of the graph of f whose arguments belong to the first, second, and third division (*i.e.* which may be written in the form $p/4$, $p/4^2$, and $p/4^3$ respectively). Neighboring points of every division under consideration have been joined together by lines (up to the third division).

Theorem 1

Suppose the numbers a and b are in $[0,1]$, and meet the condition $0 < b-a \le 1/4^n$. Then $|f(b)-f(a)| \le 4/2^n$. Thus f is (even uniformly) continuous on the interval. *Proof.* We distinguish two cases. 1) There

exists a quotient[1] $V_n = p/4^n$ with $0 \le V_n \le a < b \le V_n + 1/4^n \le 1$. (Thus a and b both lie in the same interval of the 4^n partition.) Then

$$a = V_n + \sum_{\nu=n+1}^{\infty} a_\nu/4^\nu \quad \text{and} \quad b = V_n + \sum_{\nu=n+1}^{\infty} b_\nu/4^\nu$$

---

[1] *Viererbruch,* apparently means a rational number with denominator a power of 4.–*Ed.*

and respectively

$$f(b) - f(a) = \sum_{\nu=n+1}^{\infty} B_\nu (-1)^{N_\nu^*}/2^\nu - \sum_{\nu=n+1}^{\infty} A_\nu (-1)^{N_\nu}/2^\nu.$$

One can therefore estimate:

$$|f(b) - f(a)| \le 2 \sum_{\nu=n+1}^{\infty} 1/2^\nu = 2\,(1/2)^n.$$

2) If case 1 does not apply, then there exists $V_n = p/4^n$ with $0 \le V_n - 1/4^n < a < V_n < b < V_n + 1/4^n \le 1$. (Thus a and b lie in neighboring intervals of the 4^n partition.) Since

$$|f(b) - f(a)| \le |f(b) - f(V_n)| + |f(V_n) - f(a)|$$

through the use of the result in case 1 on both $|f(b) - f(V_n)|$ and $|f(V_n) - f(a)|$ respectively, one gets

$$|f(b) - f(a)| \le 4\,(1/2)^n.$$

Theorem 2

f is non-differentiable at each point in the interval. For the proof we use two lemmas.

Lemma a: Suppose $0 \le V_n$ and $V_n^* = V_n + 1/4^n \le 1$. Then $|f(V_n^*) - f(V_n)| = 1/2^n$.

Proof. Let $V_n^* = V_n + \sum_{\nu=n+1}^{\infty} 3/4^\nu$. From this it follows

$$f(V_n^*) = f(V_n) + \sum_{\nu=n+1}^{\infty} (-1)^{N_{n+1}^*}/2^\nu$$

and

$$|f(V_n^*) - f(V_n)| = \sum_{\nu=n+1}^{\infty} 1/2^\nu = 1/2^n.$$

Lemma b: If $A \le x \le B$ with $A < B$, then at least one of the following inequalities holds:

$$\left| \frac{f(B) - f(x)}{B - x} \right| \ge \left| \frac{f(B) - f(A)}{B - A} \right| \quad \text{or} \quad \left| \frac{f(x) - f(A)}{x - A} \right| \ge \left| \frac{f(B) - f(A)}{B - A} \right|$$

Proof. For $x = A$ or $x = B$ the statement is evident. We can thus assume for further consideration that $A < x < B$. From the simple identity

$$(B - A)\frac{f(B) - f(A)}{B - A} = (B - x)\frac{f(B) - f(x)}{B - x} + (x - A)\frac{f(x) - f(A)}{x - A}$$

one gets

(*)

$$(B - A)\left|\frac{f(B) - f(A)}{B - A}\right| \leq (B - x)\left|\frac{f(B) - f(x)}{B - x}\right| + (x - A)\left|\frac{f(x) - f(A)}{x - A}\right|.$$

If the following held simultaneously:

$$\left|\frac{f(B) - f(x)}{B - x}\right| < \left|\frac{f(B) - f(A)}{B - A}\right| \text{ and } \left|\frac{f(x) - f(A)}{x - A}\right| < \left|\frac{f(B) - f(A)}{B - A}\right|,$$

then it would follow that

$$(B - x)\left|\frac{f(B) - f(x)}{B - x}\right| + (x - A)\left|\frac{f(x) - f(A)}{x - A}\right| < (B - A)\left|\frac{f(b) - f(A)}{B - A}\right|,$$

a contradiction to (*). Now we can prove Theorem 2.

Proof. For each x with $0 \leq x \leq 1$ and for every natural number n there exists at least one V_n where $0 \leq V_n \leq x \leq V_n^* = V_n + 1/4^n \leq 1$. From our two lemmas it follows for at least one of the two fractions V_n or V_n^* (we denote it by K_n) that

$$\left|\frac{f(x) - f(K_n)}{x - K_n}\right| \geq \left|\frac{f(V_n^*) - f(V_n)}{V_n^* - V_n}\right| = \frac{1/2^n}{1/4^n} = 2^n.$$

For the sequence K_n, of course $\lim_{n\to\infty} K_n = x$, however the corresponding sequence of difference quotients

$$D_n = \frac{f(K_n) - f(x)}{K_n - x}$$

does not converge.

Remarks.

We have only used elementary mathematical techniques. Our example can therefore perhaps even be understood by gifted boys.[2]

---

[2]*Primaner*, pupils of a "Prima", the final year of secondary school, 18 or 19 years old.–*Au.*

The values of the function can easily be calculated for all rational arguments. These function values are, as one can easily see, likewise all rational. Through simple modifications of our definition, infinitely many other continuous and non-differentiable functions can be given. One can for example go from the decimal system. In the general case one writes the argument in a base n expansion, where $n = m + 2k$ and $m > 1$, $k > 0$. One gets the values of the function as a base m expansion. The interested reader can work through generalization of the definition and the proof by himself. Here we have not handled the general case, since the proof would become complicated and bring no essential mathematical insight.

References.

There were, around the turn of the century, a large number of examples of continuous but not differentiable functions published, which would be impossible to mention. Therefore we must content ourselves with a choice of those works which in our opinion would be the most interesting to the reader.

The famous example of Weierstrass was first published by P. du Bois-Reymond in his work: *Versuch einer Klassifikation der willkürlichen Funktionen reeller Argumente* [1].

Knopp gives an extensive overview of the literature related to our topic up to 1918 in his work: *Ein einfaches Verfahren zur Bildung stetiger nirgends differenzierbarer Funktionen* [2]. One very beautiful example originated from v. d. Waerden: *Ein einfaches Beispiel einer nicht-differenzierbaren stetigen Funktion* [3]. Recently, I have discovered in the old *Semester Reports* an example which in certain hindsight is a forerunner of our examples; Huga Scherer: *Kronstruktion einer überall stetigen, nirgends differenzierbaren Funktion* (Sommer 1938). The contribution of K. Zeller is also interesting: *Kondensation von Singularitäten* [5]. A specific continuous non-differentiable function is not constructed there, but the existence of such functions is discussed.

The graph (see Figure 2) of the function described in this selection is a **self-affine** set in the plane. Consider these four affine transformations on $\mathbb{R}^2$, where points $(x, y) \in \mathbb{R}^2$ are identified with 2×1 column matrices $\begin{bmatrix} x \\ y \end{bmatrix}$.

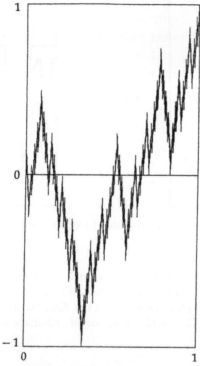

Figure 2.   Kiesswetter's fractal.

$$f_1\begin{bmatrix} x \\ y \end{bmatrix} = \begin{bmatrix} 1/4 & 0 \\ 0 & -1/2 \end{bmatrix}\begin{bmatrix} x \\ y \end{bmatrix}$$

$$f_2\begin{bmatrix} x \\ y \end{bmatrix} = \begin{bmatrix} 1/4 & 0 \\ 0 & 1/2 \end{bmatrix}\begin{bmatrix} x \\ y \end{bmatrix} + \begin{bmatrix} 1/4 \\ -1/2 \end{bmatrix}$$

$$f_3\begin{bmatrix} x \\ y \end{bmatrix} = \begin{bmatrix} 1/4 & 0 \\ 0 & 1/2 \end{bmatrix}\begin{bmatrix} x \\ y \end{bmatrix} + \begin{bmatrix} 1/4 \\ 0 \end{bmatrix}$$

$$f_4\begin{bmatrix} x \\ y \end{bmatrix} = \begin{bmatrix} 1/4 & 0 \\ 0 & 1/2 \end{bmatrix}\begin{bmatrix} x \\ y \end{bmatrix} + \begin{bmatrix} 3/4 \\ 1/2 \end{bmatrix}.$$

Each of the transformations shrinks horizontally with ratio $1/4$ and vertically with ratio $1/2$. The first transformation also reflects top-to-bottom. (See Figure 3) The compact, nonempty set K is an **invariant set** of the iterated function system f_1, f_2, f_3, f_4 iff

$$K = f_1[K] \cup f_2[K] \cup f_3[K] \cup f_4[K].$$

By an application of the contraction mapping theorem ([5], 3.1(3)(i), [12], (4.1.3), [13], Theorem 9.1, [6], p. 82), there is a unique nonempty compact invariant set. This set is the graph of Kiesswetter's function.

By Theorem 1, the function f satisfies a Lipschitz condition

$$|f(x) - f(y)| \le C|x - y|^{1/2};$$

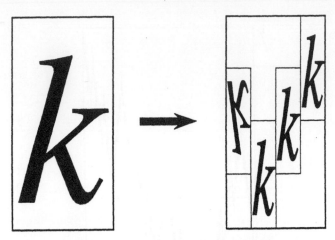

Figure 3. The iterated function system.

the choices like $x = k/4^n$, $y = (k+1)/4^n$ show that $1/2$ is the best exponent possible here. By the theorem of Selection 11 (Besicovitch and Ursell), the graph has Hausdorff dimension $\leq 3/2$. In [11] or [7] it is seen that it has dimension exactly $3/2$.

Recent work on the Hausdorff dimension of self-affine sets includes: [8], [14], [17], [18], [20]. —Ed.

Bibliography

[1] P. du Bois-Reymond, *Versuch einer Klassifikation der willkürlichen Funktionen reeler Argumente nach ihren Änderungen in den kleinsten Intervallen,* J. Reine Angew. Math. **79** (1875), 21–37.

[2] K. Knopp, *Ein einfaches Verfahren zur Bildung stetiger nirgends differenzierbarer Funktionen,* Math. Zeitschrift **2** (1918), 1–26.

[3] B. L. van der Waerden, *Ein einfaches Beispiel einer nicht-differenzierbaren stetigen Funktion,* Math. Zeitschrift **32** (1930), 474–475.

[4] H. Scherer, *Konstruktion einer überall stetigen, nirgends differenzierbaren Funktion,* Math.-Phys. Semesterber. (1938).

[5] K. Zeller, *Kondensation von Singularitäten,* Math.-Phys. Semesterber. **3** (1953), 207–213.

[6] M. F. Barnsley, *Fractals Everywhere,* Academic Press, 1988.

[7] T. Bedford, *Crinkly Curves, Markov Partitions and Dimension*, Ph.D. Thesis, University of Warwick, 1984.

[8] T. Bedford, *On Weierstrass-like functions and random recurrent sets*, Math. Proc. Cambr. Phil. Soc. **106** (1989), 325–342.

[9] Bernard Bolzano, *Funktionenlehre*, Herausgegeben und mit Anmerkungen versehen von K. Rychlik, Prague, 1930.

[10] Bernard Bolzano, *Paradoxes of the Infinite*, With a historical introduction by Donald A. Steele, Yale University Press, 1950.

[11] G. A. Edgar, *Kiesswetter's fractal has Hausdorff dimension 3/2*, Real Analysis Exchange **41** (1989), 215–223.

[12] G. A. Edgar, *Measure, Topology, and Fractal Geometry*, Springer-Verlag, New York, 1990.

[13] K. Falconer, *Fractal Geometry: Mathematical Foundations and Applications*, Wiley, 1990.

[14] K. J. Falconer, *The Hausdorff dimension of self-affine fractals*, Math. Proc. Cambr. Phil. Soc. **103** (1988), 339–350.

[15] M. Jaček, *Über den wissenschaftlichen Nachlass Bernhard Bolzanos*, Jahresberichte der deutschen Mathematikervereinigung **31** (1922), 109–110.

[16] K. Kießwetter, *Theoriebildung bei der Untersuchung nichttransitiver Würfelstrukturen: Ergebnisse aus den Förderveranstaltungen nach dem "Hamburger Modell" für mathematisch besonders befähigte Oberstufenschüler*, Mitteilungen Math. Ges. Hamburg **11** (1989), 745–762.

[17] C. McMullen, *The Hausdorff dimension of general Sierpiński carpets*, Nagoya Math. J. **96** (1984), 1–9.

[18] F. Przytycki and M. Urbanski, *On the Hausdorff dimension of some fractal sets*, Studia Math. **93** (1989), 155–186.

[19] Steve Russ, *Bolzano's analytic programme*, Math. Intelligencer **14** (1992), no. 3, 45–53.

[20] M. Urbanski, *The probability dimension of self-affine functions* (to appear).

Could any book on fractals fail to include a contribution by Benoit Mandelbrot? The following selection was chosen to close the book. It is one of the first hints that these weird sets with fractional dimension are not merely another chapter from mathematics (interesting as that may be to mathematicians), but also a chapter or two in the applications of mathematics to various branches of science. Mandelbrot's manifesto on this topic is [8].

Benoit B. Mandelbrot was born in Warsaw in 1924. He holds a doctorate from the University of Paris. Although officially Paul Lévy had no doctoral students, Mandelbrot often says he considers himself a student of Lévy. Mandelbrot held various research and teaching positions, but then in 1958 went to the I.B.M. Thomas J. Watson Research Center as an I.B.M. Fellow. Some[1] may question whether to consider Mandelbrot a mathematician. But his prominent place in the history of Twentieth Century science cannot be doubted. An interview with Mandelbrot is in *Mathematical People* [5].

---

[1] Perhaps including Mandelbrot himself: "Science would be ruined if (like sports) it were to put competition above everything else, and if it were to clarify the rules of competition by withdrawing entirely into narrowly defined specialties. The rare scholars who are nomads-by-choice are essential to the intellectual welfare of the settled disciplines."—from Mandelbrot's entry in *Who's Who*.

NINETEEN

How Long Is the Coast of Britain? Statistical Self-Similarity and Fractional Dimension

Benoit Mandelbrot

Geographical curves are so involved in their detail that their lengths are often infinite or, rather, undefinable. However, many are statistically "self-similar," meaning that each portion can be considered a reduced-scale image of the whole. In that case, the degree of complication can be described by a quantity D that has many properties of a "dimension," though it is fractional; that is, it exceeds the value unity associated with the ordinary, rectifiable, curves.

Seacoast shapes are examples of highly involved curves such that each of their portion can—in a statistical sense—be considered a reduced-scale image of the whole. This property will be referred to as "statistical self-similarity." To speak of a length for such figures is usually meaningless. Similarly (1), "the left bank of the Vistula, when measured with increased precision, would furnish lengths ten, hundred or even thousand times as great as the length read off the school map." More generally, geographical curves can be considered as superpositions of features of widely scattered characteristic size; as ever finer features are taken account of, the measured total length increases, and there is usually no clear cut gap between the realm of geography and details with which geography need not be concerned.

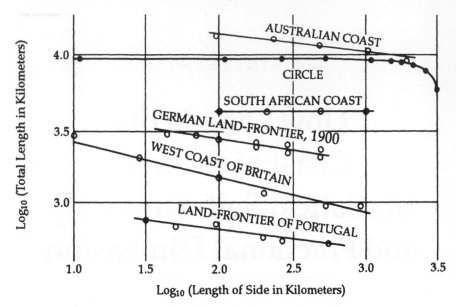

Figure 1. Richardson's data concerning measurements of geographical curves by way of polygons which have equal sides and have their corners on the curve. For the circle, the total length tends to a limit as the side goes to zero. In all other cases, it increases as the side becomes shorter, the slope of the doubly logarithmic graph being in absolute value equal to $D - 1$. (Reproduced from 2, Fig. 17, by permission).

Quantities other than length are thus needed to discriminate between various degrees of complication for a geographical curve. When a curve is self-similar, it is characterized by an exponent of similarity, D, which possesses many properties of a dimension, though it is usually a fraction greater than the dimension 1 commonly attributed to curves. We shall reexamine in this light some empirical observations by Richardson (2). I propose to interpret them as implying, for example, that the dimension of the west coast of Great Britain is $D = 1.25$. Thus, the so far esoteric concept of "random figure of fractional dimension" is shown to have simple and concrete applications and great usefulness.

Self-similarity methods are a potent tool in the study of chance phenomena, including geostatistics, as well as economics (3) and physics (4). In fact, many noises have dimensions D contained between 0 and 1, so that the scientist ought to consider dimension as a continuous quantity ranging from 0 to infinity.

Returning to the claim made in the first paragraph, let us review the methods used when attempting to measure the length of a seacoast. Since a geographer is unconcerned with minute details, he may choose a positive scale G as a lower limit to the length of geographically meaningful features. Then, to evaluate the length of a coast between two of its points A and B, he may draw the shortest inland curve joining A and B while staying within a distance G of the sea. Alternatively, he may draw the shortest line made of straight segments of length at most G, whose vertices are points of the coast which include A and B. There are many other possible definitions. In practice, of course, one must be content with approximations to shortest paths. We shall suppose that measurements are made by walking a pair of dividers along a map so as to count the number of equal sides of length G of an open polygon whose corners lie on the curve. If G is small enough, it does not matter whether one starts from A or B. Thus, one obtains an estimate of the length to beOB called $L(G)$.

Unfortunately, geographers will disagree about the value of G, while $L(G)$ depends greatly upon G. Consequently, it is necessary to know $L(G)$ for several values of G. Better still, it would be nice to have an analytic formula linking $L(G)$ with G. Such a formula, of an entirely empirical character, was proposed by Lewis. F. Richardson (2) but unfortunately attracted no attention. The formula is $L(G) = M G^{1-D}$, where M is a positive constant and D is a constant at least equal to unity. This D, a "characteristic of the frontier, may be expected to have some positive correlation with one's immediate visual perception of the irregularity of the frontier. At one extreme, $D = 1.00$ for a frontier that looks straight on the map. For the other extreme, the west coast of Britain was selected because it looks like one of the most irregular in the world; it was found to give $D = 1.25$. Three other frontiers which, judging by their appearance on the map were more like the average of the world in irregularity, gave $D = 1.15$ for the land frontier of Germany in about A.D. 1899; $D = 1.14$ for the land frontier between Spain and Portugal and $D = 1.13$ for the Australian coast. A coast selected as looking one of the smoothest in the atlas, was that of South Africa and for it, $D = 1.02$."

Richardson's empirical finding is in marked contrast with the ordinary behavior of smooth curves, which are endowed with a well-defined length and are said to be "rectifiable." Thus, to quote Steinhaus (1) again, "a statement nearly adequate to reality would be to call most arcs encountered in nature not rectifiable. This statement is contrary to the belief that not rectifiable arcs are an invention of mathematicians and that natural arcs are rectifiable: it is the opposite that is true."

I interpret Richardson's relation as contrary to the belief that curves of dimension greater than one are an invention of mathematicians. For that, it is necessary to review an elementary feature of the concept of dimension and to show how it naturally leads to the consideration of fractional dimensions.

To begin, a straight line has dimension one. Hence, for every positive integer N, the segment $(0 \leq x < X)$ can be exactly decomposed into N nonoverlapping segments of the form $[(n - 1)X/N \leq x < nX/N]$, where n runs from 1 to N. Each of these parts is deducible from the whole by a similarity ratio $r(N) = 1/N$. Similarly, a plane has dimension two. Hence, for every perfect square N, the rectangle $(0 \leq x < X; 0 \leq y < Y)$ can be decomposed exactly into N nonoverlapping rectangles of the form $[(k - 1)X/\sqrt{N} \leq x < kX/\sqrt{N}; (h - 1)Y/\sqrt{N} \leq y < hY/\sqrt{N}]$, where k and h run from 1 to $\sqrt{N}$. Each of these parts is deducible from the whole by a similarity of ratio $r(N) = 1/\sqrt{N}$. More generally, whenever $N^{1/D}$ is a positive integer, a D-dimensional rectangular parallelepiped can be decomposed into N parallelepipeds deducible from the whole by a similarity of ratio $r(N) = 1/N^{1/D}$. Thus, the dimension D is characterized by the relation $D = -\log N / \log r(N)$.

This last property of the quantity D means that it can also be evaluated for more general figures that can be exactly decomposed into N parts such that each of the parts is deducible from the whole by a similarity of ratio $r(N)$, or perhaps by a similarity followed by a rotation and even a symmetry. If such figures exist, they may be said to have $D = -\log N / \log r(N)$ for dimension (5). To show that such figures exist, it suffices to exhibit a few obvious variants of von Koch's continuous nondifferentiable curve. Each of these curves is constructed as a limit. Step 0 is to draw the segment $(0, 1)$. Step 1 is to draw either of the kinked curves of Fig. 2, each made up of N intervals superposable upon the segment $(0, 1/4)$.

Step 2 is to replace each of the N segments used in step 1 by a kinked curve obtained by reducing the curve of step 1 in the ratio $r(N) = 1/4$. One obtains altogether N^2 segments of length 1/16. Each repetition of the same process adds further detail; as the number of steps

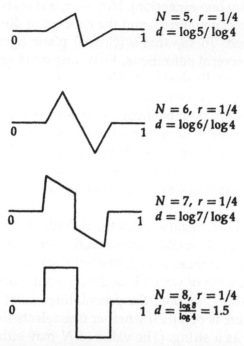

$N = 5, r = 1/4$
$d = \log 5/\log 4$

$N = 6, r = 1/4$
$d = \log 6/\log 4$

$N = 7, r = 1/4$
$d = \log 7/\log 4$

$N = 8, r = 1/4$
$d = \frac{\log 8}{\log 4} = 1.5$

Figure 2. Nonrectifiable self-similar curves can be obtained as follows. Step 1: Choose any of the above drawings. Step 2: Replace ach of its N legs by a curve deduced from the whole drawing through a similarity ratio 1/4. One is left with a curve made of N^2 legs of length $(1/4)^2$. Step 3: replace each leg by a curve obtained from the whole drawing through a similarity ratio $(1/4)^2$. The desired self-similar curve is approached by an infinite sequence of these steps.

grows to infinity, our kinky curves tend toward continuous limits and it is obvious by inspection that these limits are self-similar, since they are exactly decomposable into N parts deducible from the whole by a similarity of ratio $r(N) = 1/4$ followed by translation. Thus, given N, the limit curve can be said to have dimension $D = -\log N/\log r(N) = \log N/\log 4$. Since N is greater than 4 in our examples, the corresponding dimensions all exceed unity. Let us now consider length: at step number s, our approximation is made of N^s segments of length $G = (1/4)^s$, so that $L = (N/4)^s = G^{1-D}$. Thus, the length of the limit curve is infinite, even though it is a "line." (Note that it is not excluded for a plane curve to have dimension equal to 2. An example is Peano's curve, which fills up a square.)

Practical applications of this notion of dimension requires further consideration, because self-similar figures are seldom encountered in nature (crystals are one exception). However, a statistical form of self-similarity is often encountered, and the concept of dimension may be further generalized. To say that a (closed) plane figure is chosen at random implies several definitions. First, one must select a family of possible figures, usually designated by Ω. When this family contains a finite number of members, the rule of random choice is specified by attributing to each possible figure a well-defined probability of being chosen. However, Ω is in general infinite and each figure has a zero probability of being chosen. But positive probabilities can be attached to appropriately defined "events" (such as the event that the chosen figure differs little—in some specified sense—from some specified figure).

For the family Ω, together with the definition of events and their probabilities, to be self-similar, two conditions are needed. First, each of the possible figures must be constructible by somehow stringing together N figures, each of which is deduced from a possible figure by a similarity of ratio r; second, the probabilities must be so specified that the same value is obtained whether one selects the overall figure at one swoop or as a string. (The value of N may either be arbitrary, or chosen from some specific sequence, such as the perfect squares relative to nonrandom rectangles, or the integral powers of 4, 5, 6, or 7 encountered in the curves built as in Fig. 2.) In case that the value of r is specified by choosing N, one can consider $-\log N/\log r$ a similarity dimension. More usually, however, given r, N will take different values for different figures of Ω. As one considers points "sufficiently far" from each other, the details on a "sufficiently fine" scale may become asymptotically independent, in such a way that $-\log N/\log r$ almost surely tends to some limit as r tends to zero. In that case, this limit may be considered a similarity dimension. Under wide conditions, the length of approximating polygons will asymptotically behave like $L(G) \sim G^{1-D}$.

To specify the mathematical conditions for the existence of a similarity dimension is not a fully solved problem. In fact, even the idea that a geographical curve is random raises a number of conceptual problems familiar in other applications of randomness. Therefore, to return to Richardson's empirical law, the most that can be said with perfect safety is that it is compatible with the idea that geographical curves are random self-similar figures of fractional dimension D.

Empirical scientists having to be content with less than perfect induc-
tions. I favor the more positive interpretation stated at the beginning
of this report.

Benoit Mandelbrot

*International Business Machines,
Thomas J. Watson Research Center,
Yorktown Heights, New York 10598*

Statistical self-similarity, much more than strict self-similarity, can be used to
counterfeit real-world objects with very natural-looking results. Hutchinson [7] evaluated the Hausdorff dimension for statistically self-similar sets: see [6] for another kind
of random self-similarity. *–Ed.*

Bibliography

[1] H. Steinhaus, *Length, shape, and area,* Colloquium Math. **3** (1954),
1–13.

[2] L. F. Richardson, *The problem of contiguity: an appendix of statistics of
deadly quarrels,* General Systems Yearbook **6** (1961), 139–187.

[3] B. Mandelbrot, *The variation of certain speculative prices,* J. Business
(Chicago) **36** (1963), 394–419.

[4] B. Mandelbrot, *Self similar error clusters in communications systems and
the concept of conditional stationarity,* IEEE Trans. Commun. Technol.
13 (1965), 71–90.

[5] D. J. Albers and G. L. Alexanderson, *Mathematical People,*
Birkhäuser. Boston, 1985.

[6] S. Graf, R. D. Mauldin, and S. Williams, *The Exact Hausdorff Dimension in Random Recursive Constructions,* Memoirs of the American Mathematical Society **71**, Providence, RI, 1988.

[7] J. E. Hutchinson, *Fractals and self similarity*, Indiana Univ. Math. J. **30** (1981), 713–747.

[8] B. Mandelbrot, *The Fractal Geometry of Nature*, Freeman. San Francisco, 1982.

Index

Permissions and Acknowledgments

Blake and Fractals. Source: *Mathematics Magazine* 63 (1990) 280. Copyright © 1990 by the Mathematical Association of America. Included here by permission of the Mathematical Association of America and J. D. Memory.

On Continuous Functions of a Real Argument that do not have a Well-Defined Differential Quotient. Translated by: B. SAWHILL Original title: Über continuirliche Functionen eines reellen Arguments, die für keinen Werth des Letzteren einen bestimmten Differentialquotienten besitzen. Source: *Mathematische Werke*, Band II (Mayer & Müller, Berlin, 1895), pp. 71–74.

On the Power of Perfect Sets of Points. Translated by: ILAN VARDI. Original title: De la puissance des ensembles parfaits de points. Source: *Acta Mathematica* 4 (1884) 381–392; reprinted in *Gesammelte Abhandlungen*, pp. 252–260.
Included here by permission of the Institut Mittag-Leffler, Djursholm, Sweden.

On a Continuous Curve without Tangent Constructible from Elementary Geometry. Translated by: ILAN VARDI. Original title: Sur une courbe continue sans tangente obtenue par une construction géométrique élémentaire. Source: *Arkiv för Matematik, Astronomi och Fysik* 1 (1904) 681–702.
Included here by permission of the Kungliga Vetenskapsakademien, Stockholm.

On the Linear Measure of Point Sets—a Generalization of the Concept of Length. Translated by: B. SAWHILL. Original title: Über das lineare Maß von Punktmengen—eine Verallgemeinerung des Längenbegriffs. Source: *Nachrichten der K. Gesellschaft der Wissenschaften zu Göttingen, Mathematisch-physikalische Klasse 1914*, pp. 404–426; reprinted in *Gesammelte Mathematische Schriften*, Band IV, pp. 249–277.
Included here by permission of the Akademie der Wissenschaften, Göttingen.

Dimension and Outer Measure. Translated by: B. SAWHILL. Original title: Dimension und äußeres Maß. Source: *Mathematische Annalen* 79 (1918) 157–179. Included here by permission of Springer-Verlag, Heidelberg.

General Spaces and Cartesian Spaces. Translated by: G. A. EDGAR. Original title: Allgemeine Räume und Cartesische Räume. Source: *Proceedings of the Section of Sciences, Koniklijke Akademie van Wetenschappen te Amsterdam* 29 (1926) 476–482, 1125–1128. Included here by permission of Koninklijke Nederlandse Akademie van Wetenschappen.

Improper Sets and Dimension Numbers (excerpt). Translated by: ILAN VARDI. Original title: Ensembles impropres et nombre dimensionnel. Source: *Bulletin des Sciences Mathématiques* 52 (1928) 320–344, 361–376.

On a Metric Property of Dimension. Translated by: ILAN VARDI. Original title: Sur une propriété métrique de la dimension. Source: *Annals of Mathematics* 33 (1932) 156–162. Included here by permission of the *Annals of Mathematics*.

On the Sum of Digits of Real Numbers Represented in the Dyadic System. Source: *Mathematische Annalen* 110 (1934) 321–330. Included here by permission of Springer-Verlag, Heidelberg.

On Rational Approximations to Real Numbers. Source: *Journal of the London Mathematical Society* 9 (1934) 126–131. Included here by permission of the London Mathematical Society.

On Dimensional Numbers of Some Continuous Curves. Source: *Journal of the London Mathematical Society* 12 (1937) 18–25. Included here by permission of the London Mathematical Society.

Plane or Space Curves and Surfaces Consisting of Parts Similar to the Whole. Translated by: ILAN VARDI. Original title: Les courbes planes ou gauches et les surfaces composées de parties semblables au tout. Source: *Journal de l'École Polytechnique* 81(1938) 227–247, 249–291; reprinted in *Oeuvres de Paul Lévy*, t. II, pp. 331–394.

Additive Functions of Intervals and Hausdorff Measure. Source: *Proceedings of the Cambridge Philosophical Society* **42** (1946) 15–23. Included here by permission of Cambridge University Press.

The Dimension of Cartesian Product Sets. Source: *Proceedings of the Cambridge Philosophical Society* **50** (1954) 198–202. Included here by permission of Cambridge University Press.

On the Complementary Intervals of a Linear Closed Set of Zero Lebesgue Measure. Source: *Journal of the London Mathematical Society* **29** (1954) 449–459. Included here by permission of the London Mathematical Society and S. J. Taylor.

On some Curves Defined by Functional Equations. Translated by: ILAN VARDI. Original title: Sur quelques courbes définies par des equations fonctionnelles. Source: *Rendiconti del Seminario Matematico dell'Università e del Politecnico di Torino* **16** (1957) 101–113. Reprinted in: *Oeuvres Mathématiques* (Geneva, 1981), pp. 716–728. Included here by permission of the Seminario Matematico.

ε-Entropy and ε-Capacity of Sets in Functional Spaces (exerpt). Translated by: EDWIN HEWITT. Original title: ε-энтропия и ε-емкость множеств в функциональных пространствах. Source: *Успехи Математических Наук* **14** (1959) no. 2, 3–86; translation published in *American Mathematical Society Translations* Series 2, Volume 17, pp. 277–364. Copyright © 1961 by the American Mathematical Society. Included here by arrangement with the American Mathematical Society.

A simple Example of a Function which is Everywhere Continuous and Nowhere Differentiable. Translated by: ERIC OLSON. Original title: Ein einfaches Beispiel für eine Funktion, welche überall stetig und nicht differenzierbar ist. Source: *Mathematisch-Physikalische Semesterberichte zur Pflege des Zusammenhangs von Schule und Universität* **13** (1966) 216–221. Included here by permission of Karl Kießwetter.

Printed and bound by CPI Group (UK) Ltd, Croydon, CR0 4YY

23/10/2024

01778263-0004